N. P. Kovalenko, Yu. P. Krasny, U. Krey

Physics of Amorphous Metals

N. P. Kovalenko, Yu. P. Krasny, U. Krey

Physics of Amorphous Metals

WILEY-VCH

Berlin – Weinheim – New York – Chichester – Brisbane – Singapore – Toronto

The Authors of this Volume

Prof. Dr. N. P. Kovalenko
Physics Department
Odessa State University
Ukraine
e-mail: npk@paco.net

Prof. Dr. Yu. P. Krasny
Mathematics Department
University of Opole
Poland
e-mail: krasnyj@math.uni.opole.pl

Prof. Dr. U. Krey
Physics Department II
University of Regensburg
Germany
e-mail: uwe.krey@physik.uni-regensburg.de

Library of Congress Card No.: applied for

A catalogue record for this book is available from the British Library.

Die Deutsche Bibliothek –
CIP Cataloguing-in-Publication-Data
A catalogue record for this publication is available from Die Deutsche Bibliothek.

printed in the
Federal Republic of Germany
printed on acid-free paper

Composition K+V Fotosatz GmbH, Beerfelden, Germany
Printing Strauss Offsetdruck GmbH, Mörlenbach, Germany
Binding J. Schäffer GmbH & Co. KG, Grünstadt, Germany

ISBN 3-527-40315-9

Preface

This book deals with amorphous metals, which are systems that nature does not form very often, but which have interesting properties. Usually, metals are polycrystalline or even single-crystalline, i.e. if one starts at high temperatures in the liquid state with one metallic compound only and reduces the temperature slowly below the melting point, there will be *crystallization*. But it has been found experimentally that in cases where the melt is an *alloy* of either (i) two or more metallic compounds, e.g. Cu and Zr, or Mg and Ca, or (ii) of one or more metallic compounds, e.g. 80% Fe, plus one or more so-called *glass formers* such as B, P, Si, C or other so-called *metalloids*, then a sufficiently rapid quench from the temperature of the liquid state to much lower temperatures, which may be in the region of room temperature, typically leads to so-called *amorphous metals*, with a metastable structure of very long lifetime – months, years or decades. These amorphous metals are characterized at least approximately by the fixed 'glassy' structure of a metallic liquid, frozen from just above the melting temperature. Therefore, amorphous metals are also called 'glassy metals', 'metallic glasses' or 'metglasses', and should be distinguished from amorphous semiconductors, e.g. amorphous Si or amorphous Ge, not only because 'metglasses' behave as metals and are often magnetic or even supraconducting, but also because of their different structure: usually, metglasses have a very high coordination number of ≈ 12, as densely packed crystalline metals as Ni, Cu or Co would have, although in those crystalline systems the density of the systems is still higher (typically 14%, see below). In contrast, amorphous Si or Ge are only tetrahedrally coordinated (i.e. the coordination number is only 4), which is of course closely related to their different 'semiconducting' behavior.

The book is centered on the theoretical description, understanding and derivation of the properties of glassy metals. However, advantages and disadvantages of the systems with respect to applications will become clear. Concerning these points one should note that the systems are not simply typical *metals*, but due to the structural properties their metallic resistivity is roughly one or two orders of magnitude higher than that of the typical crystalline counterparts, namely of the same order of magnitude as in the liquid state just above the melting point. At the same time, amorphous metals with Fe, Co and Ni are usually *magnetic* and, in connection with the high resistivity, the material is interesting for voltage transformers, because of their lower electrical losses compared with crystalline Fe–Si transformer material. Thirdly, it is advantageous that the typical pinning centers for magnetic domain walls of crystalline metallic magnets, namely dislocations and grain boundaries, are missing (although a continuous distribution of sources of internal stresses also exists in the amorphous metals). This is the reason for the *extremely soft-magnetic* properties of some of the glassy magnetic materials prepared by industry, which are used for shielding external magnetic fields very efficiently.

However, the applications of the systems are not at the center of this book, which instead stresses the theoretical description of the relevant properties, as already mentioned. This description introduces the relevant theoretical techniques in a self-consistent way, and therein it goes beyond the introductory level. However, we try to keep the applications in mind, whenever possible, and mention the advantageous and disadvantageous properties. Furthermore, although our book emphasizes the *analytical* methods in the theory of amorphous systems and goes as far as to the Eliashberg and Nambu-Gorkov equations in the theory of amorphous superconductivity, we also discuss some of the more recent *numerical* work, i.e. computer simulation methods based on the so-called 'molecular dynamics' and similar techniques, and draw qualitative and semiquantitative conclusions which may be useful not only for the theorist.

Odessa/Opole/Regensburg, January 2001

N. P. Kovalenko
Yu. P. Krasny
U. Krey

Contents

1 On the structure of amorphous metals *1*

1.1 Introduction: preparation of amorphous metals, and simple models *1*

1.2 The Radial Distribution Function and the Structure Function *3*

1.3 Structural models of glassy metals *7*

2 The pseudopotential method *15*

2.1 An effective Hamiltonian for the electron-ion system of elemental metals *15*

2.2 Pseudopotential theory *18*

2.3 Model pseudopotentials *25*

2.4 Microscopic theory of the homogeneous electron gas *29*

2.5 The effective interaction between ions in liquid or amorphous metals *37*

2.6 The effective two-particle ion-ion interaction (to second order in the pseudopotential) *47*

2.7 Effective ion-ion pair interaction to cubic order in the pseudopotential *52*

2.8 Many-particle interactions in metallic hydrogen *58*

2.9 Computer calculations of the electronic structure in metallic amorphous alloys *69*

3 Atomic properties of amorphous metals: ow-energy excitations *77*

3.1 Experiments on the atomic dynamics in glasses *77*

3.2 A tunnelling model *80*

3.3 A quasi-phonon model for amorphous metals; heat capacity at moderately low temperatures *99*

3.4 The quasi-phonon contribution to the heat conductivity and sound absorption at moderately low temperatures *111*

3.5 Beyond the quasi-phonon approximation *129*

4 Magnetic properties of amorphous metals *135*

4.1 Review of experimental results *135*

4.2 Thermodynamic properties of amorphous ferromagnets near the Curie temperature *139*

4.3 The spectrum of quasi-magnon excitations in amorphous ferromagnets *150*
4.4 Low-temperature magnetic behavior of amorphous ferromagnets *156*
4.5 Beyond the quasi-magnon approach: computer simulations *167*
4.6 The thermodynamics of amorphous ferromagnets *169*
4.7 Itinerant magnetism and itinerant spin-glass behavior in amorphous alloys *171*

5 Superconductivity of glassy metals *179*
5.1 The Eliashberg equation for amorphous metals *179*
5.2 The electron-phonon coupling constant and the superconducting transition temperature for simple amorphous metals *198*
5.3 Superconducting properties of binary alloys of simple amorphous metals *205*

6 Conclusions *215*

7 Appendices *217*
Appendix A Calculation of the free energy of amorphous metals *217*
Appendix B Calculation of the free energy of amorphous ferromagnets *220*
Appendix C Derivation of the Eliashberg equation for amorphous metals *238*
Appendix D Simplification of the Eliashberg equation *256*

References *269*

Index *275*

1 On the Structure of Amorphous Metals

1.1 Introduction: Preparation of Amorphous Metals, and Simple Models

Most solids exist in the crystal state. However, at room temperature, there are also *amorphous solids* of essentially two kinds: (i) semiconducting systems, which usually exist as tetrahedrally coordinated networks, and (ii) amorphous *metallic* alloys, which appear in a state that is essentially similar to a frozen picture of a dense metallic liquid composed of at least two different alloy components. (Pure metals apparently do not form a stable or metastable amorphous state, except as films on very cold substrates, e.g. at He temperatures and apparently only with small amounts of impurities (Felsch, 1969, 1970a,b; Leung and Wright, 1974).)

Systems of type (i), e.g. amorphous Si or amorphous Ge, are not the focus of this book. Instead, we concentrate on systems of type (ii), which are often called 'glassy metals', metallic glasses ('metglasses'), or 'amorphous metals'. There are a vast number of such systems, which can be:

(a) composed of at least two different metals, e.g. amorphous $Cu_{1-x}Zr_x$ alloys, which can be prepared in a vast concentration range from $x \approx 0.35$ to $x \approx 0.75$ (or hydrogenated $Fe_{1-x}Zr_xH_y$, or $Mg_{1-x}Ca_x$ alloys, and many other species), or
(b) composed of at least one metallic and one non-metallic 'metalloid' compound, as e.g. amorphous $Fe_{1-x}B_x$, which exists in the composition range between $x=0.15$ and 0.33. These systems are stable in the amorphous state at room temperature, or at least metastable, with lifetimes of up to years or decades, and they can be prepared in different ways.

Some preparation methods are mentioned here, namely

- the so-called 'splat-cooling' technique, in which an 'anvil' with a droplet of the melt on it is hit with a kind of piston – the droplet is quenched immediately to a flat amorphous film;
- the so-called 'melt spinning' method, which is generally applicable to industrial and technical mass production, but is also commonly used in research institutes. The melt is continuously produced by induction heating and dropped in a continuous flow onto a rotating Cu wheel of typically 1 m diameter, such that a continuous ribbon of the amorphous metal is tangentially ejected from the

wheel with a high velocity of typically 2 m/s. The thickness of the ribbons produced in this way is typically 50 μm, and they are up to 2 cm wide;
- the so-called 'solid-state reaction' technique, in which after having produced powders of the alloy components by milling techniques, one can even prepare bulk amorphous metals (Samwer *et al.*, 1994). This technique has apparently not yet found widespread technical applications, whereas the *amorphous magnetic ribbons* produced by melt spinning are actually used for the mass production of active and passive magnetic sensor materials (Hilzinger, 1990).

Other techniques that also deserve mention are the methods of *sputtering* and the method of *laser glazing*.

In addition, large-scale applications of amorphous material, which would have led to a replacement of the conventional FeSi transformer magnets, looked promising in the 1980s, but to date the conventional systems have won the economic competition. In fact, in addition to the difficulty of preparing bulk amorphous metals directly, one of their main shortcomings is the '*aging*' phenomenon, i.e. the fact that the material properties of amorphous metals may deteriorate significantly after cycle times of typically several months or years. As a consequence, the ultra-soft magnetic properties of the shielding material can only be kept for long periods of time if the system is treated very cautiously, both thermally, magnetically and mechanically.

Now let us consider the stability of the amorphous state and questions of model formation. Two of the important physical questions related to these problems concern the 'glass forming ability', and the stability against crystallization. As already mentioned, and as is also known from the conventional non-metallic window glasses, the amorphous state is typically only metastable, and we observe the above-mentioned aging properties. These are of course preparation-dependent, and certain 'annealing procedures', which usually follow the production process, are essential for the quality and stability of industrial products (Hilzinger, 1990). The question therefore naturally arises as to when and why amorphous systems are formed at all.

There are various theoretical aspects to this question, which are intimately related to the problem of modeling the atomic structure of amorphous metals. For example:

(i) the role of deep eutectics

For liquid alloys at a given temperature there exists usually a so-called *miscibility gap*, i.e. a range Δx of concentrations, where the liquid alloy $A_{1-x}B_x$ coexists with separate A resp. B liquids. Considering liquid–vapor phase transitions as an analogon to the demixing transition, the analogon to the miscibility gap is the coexistence region between a single-component liquid and the corresponding vapor phase, and the density range of coexistence, $\Delta\rho$, plays a role corresponding to that of Δx. Outside the miscibility gap, a liquid alloy phase is not thermodynamically stable. With decreasing temperature of the melt, Δx becomes gradually smaller and smaller, until at a low, so-called 'eutectic temperature', which may approach the range of room temperature, a single so-called *eutectic composition* x_t remains. In this eutectic concentration region the glass-forming ability after a rapid quench is particularly high, which seems natural, but is not easy to formulate quantitatively.

(ii) hard-sphere models

There are also '*steric*' aspects (i.e. in contrast to energetic ones) that favor amorphization. One of the first models to describe amorphous alloys was the so-called *Bernal model*, i.e. a dense random packing of hard spheres (Bernal, 1959, 1960): If we try to compress a large set of hard spheres, i.e. taken from ball bearings, into a box, e.g. 80% of the atoms resembling Fe in its atomic radius, and 20% with a radius corresponding to B, then we usually end up with a metastable state. Finney and coworkers have produced such models by hand (Finney, 1970) and found that the packing fraction $f = N(4\pi r^3/3)/V$ of such models, if we are dealing with fictitious one-component cases, are somewhat lower (typically $\approx 14\%$) than those of crystalline face-centered or hexagonal crystals, namely 0.6366 ± 0.0004 instead of 0.74 (Finney, 1970). Here r is the hard-core radius of the atoms, N the number of balls, and V the volume of the box considered. Furthermore, it has been found that in such models there is a sufficient number of small holes, into which 20% small 'B spheres' would fit. More details will be discussed in the following section.

1.2 The Radial Distribution Function and the Structure Function

A more complete picture of the structure of amorphous systems can only be obtained by the set of averaged n-point correlation functions for $n=2, 3, 4, \ldots$, namely

$$F_n(\mathbf{R}_1, \mathbf{R}_2, \ldots, \mathbf{R}_n) = \overline{\delta(\mathbf{r}_1 - \mathbf{R}_1)\delta(\mathbf{r}_2 - \mathbf{R}_2)\ldots\delta(\mathbf{r}_n - \mathbf{R}_n)} \tag{1.2.1}$$

Here, the $\mathbf{r}_i$ denote the positions of the n atoms, $\delta(\mathbf{r})$ is Dirac's Delta function and the overbar denotes an average over an ensemble of amorphous systems with the same statistical properties as the considered one. We assume in the following that this ensemble can be obtained by 'self-averaging', i.e. in the so-called *thermodynamic limit* of infinite system size, $V \to \infty$, the system can be divided into infinite blocks, which represent the samples of the ensemble. Thus we also assume *statistical homogeneity* of the system, i.e. the averages and averaged correlation functions considered are assumed to be translationally invariant.

The most important of the averaged correlation functions in Eq. (1.2.1) is the so-called *Radial Distribution Function* $\rho(\mathbf{r})$, and the so-called *Partial Radial Distributions* $\rho_{\alpha\beta}(\mathbf{r})$ (see below), which can be derived from Eq. (1.2.1) for $r := |\mathbf{r}_1 - \mathbf{r}_2|$ with $n = 2$. These functions appear throughout the following chapters, when the thermodynamics of vibrational, magnetic, electronic and superconducting excitations in amorphous metals are considered, and this is why we emphasise them here.

In fact, $\rho(\mathbf{r})$ and $\rho_{\alpha\beta}(\mathbf{r})$ can most easily be determined for computer models of amorphous systems, just by counting the numbers of atoms of kind β that are at a distance between r and $r + dr$ from a fixed atom of kind α, and averaging over all atoms α, see below. Experimentally, their determination is somewhat more involved and involves diffraction experiments utilizing X-ray diffraction, electron diffraction or neutron diffraction, or combinations thereof. Namely, for the different diffraction sources, the *scattering amplitudes* f_α (see below) are different, such that from the differences of the scattering cross-sections $d\sigma/d\Omega$ obtained by three different experiments we can usually determine the partial radial distribution functions $\rho_{\alpha\beta}(\mathbf{r})$ for binary amorphous alloys, i.e. with $\alpha, \beta = 1, 2$. (With neutron scattering, we can also use amorphous alloys of the same composition, but with different *isotopes* of the compounds considered, because these isotopes sometimes also have very different neutron scattering lengths.)

The description of the diffraction experiments, and their relation to the radial distribution functions, is as follows.

Let a monochromous beam of X-ray photons, or electrons, or neutrons, impinge onto our amorphous sample of volume V and particle number N. The monochromous beam is assumed to have a current-density j_0 $(= dN_0/(dt\,\mathrm{cm}^2))$ and an incoming wavevector $\mathbf{k_i} = \mathbf{k}\mathbf{e_z}$, where $\lambda = 2\pi/k$ is the wavelength or de Broglie wavelength of the particles and $\mathbf{e_z}$ the direction of the flow (the z-direction without restriction). The particles are scattered elastically, and a counter, which is at a very large distance r from the sample, counts the number N_s of scattered particles with polar scattering angles between θ and $\theta + d\theta$, and azimuthal ones between ϕ and $\phi + d\phi$. The solid angle $d\Omega$ covered by the counter is thus given by $d\Omega = \sin\theta\, d\theta\, d\phi$. From the number of counts per second, dN_s/dt, by the counter, we can thus determine *experimentally* the so-called differential cross-section:

$$(d\sigma/d\Omega) = (dN_s/dt)/(j_0 d\Omega) \tag{1.2.2}$$

Theoretically, $d\sigma/d\Omega$ is given by the following expression:

$$(d\sigma/d\Omega)(\mathbf{q}) = \overline{|\sum_{i=1}^{N} f_i \exp[i\mathbf{q}(\mathbf{r} - \mathbf{r}_i)]|^2} \tag{1.2.3}$$

Here, as already mentioned, $\mathbf{r}$ is the position of the counter and $\mathbf{r}_i$ are the positions of the scattering atoms in the amorphous sample; $|\mathbf{r}|$ is always >> $|\mathbf{r}_i|$, i.e. the center of mass of the amorphous sample can be taken as the origin. The vector $\mathbf{q} = \mathbf{k}_f - \mathbf{k}_i$ is the difference between the wavenumbers of the scattered particle before and after the scattering. Finally, for the magnitude we have $|\mathbf{q}| = (4\pi/\lambda)\sin(\theta/2)$, and the f_i are the complex scattering amplitudes.

Now, by evaluation of Eq. (1.2.3), it follows that for an amorphous alloy with $\alpha = 1, 2, \ldots, n$ different alloy components:

$$(d\sigma/d\Omega) = N \sum_{\alpha,\beta=1}^{n} c_\alpha c_\beta f_\alpha^* f_\beta \int d^3r \exp(i\mathbf{q}\,\mathbf{r})\rho_{\alpha\beta}(r) \tag{1.2.4}$$

Here the product $c_\beta 4\pi r^2 dr\rho_{\alpha\beta}(r)$ is the expectation value of the number of β atoms that have a distance between r and $r + dr$ from a given α atom. Thus, for $r \to \infty$, $\rho_{\alpha\beta}(r)$ converges to $\rho_0 := (N/V) =: 1/\upsilon_0$, i.e. the reciprocal of the specific volume, independent from α and β, and for the Fourier transform of $\rho_{\alpha\beta}(r)$, we obtain

$$S_{\alpha\beta}(q) = \int d^3r \exp(\mathbf{iq} \cdot \mathbf{r})\rho_{\alpha\beta}(r) = [(2\pi)^3/\upsilon_0]\delta(\mathbf{q}) + a_{\alpha\beta}(q) \tag{1.2.5}$$

Here $a_{\alpha\beta}(q)$ is the so-called *Partial Structure Function*, which is often used in the following chapters. According to Eq. (1.2.4) we thus have

$$(d\sigma/d\Omega) = N\Sigma_{\alpha\beta} c_\alpha c_\beta f_\alpha^* f_\beta S_{\alpha\beta}(q) \tag{1.2.6}$$

In this way, from the experiments, $S_{\alpha\beta}(q)$ and thus $\rho_{\alpha\beta}(r)$ can be determined. By the spherical symmetry of both quantities, we can reduce the necessary integrations to one-dimensional ones from 0 to ∞, namely

$$a_{\alpha\beta}(q) = 4\pi \int_0^\infty r^2 dr[\sin(qr)/(qr)][\rho_{\alpha\beta}(r) - \rho_0] \tag{1.2.7}$$

$$\rho_{\alpha\beta}(r) = \rho_0 + (2\pi^2)^{-1} \int_0^\infty q^2 dq[\sin(qr)/(qr)]a_{\alpha\beta}(q) \tag{1.2.8}$$

As can be seen from Eq. (1.2.8), it is sometimes useful to plot the following 'reduced radial distribution' $G_{\alpha\beta}(r)$, which is used in some of the following figures:

$$G_{\alpha\beta}(r) := 4\pi r[\rho_{\alpha\beta}(r)\rho_0] \tag{1.2.9}$$

which for $q \neq 0$ is related to $a_{\alpha\beta}(q)$ by

$$a_{\alpha\beta}(q) = (1/q) \int_0^\infty dr G_{\alpha\beta}(r) \sin(qr) \tag{1.2.10}$$

and

$$G_{\alpha\beta}(r) = (2/\pi) \int_0^\infty dq[q a_{\alpha\beta}(q)] \sin(qr) \tag{1.2.11}$$

Figures 1.2.1, 1.2.2 and 1.2.3 present the partial distribution functions $\rho(r)$ and the structure functions $S(q)$, which von Heimendahl (1979) calculated from a computer simulation for the structure of a two-component metallic glass, namely for $Mg_{70}Zn_{30}$. The calculation produced a 'relaxed hard-core model' of 800 atoms; such models are often more simply called 'soft-core models'. In the calculation,

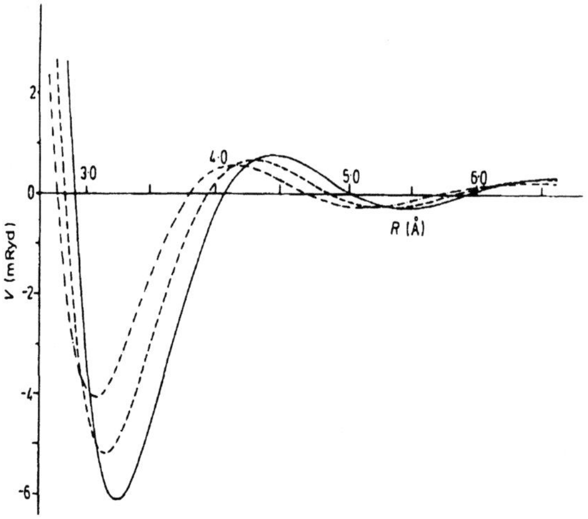

Fig. 1.2.1 The pair potentials used in the computer simulation of von Heimendahl (1979). This simulation produced a 'soft-core model' for $Mg_{70}Zn_{30}$ with 800 atoms in a box with periodic boundary conditions. In the calculation, three effective pair potentials were used, calculated by J. Hafner (Hafner, 1976, 1977a,b) from the general non-local 'pseudopotential theory' described in Chapter 2. In this figure, the 'deepest', 'second deepest' and 'most shallow' lines, respectively, represent the pair potentials for MgMg, Mg-Zn and Zn-Zn

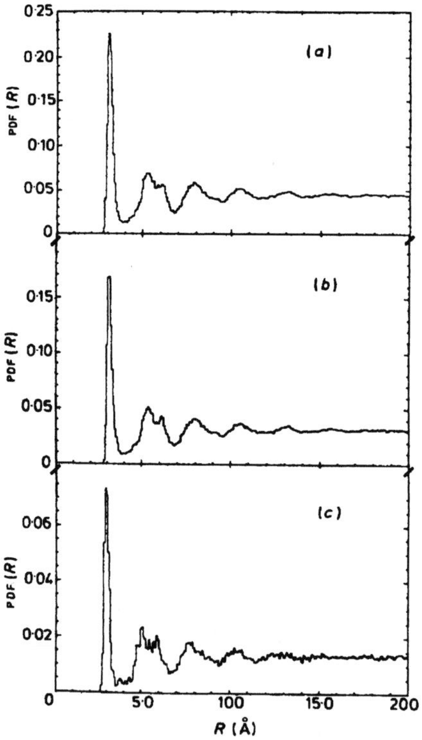

Fig. 1.2.2 The total 'radial distribution function' $\rho(r)$ for the amorphous model system $Mg_{70}Zn_{30}$ according to the computer simulation of von Heimendahl (1979). Parts b and c represent the partial radial distribution functions corresponding to the Mg-Mg and Zn-Zn correlations respectively

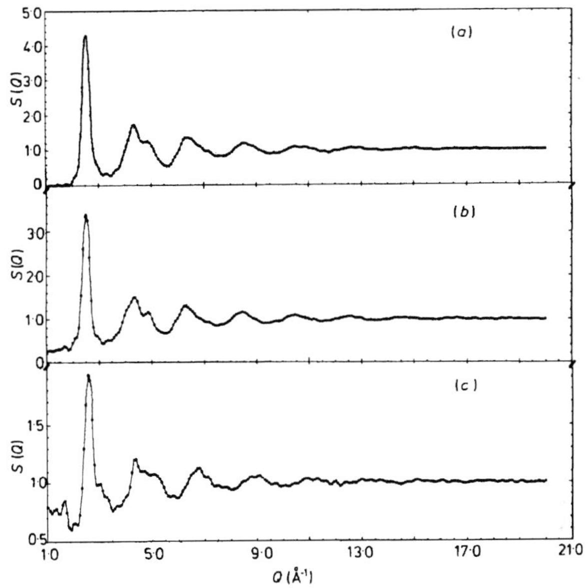

Fig. 1.2.3 The corresponding 'structure function' $S(q)$ to Fig. 1.20.2 for the amorphous model system $Mg_{70}Zn_{30}$ according to the computer simulation of von Heimendahl (1979). Parts b and c represent the partial radial distribution functions and partial structure functions corresponding to the Mg-Mg and Zn-Zn correlations respectively

three effective pair potentials were used, calculated by J. Hafner (Hafner, 1976, 1977a,b) from the general 'non-local pseudopotential theory' described in Chapter 2. Figure 1.2.1 shows these pair potentials: in this figure, the 'deepest', 'second deepest' and 'most shallow' lines, respectively, present the pair potentials for Mg-Mg, Mg-Zn and Zn-Zn. Figure 1.2.2 depicts the total radial distribution (part a) and the partial distribution functions $\rho_{\text{Mg-Mg}}(r)$, part b, and $\rho_{\text{Zn-Zn}}(r)$, part c. Finally Fig. 1.2.3 (a, b, c) shows the corresponding 'structure functions' $a_{\alpha\beta}(q)$ (= $S_{\alpha\beta}(q)$ for $q \neq 0$).

1.3 Structural Models of Glassy Metals

As already mentioned in Section 1.1, there are many models constructed so as to yield a sufficient description of metallic glasses. The earliest models were in fact constructed 'by hand', i.e. the Finney model (Finney, 1970), one of the most prominent DRPHS models (DRPHS = *D*ense *R*andom *P*ackings of *H*ard *S*pheres).

Figure 1.3.1 shows the total 'reduced radial distribution function' $G(r)$ for Finney's hard-sphere model, compared with experimental results.

Here the first pronounced peak corresponding to the nearest-neighbor shell and the hard-core condition is most clearly visible, as is the following pronounced

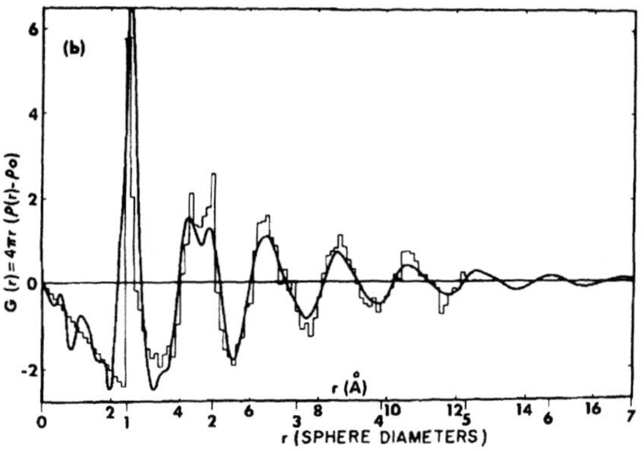

Fig. 1.3.1 Comparison of the reduced radial distribution function $G(r)$ obtained from Finney's original ball bearing model (the histogram) with experiments for amorphous $Ni_{76}P_{24}$ (Cargill, 1975, p. 304)

minimum, followed by a second, less pronounced 'split' maximum, corresponding to the shell of second- and third-nearest neighbors. However, the shortcomings of the DRPHS model can also be clearly seen by the comparison with experiment. Namely, in the DRPHS model, because of the hard-core condition, there are no Fe atoms with distances less than $2r_0$, where r_0 is the hard-core radius corresponding to Fe, namely $r_0 = 2.86$ Å, whereas in reality (i.e. for the solid line in Fig. 1.3.1, and also in Fig. 1.3.2) the first peak, corresponding to the shell of 'nearest neighbors', although being quite pronounced, is clearly somewhat 'smeared' over a small but finite range, and also the two smaller peaks in the following shell corresponding to second- and third-nearest neighbors are different in reality: in the DRPHS model, the second of these two peaks is clearly more pronounced, which is the opposite of reality (see Fig. 1.3.2).

Furthermore, in the liquid state of metallic glasses, as opposed to the amorphous one, there is no splitting at all (Fig. 1.3.3), but otherwise the radial distributions for liquid metals just above the melting point are quite similar to those of the amorphous state, which justifies the 'frozen liquid' picture mentioned above.

These observations give a clear hint that hard-core models are oversimplified and should be used with care. At least the positions of the atoms in such models should be energetically *'relaxed'* in realistic inter-ionic potentials, e.g. the self-consistent pseudopotentials discussed in the next chapter. The improved models based on such procedures could be termed DRPSS models (i.e. Dense Random Packings of 'Soft Spheres'), or simply soft-sphere models.

For liquids there is a well-known analytical treatment of the radial distribution function in hard-sphere approximation, the *Percus-Yevick* approximation (Percus and Yevick, 1958). We do not describe this here, because there are several comprehensive reviews on it available (e.g. Kovalenko and Fisher, 1973). Results from this approximation are shown in Fig. 1.3.4, which should be compared with those of Fig. 1.3.3.

Because the agreement is reasonable, even with the results for the amorphous metals in Fig. 1.3.2, and because there are further approximations made in the

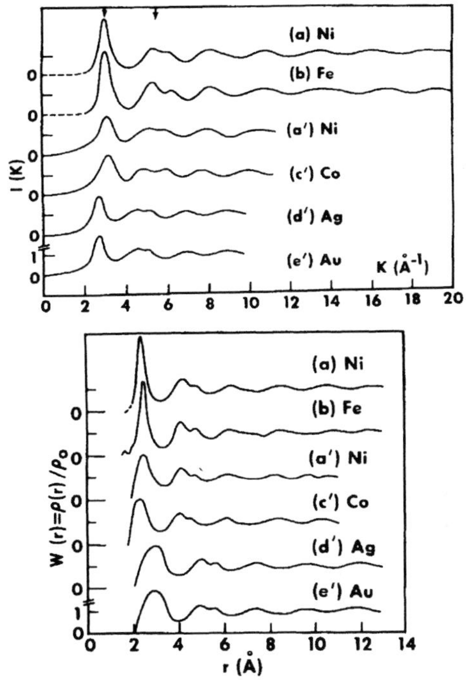

Fig. 1.3.2 Interference function $I(k)$, $\equiv S(q)$ in the text (upper part), and radial distribution function $W(r) = \rho(r)/\rho_0(r)$ (lower part) for amorphous films of different metals, after Cargill (1975), p. 262. The results for amorphous Ni and Fe (a,b) are taken from Ichikawa (1973), and for the amorphous films of Ni, Co, Ag and Au (a',c',d',e') from Davies and Grundy (1971, 1972)

analytical treatment of the excitations discussed in later chapters, the 'liquid-like' Percus-Yevick radial distribution function will be used frequently below.

As just mentioned, to obtain more realistic models of metallic glasses as a prerequisite for calculations of vibrational, spin, electronic or superconducting excita-

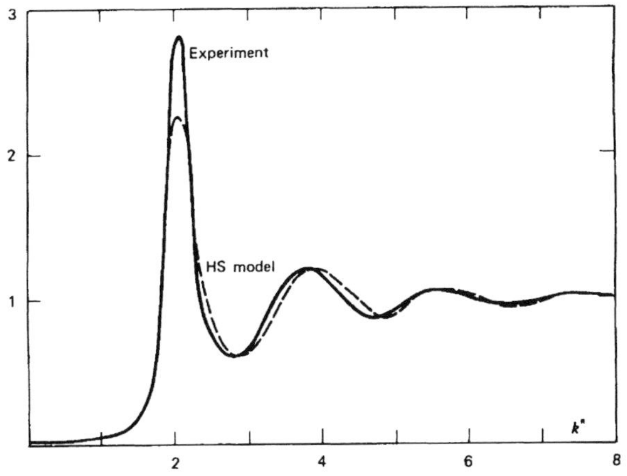

Fig. 1.3.3 The 'structure function' $S(q)$ of liquid Na is compared with results from a 'molecular dynamics' simulation (hard-sphere model) corresponding to high densities, i.e. with a filling factor of the hard spheres of $\eta = 45\%$ of the volume (after Balescu, 1975, p. 288)

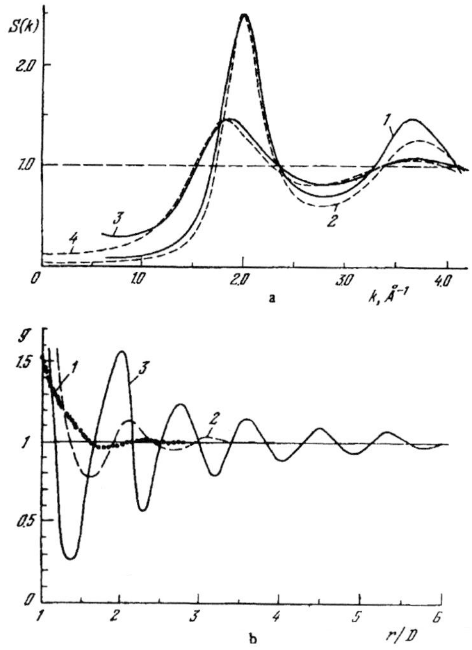

Fig. 1.3.4 Upper plot: Structure function for liquid Ar from experiment (dashed) and from the Percus–Yevick (PY) theory with a hard-sphere diameter $D = 3.44$ Å. Lower plot: Different radial distributions for hard-sphere liquids according to the PY theory at different densities (solid lines) $nD^3 = 0.3$ (1), 0.7 (2) and 1.1 (3) (after Kovalenko and Fisher, 1973)

tions, we should use realistic self-consistent inter-ionic potentials, which depend on the *electronic structure* of the system considered and thus in principle should be determined self-consistently (see Chapter 2), because the electronic state is influenced by the 'amorphous state' of the positions of the ions themselves. Hausleitner and Hafner (1993) have performed such self-consistent structure calculations, although using a simplified semi-empirical tight-binding scheme for the treatment of the d-electrons. In their calculations, the liquid state itself, from which the final amorphous structure was obtained by rapid quenching and subsequent relaxation, was also simulated right from the beginning by ambitious computer simulations of the so-called 'molecular dynamics' type (Hausleitner and Hafner, 1991). However, a somewhat less ambitious procedure seems reasonable. In fact, one of the present authors has used, with two coworkers (Krey *et al.*, 1987), a procedure, described below, to produce satisfactory models of amorphous Cu_{1-x} Zr_x alloys, as well as Fe_{1-x} B_x and $Fe_{1-x}B_x$ and $Fe_{1-x}Zr_xH_y$ (where H means hydrogen, and the concentration was either $y=0$ or $y=2x$, with a wide concentration region of x, over which these systems are stable at room temperature). These systems have interesting electronic and magnetic properties, and the model generation described in the following was used as a pre-requisite to study those properties. The models were constructed as follows.

Into a cubic box containing 10^3 bcc unit cells with two sites per cell, 2000 Cu-rsp. Zr-atoms were randomly placed on the 2000 sites of the bcc lattice. Then the system was 'thermalized' to high temperatures by random displacements of the atoms by typically 30% of the lattice constant, and afterwards the system was quenched to $T = 0$ K and subsequently energetically relaxed, where standard Lennard–Jones potentials have been used, i.e. of the type

$$U_{ij}(r) = 4(\varepsilon_i\varepsilon_j)^{1/2}\{[(\sigma_i + \sigma_j)/(2r)]^{12} - [(\sigma_i + \sigma_j)/(2r)]^6\} \qquad (1.3.1)$$

Note that these potentials are 'additive' in σ_i and σ_j, which is no longer true for the self-consistent transition metal-metalloid potentials in the calculations of Hausleitner and Hafner (1993).

In Eq. (1.3.1), for the alloys considered, the standard Lennard-Jones 'soft-core radii' σ_i and the energy parameters ε_i had simply to be taken from the book by Kittel (1971). The energy relaxation was started with a somewhat enlarged edge length of the box. Periodic boundary conditions were used, and in the course of the relaxation, the cubic box was gradually compressed, until finally after a large number of 'Monte Carlo sweeps' through the whole sample, the system was stuck in a metastable state, i.e. the energy could not be reduced further by small deviations of the atoms. The resulting structures compared favorably with experiments concerning the radial distribution functions (see Krey *et al.*, 1987), and so the structures calculated in this way were taken as the basis for calculations of the electronic and magnetic states of the system (see, for example, Krompiewski *et al.*, 1989 a, b, 1992; Krey *et al.* 1990, 1992). Concerning the magnetic states, we would like to stress that an *itinerant approach* was used (i.e. in the crystal we would talk about 'band magnetism'), which appears more realistic, but is also more complicated than the conventional Heisenberg model (see Chapters 4 and 5). In particular, we will find that not only the *directions* but also the *magnitudes* of the atomic magnetic moments can differ considerably from site to site in the amorphous systems considered, and that not only ferromagnetic states arise, but also states that are called *speromagnetic* (or spin-glass states) according to the terminology of J. M. Coey (Moorjani and Coey, 1984), i.e. in these states the spins are far from being aligned.

P. Kizler, a coworker of S. Steeb in Stuttgart, has performed a detailed experimental and computational comparison of different structural models for typical metallic glasses obtained by various methods, such as the one sketched in Fig. 1.3.5 (Kizler *et al.*, 1988). The essential conclusion of his systematic survey was that, using the available experimental methods, the models can hardly be distinguished in quality, although the agreement is far from overwhelming.

Kizler also looked for *three-body correlations* (Kizler, 1993), i.e. properties that go beyond the description of the radial distribution function, which is only a two-body correlation. By means of the EXAFS method (EXAFS = Extended X-ray Adsorption Fine Structure), using experimental spectroscopy compared with extensive simulations, Kizler was also able to make some statements on the angle between a B atom and two nearest-neighbor Fe atoms, or vice versa, in amorphous Fe–B alloys (Kizler, 1991, 1993). Again, however, the computer models turned out not to be significantly different to these correlations, although now one of the computer models – that of Dubois *et al.* (1985) – turned out to be slightly preferable.

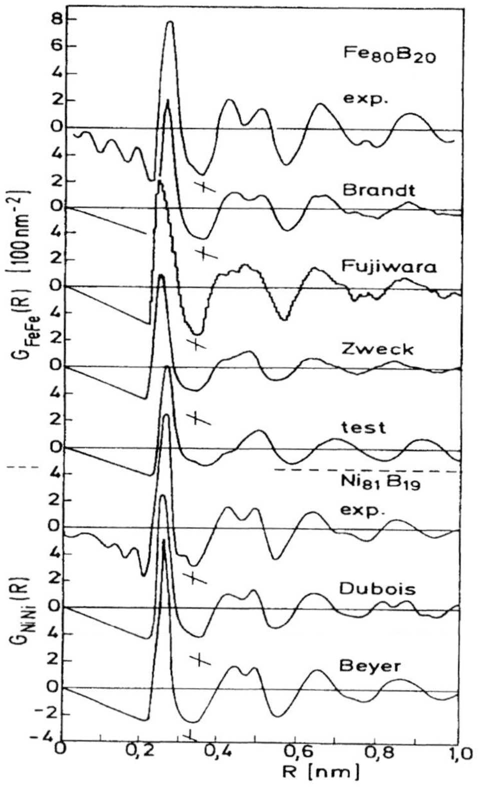

Fig. 1.3.5 Comparing radial distribution functions from different computer simulations with experiments, (a) for $Fe_{80}B_{20}$, and (b) for $Ni_{81}B_{19}$. Only the 'test' simulation, which is hardly relaxed, is very insufficient (Kizler, 1988, Fig. 1)

Of course, with a great deal of effort in the potential parameters and procedures it is possible to get a practically perfect fit, as in Fig. 1.3.6, which is taken from Fig. 4-33 in the review of Lamparter and Steeb (1993).

In conclusion, we must keep in mind that, within experimental accuracy, the various computer models, using reasonable sizes and construction principles, can hardly be distinguished experimentally and thus can also be used as a reasonable basis for calculations of excitations in amorphous systems.

This statement, used with some caution, should be kept in mind in the following chapters, where in Chapter 2 we discuss the general theory of electrons plus ions in a liquid or solid, leading to the construction and properties of the relevant *pseudopotentials*. This is followed by a treatment of *many-particle interactions* in metallic hydrogen. Computer calculations of the *electronic structure* of amorphous metals are treated at the end of Chapter 2.

In Chapter 3, harmonic *phonons* and extremely anharmonic so-called *tunneling states* in amorphous systems are treated, and the *ultrasound adsorption* is calculated. At the end of Chapter 3 we discuss the numerical simulation of harmonic vibrational excitations in amorphous systems, which is based on implementation of Green's function techniques and the method of 'continued fractions', where the

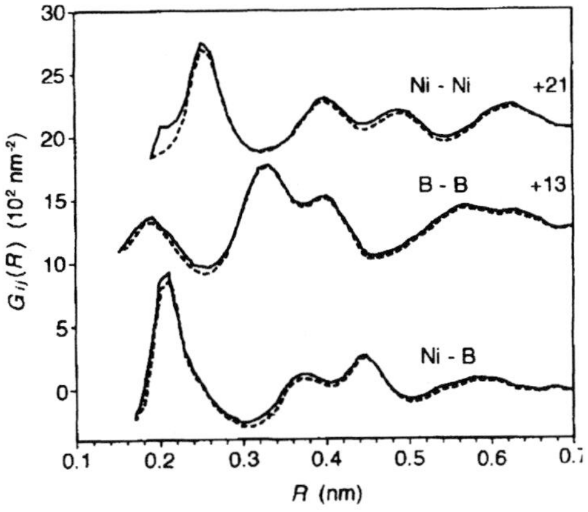

Fig. 1.3.6 Partial radial distribution functions $\rho_{\alpha\beta}(r)$ for $\alpha, \beta = \text{Ni, B}$ according to model calculations of Pusztai (1991) (solid lines) and experiments from Ishmaev *et al.* 1987. (Lamparter and Steeb 1993, Fig. 4-33)

'line broadening' of the vibrational excitations observed in 'inelastic neutron scattering' experiments plays a major role.

Chapter 4 treats *amorphous ferromagnetism,* mainly (i) in the already mentioned *Heisenberg model,* both in the mean-field approach (which works near T_C) and in the magnon approximation at low temperatures. At the end of Chapter 4, we also include a treatment of the *itinerant magnetism* of amorphous metals, i.e. here the origin and the strongly inhomogeneous properties of the magnetism in amorphous metals find its natural explanation. This includes a discussion of *itinerant spin glass behavior* in amorphous Fe, and in iron-rich amorphous alloys.

Chapter 5 treats *superconductivity* in amorphous metals, i.e. the so-called *Eliashberg theory* of strongly coupled superconductors is extended to amorphous systems, and it is found that typically the amorphous systems have a higher transition temperature than the corresponding crystalline compounds.

Finally, in Chapter 6, some remarks are made on *aging* in glassy metals, which is an important aspect for many applications.

It should be mentioned here that in large parts of this book we stress the powerful 'diagrammatic and analytical theories' promoted by the famous Russian school of theoretical physics (Abrikosov *et al.*, 1965), which however are usually only approximative for the amorphous systems and must be combined with thermodynamic perturbation theory, leading to typical quasi-crystalline approximations (e.g. quasi-phonons, quasi-magnons, etc., see below). But in some chapters, e.g. Chapter 4, we also deal with the results of more recent numerical methods, which are largely complementary to the analytic approach.

2
The Pseudopotential Method

2.1
An Effective Hamiltonian for the Electron-Ion System of Elemental Metals

Simple crystalline or amorphous metals are good conductors of electric current, because the external electron shells of their atoms overlap. The electrons from these shells, called valence or conduction electrons, are *collectivized*; they easily move from one atom to another and thus form a quasi-free electron subsystem. Internal electron shells belonging to the ion core, however, hardly overlap at all and thus retain the properties of isolated ions. Being comparatively small in size (they occupy about 10% of the volume), the ions interact with one another by means of (screened) Coulomb forces. The mass of ions is much greater than the mass of valence electrons, therefore we may approximately assume that light electrons move in the field of motionless ions, while heavy ions shift very slowly in the electron liquid made up of 'spread' conduction electrons ('spread' due to their swift motion). This is called the *adiabatic approximation* or Born-Oppenheimer approximation.

The interaction of valence electrons with ions is purely Coulombian in the main part of the metal. Within an ion, however, the true wave function of the electron is strongly affected by oscillations because it has to be orthogonal to the wave functions of the electrons in the core of the ion. This leads to a sharp reduction in the effective interaction of the electrons with the ions and, therefore, to a much smaller amplitude of electron scattering from the ions. Such a reduction in the effective interaction is equivalent to the emergence of an effective repulsion potential within an ion that largely compensates the self-consistent potential of the ion core. Thus it is possible to substitute the true potential by some effective *pseudopotential* that is non-local in the general case and gives a 'close-to-real' value of the scattering amplitude, and also describes the behavior of the wave function of the electron in the bulk of the volume closely to a plane-wave behavior. The effective reduction in the interaction within an ion introduces a small parameter into the theory of simple metals: this parameter is the ratio of the pseudopotential's Fourier component to the Fermi energy.

Because ions and conduction electrons are thus only loosely connected, their properties can be determined almost independently of each other. For example, the degeneration temperatures of ion and electron gases are very precisely defined by the relations:

$$T_{\text{i}} = \frac{\hbar^2}{2k_{\text{B}}M}\left(\frac{3\pi^2 N}{V}\right)^{2/3}, \quad T_{\text{e}} = \frac{\hbar^2}{2k_{\text{B}}m}\left(\frac{3\pi^2 ZN}{V}\right)^{2/3} \tag{2.1.1}$$

where N is the number of ions in the metal, V is the volume of the metal, Z is the valency of atoms in the metal, M and m are the masses of the ion and the electron, respectively, $\hbar$ is Planck's constant (divided, as usual, by 2π) and k_{B} is Boltzmann's constant. As is easy to estimate, these temperatures have the following order of magnitude for simple metals: $T_{\text{i}} \sim 1$–1.5 K; $T_{\text{e}} \sim 10^4$–10^5 K. In the following, we will be interested in temperatures satisfying the inequality $T_{\text{i}} \ll T \ll T_{\text{e}}$. This temperature interval is fairly wide and includes the whole domain of the liquid state of the metal. The motion of the ions at such temperatures is governed by the laws of classical mechanics, so the *ion subsystem* may be considered as a classical liquid. The motion of the *electrons*, however, has to be described by means of quantum mechanics, i.e. the electron gas should be considered as a strongly degenerate Fermi gas. This understanding of the structure of simple metals allows us to choose the Hamiltonian of the system in the following form:

$$\hat{H} = \sum_{\nu} \frac{\mathbf{P}_{\nu}^2}{2M} + \frac{1}{2}\sum_{\substack{\nu,\nu' \\ (\nu\neq\nu')}} \frac{Z^2e^2}{|\mathbf{R}_{\nu} - \mathbf{R}_{\nu'}|} + \sum_{j} \frac{\hat{\mathbf{P}}_j^2}{2m} + \sum_{\substack{j,j' \\ (j\neq j')}} \frac{e^2}{|\mathbf{r}_j - \mathbf{r}_{j'}|} + \sum_{j,\nu} w(\mathbf{r}_j - \mathbf{R}_{\nu}) \,. \tag{2.1.2}$$

Here $w(\mathbf{r})$ is the local *pseudopotential* of the electron-ion interaction, $\mathbf{P}_{\nu}$ is the momentum of the νth ion, $\hat{\mathbf{P}}_j = -i\hbar\nabla_j$ is the momentum operator of the jth electron, $\mathbf{r}_j$ and $\mathbf{R}_{\nu}$ are the position vectors of electron and ion, respectively, and e is the elementary charge. If the electron-ion interaction is neglected in zeroth approximation, the electronic and ionic subsystems should be placed in *compensating 'background' charges* evenly spread over the whole volume, otherwise the subsystems will not be stable. This is why, before using perturbation theory, we rewrite the Hamiltonian $\hat{H}$ as follows:

$$\hat{H} = \hat{H}_i + \hat{H}_e \tag{2.1.3}$$

Here

$$\hat{H}_{\text{i}} = \sum_{\nu} \frac{\mathbf{P}_{\nu^2}}{2M} + \frac{1}{2}\sum_{\substack{\nu,\nu' \\ (\nu\neq\nu')}} \frac{Z^2e^2}{|\mathbf{R}_{\nu} - \mathbf{R}'_{\nu}|} + \sum_{\nu} \frac{ZN}{V}\int_V w(\mathbf{R}_{\nu} - \mathbf{r})\mathrm{d}^3r + \frac{1}{2}\frac{N^2Z^2}{V^2}\int \frac{e^2}{|\mathbf{r} - \mathbf{r}'|}\mathrm{d}^3r\mathrm{d}^3r' \tag{2.1.4}$$

is the Hamiltonian of the ions placed in the negative background charge evenly spread over the whole volume. The third summand in Eq. (2.1.4) describes the interaction of the ions with the negative background, while the last summand is the self-energy of the background charges. The expression

$$\hat{H}_{ie} = \sum_j \frac{\mathbf{P}_j^2}{2m} + \frac{1}{2}\sum_{\substack{j,j' \\ (j \neq j')}} \frac{e^2}{|\mathbf{r}_j - \mathbf{r}_{j'}|} - \sum_j \frac{N}{V}\int \frac{Ze^2}{|\mathbf{r}_j - \mathbf{r}|}\mathrm{d}^3r$$
$$+ \frac{1}{2}\frac{N^2Z^2}{V^2}\int \frac{e^2}{|\mathbf{r} - \mathbf{r}'|}\mathrm{d}^3r\mathrm{d}^3r' + \sum_{j,\nu}\left\{w(\mathbf{r}_j - \mathbf{R}_\nu) - \frac{1}{V}w(\mathbf{r})\mathrm{d}^3r\right\} \tag{2.1.5}$$

is the Hamiltonian of the interacting electron gas placed in the positive background; the last summand in this Hamiltonian describes the effective electron-ion interaction. In writing Eqs (2.1.1) to (2.1.5) we have taken into account that, in the limit $V \to \infty$,

$$\int_V w(\mathbf{R} - \mathbf{r})\mathrm{d}^3r = \int_V w(\mathbf{r})\mathrm{d}^3r$$
$$\int_V \frac{Z^2e^2}{|\mathbf{r}_j - \mathbf{r}|}\mathrm{d}^3r = \frac{1}{V}\int \frac{Z^2e^2}{|\mathbf{r} - \mathbf{r}'|}\mathrm{d}^3r\mathrm{d}^3r'$$

In the following it is more convenient to work with the Hamiltonian in the second-quantized form. For this purpose, let us introduce field operators for the electron gas:

$$\hat{\psi}_\sigma(\mathbf{r}) = \frac{1}{\sqrt{V}}\sum_{\mathbf{k}} a_{\mathbf{k}\sigma}\exp(i\mathbf{k}\cdot\mathbf{r}) \tag{2.1.6}$$

where $a_{\mathbf{k}\sigma}$ is the annihilation operator of electrons in a state with momentum $\hbar\mathbf{k}$ and spin σ ($a^+_{\mathbf{k}'\sigma'}$ is the creation operator) which fulfill the usual *anticommutation rules*:

$$[a_{\mathbf{k}\sigma}, a^+_{\mathbf{k}'\sigma'}]_+ \equiv a_{\mathbf{k}\sigma}a^+_{\mathbf{k}'\sigma'} + a^+_{\mathbf{k}'\sigma'}a_{\mathbf{k}\sigma} = \delta_{\mathbf{k}\mathbf{k}'}\delta_{\sigma\sigma'}$$
$$[a_{\mathbf{k}\sigma}, a_{\mathbf{k}'\sigma'}]_+ = [a^+_{\mathbf{k}\sigma}, a^+_{\mathbf{k}'\sigma'}]_+ = 0$$

Next, we can proceed to the momentum representation for the interaction potentials:

$$\frac{e^2}{|\mathbf{r}_j - \mathbf{r}_{j'}|} = \frac{1}{V}\sum_{\mathbf{q}(\neq 0)} \frac{4\pi e^2}{q^2}\exp[i\mathbf{q}\cdot(\mathbf{r}_j - \mathbf{r}_{j'})]$$
$$w(\mathbf{r}_j - \mathbf{R}_\nu) = \frac{1}{V}\sum_{\mathbf{q}(\neq 0)} w(\mathbf{q})\exp[i\mathbf{q}\cdot(\mathbf{r}_j - \mathbf{r}_{j'})] \tag{2.1.7}$$

(Here the exclusion of $\mathbf{q} = 0$ from the sum again reflects the electrical neutrality of the system as a whole.)

Then, with, $\rho(\mathbf{q}) = \frac{1}{N}\sum_{\nu} \exp(-i\mathbf{q}\cdot\mathbf{R}_\nu)$, $\upsilon_0 = V/N$, $\hat{H}_j = \hat{H}_j' + \hat{H}_j''$, Eqs (2.1.4) and (2.1.5) will look as follows:

$$\hat{H}_i = \sum_{\nu}\left(\frac{\mathbf{P}_{\nu^2}}{2M} + \frac{Z}{\upsilon_0}b\right) + \frac{1}{2}\sum_{q(\neq 0)} \frac{N}{\upsilon_0}\frac{4\pi^2 Z^2 e^2}{q^2}\left(|\rho(q)|^2 - \frac{1}{N}\right), \tag{2.1.8}$$

$$\hat{H}_{ie} = \hat{H}_0 + \hat{H}_j' + \hat{H}_j'' \equiv \hat{H}_0 + \hat{H}_j \tag{2.1.9}$$

in which $b = \lim_{\mathbf{q}\to 0}\{w(\mathbf{q}) + 4\pi Ze^2/q\}^2$ is the pseudopotential's non-Coulombian part, and the different contributions to the Hamiltonian are explained below.

If the electron-ion interaction were purely Coulombian, i.e. if $w(\mathbf{q}) = -4\pi Ze^2/q^2$ then b would automatically vanish, which follows from the electrical neutrality of the system as a whole. However, the additional interaction of conduction electrons with the electron shells of the ions makes this limit different from zero.

$$H_0 = \sum_{\sigma}\int \hat{\psi}_\sigma^+(\mathbf{r})\left\{-\frac{\hbar^2}{2m}\Delta\right\}\hat{\psi}_\sigma(\mathbf{r})\mathrm{d}^3r = \sum_{\mathbf{k}\sigma}\frac{\hbar^2 k^2}{2m}a_{\mathbf{k}\sigma}^+ a_{\mathbf{k}\sigma} \tag{2.1.10}$$

$$\begin{aligned}\hat{H}_j' &= \frac{1}{2}\sum_{\sigma,\sigma'}\int_V \mathrm{d}^3r \int_V \mathrm{d}^3r'\,\hat{\psi}_\sigma^+(\mathbf{r})\hat{\psi}_{\sigma'}^+(\mathbf{r}')\left\{\frac{e^2}{|\mathbf{r}-\mathbf{r}'|} - \frac{1}{V}\int\frac{e^2}{r}\mathrm{d}^3r\right\}\\ &\times \hat{\psi}_{\sigma'}(\mathbf{r}')\hat{\psi}_\sigma(\mathbf{r}) = \frac{1}{2}\sum_{\substack{\mathbf{k},\mathbf{k}',\mathbf{q},\\ \sigma,\sigma'(\mathbf{q}\neq 0)}}\frac{4\pi e^2}{q^2}a_{\mathbf{k}+\mathbf{q},\sigma}^+ a_{\mathbf{k}'-\mathbf{q},\sigma'}^+ a_{\mathbf{k}'\sigma}a_{\mathbf{k}\sigma}\end{aligned} \tag{2.1.11}$$

$$\begin{aligned}\hat{H}_j'' &= \sum_{\sigma}\int \hat{\psi}_\sigma^+(\mathbf{r})\int\left\{\sum_{\nu}\left[w(\mathbf{r}-\mathbf{R}_\nu) - \frac{1}{V}\int w(\mathbf{r})\mathrm{d}^3r\right]\right\}\hat{\psi}_\sigma(\mathbf{r})\mathrm{d}^3r\\ &= \sum_{\substack{\mathbf{k},\mathbf{q},\sigma\\(\mathbf{q}\neq 0)}}\frac{1}{V_0}w(\mathbf{q})\rho(\mathbf{q})a_{\mathbf{k}+\mathbf{q},\sigma}^+ a_{\mathbf{k}\sigma}\end{aligned} \tag{2.1.12}$$

2.2 Pseudopotential Theory

Section 2.1 described a model of simple liquid and amorphous metals on the basis of physical considerations, in which the interaction of the conduction electrons with the ions is described by local pseudopotentials. This model, naturally, needs substantiation: For simplicity, we may confine this problem to the motion of non-interacting electrons in the field of ions with (essentially) fixed, but disordered, positions. The behavior of such electrons is described by the Hamiltonian

$$\hat{H} = \hat{T} + \hat{V} \tag{2.2.1}$$

where $\hat{T} = -\dfrac{\hbar^2}{2m}\Delta$ is the operator of the kinetic energy of an electron, and $\hat{V}(\mathbf{r})$ is the self-consistent potential affecting the electron, as derived from the presence of the other electrons and all the ions.

The eigenfunctions of the electron satisfy the Schrödinger equation:

$$\hat{H}\psi_i = E_i\psi_i \tag{2.2.2}$$

We will make a distinction between the wave functions of localized 'core' states (or wave functions of interior shells) and the wave functions of delocalized states $\psi_{\mathbf{k}}$ (or valence and conduction wave functions). The 'core' wave functions of interior shells for simple metals are strongly localized near the ions and hardly overlap at all. Therefore, we may consider with sufficient accuracy that they coincide with the electronic wave functions of an isolated ion, while the corresponding energies have different values. These 'core' wave functions satisfy the equation

$$H\psi_a = E_a\psi_a \tag{2.2.3}$$

where a is an index defining both the ion's position and the quantum numbers of the analyzed state.

In contrast, the *delocalized* electronic states are nearly free. This implies that their wave functions $\psi_{\mathbf{k}}$ in the space between the ions are close to plane waves. Plane waves are, nevertheless, a bad basis for expanding. This is why Herring (1940) proposed expanding $\psi_{\mathbf{k}}$ by plane waves made *orthogonal* to ψ_a (see also Harrison (1966) and Abrikosov (1987)). This orthogonalization leads to a distortion of the plane waves that makes them similar to the desired functions. We may hope that an expansion by *orthogonalized plane waves* (OPW) is more effective than by ordinary plane waves.

The orthogonalized plane wave with the wavevector $\mathbf{k}$ may be represented as

$$\mathrm{OPW}_{\mathbf{k}} = V^{-1/2}\left\{\exp(i\mathbf{k}\cdot\mathbf{r}) - \sum_a \psi_a(\mathbf{r})\int \psi_a^*(\mathbf{r}')\exp(i\mathbf{k}\cdot\mathbf{r}')\mathrm{d}^3r'\right\} \tag{2.2.4}$$

By construction, this function is orthogonal to any wave function of interior shells. Indeed,

$$\int \psi_\beta^*(\mathbf{r})\mathrm{OPW}_{\mathbf{k}}\mathrm{d}^3r = V^{-1/2}\int \psi_\beta^*(\mathbf{r})\exp(i\mathbf{k}\cdot\mathbf{r})\mathrm{d}^3r$$
$$-\sum_a \delta_{a\beta}V^{-1/2}\int \psi_a^*(\mathbf{r}')\exp(i\mathbf{k}\cdot\mathbf{r}')\mathrm{d}^3r' = 0\,.$$

For further convenience, we will use the following bra- and ket-forms for the wave functions:

$$\begin{cases} \langle \mathbf{k}|^* = |\mathbf{k}\rangle = V^{-1/2}\exp(i\mathbf{k}\cdot\mathbf{r}) \\ \langle a|^* = |a\rangle = \psi_a(\mathbf{r}) \end{cases} \tag{2.2.5}$$

Then, orthogonalized plane waves may be represented as

$$\mathrm{OPW}_{\mathbf{k}} = |\mathbf{k}\rangle - \sum_a |a\rangle\langle a|\mathbf{k}\rangle \tag{2.2.6}$$

where $\langle a|\mathbf{k}\rangle = V^{-1/2}\int \psi_a^*(\mathbf{r})\exp(i\mathbf{k}\cdot\mathbf{r})\mathrm{d}^3r$.

To shorten the forms still further, let us introduce the operator

$$\hat{L} = 1 - \sum_a |a\rangle\langle a| \tag{2.2.7}$$

Then

$$\mathrm{OPW}_{\mathbf{k}} = \hat{L}|\mathbf{k}\rangle \tag{2.2.8}$$

Let us expand the wave function $\psi_{\mathbf{k}}$ of the conduction zone into a series of orthogonalized plane waves:

$$\psi_{\mathbf{k}}(\mathbf{r}) \equiv |\psi_{\mathbf{k}}\rangle = \sum_{\mathbf{q}} a_{\mathbf{q}}(\mathbf{k})\mathrm{OPW}_{\mathbf{q}} = \sum_{\mathbf{q}} a_{\mathbf{q}}(\mathbf{k})\hat{L}|\mathbf{q}\rangle \tag{2.2.9}$$

We can substitute Eq. (2.2.9) into Eq. (2.2.2) and rearrange its summands to give

$$(\hat{T} + \hat{W})|\varphi_{\mathbf{k}}\rangle = E_{\mathbf{k}}|\varphi_{\mathbf{k}}\rangle \tag{2.2.10}$$

where $\hat{W}$ is the so-called pseudopotential defined by the equality

$$\hat{W} = \hat{V} - (E_{\mathbf{k}} - \hat{H})(\hat{L} - 1) = \hat{V} + \sum_a (E_{\mathbf{k}} - E_a)|a\rangle\langle a| \tag{2.2.11}$$

and

$$\varphi_{\mathbf{k}}(\mathbf{r}) = |\varphi_{\mathbf{k}}\rangle = \sum_{\mathbf{q}} a_{\mathbf{q}}(\mathbf{k})|\mathbf{q}\rangle \tag{2.2.12}$$

is the pseudowave function. The true wave function $\psi_{\mathbf{k}}$, according to Eq. (2.2.9), is related to it as

$$|\psi_{\mathbf{k}}\rangle = \hat{L}|\varphi_{\mathbf{k}}\rangle \tag{2.2.13}$$

Let us now find the properties of the pseudopotential $\hat{W}$ obtained with the help of orthogonalized plane waves.

Because $E_{\mathbf{k}} > E_a$ for r close to an ion, the second summand in Eq. (2.2.11) is positive; it corresponds to a repulsive force from the ion and compensates the strong attraction defined by the potential $\hat{V}(\mathbf{r})$. For simple metals, therefore, the operator $\hat{W}$ may be considered as a small quantity (it is small *between* ions on account of a rapid decrease in the electron-ion interaction and, *around* an ion, due to the compensation of summands in Eq. (2.2.11)). This is one of the chief advantages of the pseudopotential $\hat{W}$ compared to the true potential V.

The pseudopotential's smallness, however, has to be paid for with other, less pleasant properties. First, as follows directly from Eq. (2.2.11), the operator $\hat{W}$ is *not local,* even if the true potential V is local. Besides, the pseudopotential evidently includes the energy values of the conduction electrons, cf. Eq. (2.2.11). This is why it depends on the solutions of the eigenvalue problem to be solved, i.e. it is not linear and has to be considered self-consistently.

Another peculiarity of the pseudopotential $\hat{W}$ is its lack of uniqueness. If to a certain pseudowave function any linear combination of 'core' wave functions is added, the modified function thus obtained satisfies Eq. (2.2.10) with the same eigenvalue, as we can easily see from a simple substitution. Moreover, when substituted into Eq. (2.2.13), this modified sum leads to the same *true* wave function as the initial pseudowave function. Thus, the solution of Eq. (2.2.10) is not unique. This non-uniqueness is ultimately determined by the fact that orthogonalized plane waves are not orthogonal to each other. The non-uniqueness would be of no importance if we were to solve the pseudopotential equation accurately, because for any of its solutions the correct eigenvalue would be obtained, and the wave function determined by Eq. (2.2.13) would be the same for any obtained pseudowave function.

In fact, however, we do not solve the pseudopotential equation accurately, but only treat it to the second order of perturbation theory. The eigenvalues obtained in this way depend on the method of solution. This is why it is important to study the nature of this non-uniqueness in more detail. A particularly fruitful study was carried out by Austin *et al.* (1962), but the reader is also referred to Güntherodt and Beck (1981), Beck and Güntherodt (1983), and March and Tosi (1976). They pointed out that if a more general form of pseudopotential, namely

$$\hat{W}|\varphi_{\mathbf{k}}\rangle' = \hat{V}(\mathbf{r})\varphi_{\mathbf{k}}(\mathbf{r}) + \sum_{a} \psi_{a}(\mathbf{r}) \int F_{a}(\mathbf{r}')\varphi_{\mathbf{k}}(\mathbf{r}')\mathrm{d}^3 r' \tag{2.2.14}$$

(in which $F_a(\mathbf{r}')$ is an arbitrary function of the coordinates and of the index a), is substituted into Eq. (2.2.10), the energy eigenvalues from this equation will coincide with the energy eigenvalues $E_{\mathbf{k}}$ in Eq. (2.2.10), where the pseudopotential has the form of Eq. (2.2.11).

Therefore, a criterion for an unambiguous choice of the pseudopotential is needed. The most natural requirement is that the function $\varphi_{\mathbf{k}}$ should have *maximum* smoothness, i.e. $\int (\nabla\varphi_{\mathbf{k}})^2 \mathrm{d}^3 r = \min.$[1] For simple metals, this requirement makes the second summand in Eq. (2.2.14) almost totally neutralize the first one

1) A recent calculation for the magnetic metals Fe, Co and Ni with so-called '*ultrasoft pseudopotentials*' has been performed by Moroni *et al.* (1997).

around the ion core. This is quite obvious. The condition of orthogonality to the 'core' wave functions of the ions' interior shells is fulfilled, if the functions $\varphi_{\mathbf{k}}$ oscillate very quickly in the vicinity of the core of the ions. This brings plenty of kinetic energy which, in accordance with quantum mechanics, equals $\frac{\hbar^2}{2m}\int(\nabla\psi_{\mathbf{k}})^2 d^3r$. The transition to smooth functions $\varphi_{\mathbf{k}}$ means that the greater part of this energy is included in the potential energy, and this is why strong neutralization takes place.

Yet another significant peculiarity of the pseudopotential $\hat{W}$ is that it can be divided into a sum of pseudopotentials of separate ions (pseudopotential factorization). This follows directly from Eq. (2.2.14). The potential V is created by all ions together with their interior shells and can be presented as the sum of spherically symmetrical potentials of separate ions

$$\hat{V}(\mathbf{r}) = \sum_j v_{\text{ion}}(\mathbf{r} - \mathbf{R}_j) \tag{2.2.15}$$

where j denumerates the ions. Then, the wave functions of interior shells, included in the second summand of Eq. (2.2.14), can be represented as

$$|a\rangle \equiv \psi_a(\mathbf{r}) = \psi_t(\mathbf{r} - \mathbf{R}_j) \tag{2.2.16}$$

where the index a has been divided into two indices, j and t, so that j again defines the number of the ion, and t denotes the set of quantum numbers that characterize the state of the core electrons. Let us stipulate that the function $F_a(\mathbf{r})$ should only depend on a through t. Then the pseudopotential depends on the position $\mathbf{R}_j$ of the ion only through the variable $(\mathbf{r} - \mathbf{R}_j)$; and because the variable $\mathbf{r}$ can only be present in the pseudopotential in the combinations $(\mathbf{r} - \mathbf{R}_j)$, the pseudopotential can be represented as the sum of potentials of separate ions, namely

$$\hat{W}(\mathbf{r}) = \sum_j \hat{w}(\mathbf{r} - \mathbf{R}_j) \tag{2.2.17}$$

in which

$$\hat{w}(\mathbf{r} - \mathbf{R}_j)\varphi_{\mathbf{k}}(\mathbf{r}) = \hat{v}(\mathbf{r} - \mathbf{R}_j)\varphi_{\mathbf{k}}(\mathbf{r}) + \sum_t \psi_t(\mathbf{r} - \mathbf{R}_j)\int F_t(\mathbf{r}')\varphi_{\mathbf{k}}(\mathbf{r}')d^3r' \tag{2.2.18}$$

It is particularly important to break the pseudopotential into a sum of pseudopotentials related to separate ions in such a way when a many-electron problem is considered.

As we said earlier, a disadvantage of the pseudopotential $\hat{W}$ obtained with the help of the OPW approach is its ambiguity and dependence on the energy of the conduction electron. This disadvantage is caused by the fact that the set of OPW states is overcomplete, and they are not orthogonal among themselves.

This disadvantage can be removed by using a basis of completely orthogonalized plane waves (COPW). Such a method was proposed by Girardeau (1971) and further developed by Bobrov and Trigger (1985 a, b), and Gurski and Gurski (1976 a, b, 1977). This is how a (COPW) basis is built.

In the plane waves basis $|\mathbf{k}\rangle$ we exclude a certain number of states $|\mathbf{k}_a\rangle$ equal to the number of localized one-electron states $|a\rangle$ of the Hamiltonian $\hat{H} = \hat{T} + \hat{V}$. The remaining plane waves are orthogonalized to the localized states and to one another. A completely orthogonalized plane wave with the wavevector $\mathbf{k} \neq \mathbf{k}_a$ is represented as

$$\tilde{\psi}_{\mathbf{k}}(\mathbf{k}) \equiv |\tilde{\mathbf{k}}\rangle = |\mathbf{k}\rangle - \sum_a C_{\mathbf{k}a}\{|a\rangle - |\mathbf{k}_a\rangle\} \tag{2.2.19}$$

The condition of orthogonality to the localized states,

$$\langle a|\tilde{\mathbf{k}}\rangle = 0 \tag{2.2.20}$$

directly leads to the system of equations defining the coefficients $C_{\mathbf{k}a}$:

$$\sum_\beta \{\delta_{a\beta} - \langle a|\mathbf{k}_\beta\rangle\} C_{\mathbf{k}\beta} = \langle a|\mathbf{k}\rangle \tag{2.2.21}$$

It should be pointed out that the coefficients satisfying Eq. (2.2.21) also meet the requirement $\langle\tilde{\mathbf{k}}_1|\tilde{\mathbf{k}}_2\rangle = \delta_{\mathbf{k}_1\mathbf{k}_2}$.

We can also show that $|\tilde{\mathbf{k}}\rangle$, together with the wave functions of localized states $|a\rangle$, form a complete orthonormalized basis.

Let us introduce the operator $\hat{L}_c$ that satisfies the equation

$$\hat{L}_c|\mathbf{k}\rangle = |\tilde{\mathbf{k}}\rangle\ , \ \hat{L}_c|\mathbf{k}_a\rangle = |\xi_a\rangle \tag{2.2.22}$$

where $\xi_a(\mathbf{r}) \equiv |\xi_a\rangle$ is an arbitrary set of functions.

It follows from Eq. (2.2.22) that $\hat{L}_c$ has the form

$$\hat{L}_c = \sum_{\mathbf{k}(\mathbf{k}\neq\mathbf{k}_a)} |\tilde{\mathbf{k}}\rangle\langle\mathbf{k}| + \sum_a |\xi_a\rangle\langle\mathbf{k}_a| \tag{2.2.23}$$

It is a linear, bounded and continuous operator; it has the conjugated operator $\hat{L}_c^+$ and inverse operator $\hat{L}_c^{-1}$ (Girardeau, 1971; Bobrov and Trigger 1985 a; Gurski and Gurski, 1976 a, 1977). If we expand a wave function of delocalized states $|\psi_{\mathbf{k}}\rangle$ into a series in the above-cited orthonormalized basis, and take into account its orthogonality to the wave functions of delocalized states, it is easy to show that

$$|\psi_{\mathbf{k}}\rangle = \sum_{\mathbf{q}(\mathbf{q}\neq\mathbf{k}_a)} a_{\mathbf{q}}(\mathbf{k})|\tilde{\mathbf{q}}\rangle = \sum_{\mathbf{q}(\mathbf{q}\neq\mathbf{k}_a)} a_{\mathbf{q}}(\mathbf{k})\hat{L}_c|\mathbf{q}\rangle \tag{2.2.24}$$

By substituting Eq. (2.2.24) into Eq. (2.2.2) we obtain

$$(\hat{T} + \hat{W}_c)|\varphi_{\mathbf{k}}\rangle = E_{\mathbf{k}}|\varphi_{\mathbf{k}}\rangle \tag{2.2.25}$$

in which

$$|\varphi_{\mathbf{k}}\rangle = \sum_{\mathbf{q}(\mathbf{q}\neq\mathbf{k}_a)} a_{\mathbf{q}}(\mathbf{k})|\mathbf{q}\rangle \tag{2.2.26}$$

is the *pseudowave* function related to the true wave function $|\psi_{\mathbf{k}}\rangle$ as

$$|\psi_{\mathbf{k}}\rangle = \hat{L}_c|\varphi_{\mathbf{k}}\rangle \tag{2.2.27}$$

while

$$\hat{W}_c = \hat{L}_c^{-1}(\hat{T}+\hat{V})\hat{L}_c - \hat{T} \tag{2.2.28}$$

is the pseudopotential obtained on the basis of completely orthogonalized plane waves.

The ambiguity of the pseudopotential $\hat{W}_c$ caused by the presence of arbitrary functions $|\xi_a\rangle$ in the operator $\hat{L}_c$ is not manifest in matrix elements $\langle\mathbf{k}|\hat{W}_c|\mathbf{p}\rangle$ given in the space of plane waves with $\mathbf{k} \neq \mathbf{k}_a$. Therefore, in accordance with Eq. (2.2.26), no term in the series of perturbation theory will depend on the choice of $\hat{W}_c$ (unlike the pseudopotential $\hat{W}$ on the OPW basis). Let us prove this. According to Eq. (2.2.23),

$$\hat{L}_c^{+} = \sum_{\mathbf{k}(\mathbf{k}\neq\mathbf{k}_n)} |\mathbf{k}\rangle\langle\tilde{\mathbf{k}}| + \sum_a |\mathbf{k}_a\rangle\langle\xi_a| \tag{2.2.29}$$

Then

$$\hat{L}_c\hat{L}_c^{+} = \sum_{\mathbf{k}(\mathbf{k}\neq\mathbf{k}_n)} |\tilde{\mathbf{k}}\rangle\langle\tilde{\mathbf{k}}| + \sum_a |\xi_a\rangle\langle\xi_a| = \hat{I} + \sum_a\{|\xi_a\rangle\langle\xi_a| - |a\rangle\langle a|\} \tag{2.2.30}$$

in which $\hat{I} = \sum_{\mathbf{k}(\mathbf{k}\neq\mathbf{k}_a)} |\tilde{\mathbf{k}}\rangle\langle\tilde{\mathbf{k}}| + \sum_a |a\rangle\langle a|$ is the identity operator. Remembering $\hat{I} = \hat{L}_c\hat{L}_c^{-1}$, we may rewrite Eq. (2.2.30) as

$$\hat{L}_c\hat{L}_c^{+} = \hat{L}_c\hat{L}_c^{-1} + \sum_a\{|\xi_a\rangle\langle\xi_a| - |a\rangle\langle a|\}$$

or

$$\hat{L}_c^{-1} = \hat{L}_c^{+} - \sum_a\{|\mathbf{k}_a\rangle\langle\xi_a| - \hat{L}_c^{-1}|a\rangle\langle a|\} \tag{2.2.31}$$

Thus,

$$\hat{L}_c^{-1}(\hat{T}+\hat{V})\hat{L}_c = \hat{L}_c^{+}(\hat{T}+\hat{V})\hat{L}_c - \sum_a |\mathbf{k}_a\rangle\langle\xi_a|\{\hat{T}+\hat{V}\}\hat{L}_c$$
$$+\sum_{a,\beta} E_a\hat{L}_c^{-1}|a\rangle\langle a|\xi_\beta\rangle\langle\mathbf{k}_\beta| \tag{2.2.32}$$

We have taken into account here that $\{\hat{T}+\hat{V}\}|a\rangle = E_a|a\rangle$. By substituting Eq. (2.2.32) into Eq. (2.2.28) we find that, if $\mathbf{k}$, $\mathbf{p} \neq \mathbf{k}_a$, the desired matrix element has the form

$$\langle \mathbf{k}|\hat{W}_c|\mathbf{p}\rangle = \langle \mathbf{k}|\hat{L}^{+}(\hat{T}+\hat{V})L|\mathbf{p}\rangle - \frac{\hbar^2 k^2}{2m}\delta_{\mathbf{kp}} = \langle \tilde{\mathbf{k}}|\hat{T}+\hat{V}|\tilde{\mathbf{p}}\rangle - \frac{\hbar^2 k^2}{2m}\delta_{\mathbf{kp}} \qquad (2.2.33)$$

and does not depend on the choice of the functions $|\xi_a\rangle$.

Similarly, it can be proved that the matrix element in question does not depend on the energy of the conduction electron either (Girardeau, 1971; Bobrov and Trigger, 1985 a; Gurski and Gurski, 1976 a, 1977).

The explicit form of the pseudopotential $\hat{W}_c$ is determined by the operators $\hat{L}_c$ and $\hat{L}_c^{-1}$. Generally, the operator $\hat{L}_c$ is not unitary, and the problem of finding $\hat{L}_c^{-1}$ is very difficult. However, it is much simpler if the functions $|\xi_a\rangle$ and $|\mathbf{k}\rangle$ form a complete orthonormalized set. For instance, if we assume that $|\xi_a\rangle = |a\rangle$, then, according to Bobrov and Trigger (1985 a)

$$\begin{aligned} \hat{L}_{\mathrm{BT}} &= \sum_{\mathbf{k}(\mathbf{k}\neq\mathbf{k}_a)} |\tilde{\mathbf{k}}\rangle\langle\mathbf{k}| + \sum_a |a\rangle\langle\mathbf{k}_a| \\ \hat{L}_{\mathrm{BT}}^{-1} = \hat{L}_{\mathrm{BT}}^{+} &= \sum_{\mathbf{k}(\mathbf{k}\neq\mathbf{k}_a)} |\mathbf{k}\rangle\langle\tilde{\mathbf{k}}| + \sum_a |\mathbf{k}_a\rangle\langle a| \end{aligned} \qquad (2.2.34)$$

The operators $\hat{L}_c^{-1}$ and $\hat{W}_c$ for crystalline metals were first obtained in an explicit form by Gurski and Gurski (1976b, 1977) at $|\xi_a\rangle = |\mathbf{k}_a\rangle$. The same papers presented an approximate expression of $\hat{W}_c$ for non-ordered systems.

In addition to problems of computation, there are also some fundamental problems. Firstly, the COPW-based pseudopotentials $\hat{W}_c$, just like the OPW-based pseudopotentials $\hat{W}$, are not local. Secondly, the choice of vectors $\mathbf{k}_a$ in the definition of the completely orthogonalized plane waves is in no way regulated by the general scheme. No serious analysis of this problem is offered by the available literature. It is often asserted, starting from physical considerations, that $|\mathbf{k}_a| \gg k_F$ (Gurski and Gurski, 1976). Finally, as follows from Eq. (2.2.28), the pseudopotential $\hat{W}_c$ is not factorized in the general form.

In conclusion, we point out that the main advantage of $\hat{W}_c$ is the fact that it is based on a complete orthogonalized set of basis functions. This advantage becomes indisputable when transition metals are studied, or when the effective Hamiltonian (2.1.2) is substantiated, etc.

2.3 Model Pseudopotentials

The previous section describes the main methods of building pseudopotentials from 'first principles'. Such methods, however, are very difficult, labor-intensive, and require plenty of simplifying approximations when they are applied to actual cases, which diminishes the accuracy of the final results. It is no wonder that such pseudopotentials often do not provide an acceptable degree of consistency between the theoretical and experimental values of a metal's physical properties.

These factors have made many researchers share the viewpoint that it is more reasonable to set a certain expression for a local model pseudopotential, and then to define its parameters requiring that experimental values are consistent with theoretical ones. This method of building model pseudopotentials was mainly developed by Heine *et al.* (1970), and Abarenkov and Heine (1964), among others.[2)]

This method presumes that the pseudopotential $\hat{W}$ can be 'decomposed', i.e. $\hat{W}(\mathbf{r}) = \sum_j w(\mathbf{r} - \mathbf{R}_j)$, where $w(\mathbf{r})$ is the pseudopotential of a separate ion. Then, the task of defining $\hat{W}$ is reduced to finding an electron's pseudopotential in a free ion's field. Abarenkov and Heine (1964) proposed defining a model pseudopotential for an ion in the following way. A sphere of radius r_0 is drawn around the ion. Outside the sphere the potential affecting an electron is just $-Ze^2/r$. The form of electronic wave functions beyond the sphere depends on the azimuthal and magnetic quantum numbers, the energy E (or the main quantum number), and on the values of the logarithmic derivative of the wave function at the distance r_0. If these four parameters are given, the wave function beyond the sphere is completely determined (up to a normalization factor). Let us now consider a model potential, which is equal to $-Ze^2/r$ beyond the sphere and to a constant $A_l(E)$ within the sphere; for the given eigenvalue, the constant is chosen such that the corresponding solution within the sphere, $j_l(\hat{e}r)$, with, $\hat{e} = \{2m(E + A_l(E))/\hbar^2\}^{1/2}$, would have the same value of the logarithmic derivative just inside the surface of the sphere as the above-mentioned solution outside the sphere. The model potential determined in such a way for a free ion will correspond beyond the sphere to a wave function of the same form as the wave function obtained with the true potential, and with the same eigenvalue E. We should point out that the value of $A_l(E)$ differs for various eigenfunctions.

The model potential that goes with $A_l(E)$ can also have its energy eigenvalues different from E; however, this does not interest us. We are going to set E, fix $A_l(E)$, and then look for a solution that corresponds to the eigenvalue E. At this stage, $A_l(E)$ is determined at a given l for only such values of E corresponding to the energy eigenvalues of the ion at the same value of l.

Thus, we have built a model pseudopotential (i.e. depending on angular momentum and energy) for a free ion that corresponds to the same energy eigenvalues as the true potential, and with eigenfunctions which coincide with the true wave functions beyond the domain of the interior atomic shells. Both eigenfunctions can be normalized in different ways, but this has no effect on their energy. The general model pseudopotential of Abarenko and Heine (1964) thus has the form

$$w_{\mathrm{HA}}(\mathbf{r}) = \begin{cases} -\sum\limits_l A_l(E)\hat{P}_l & \text{when} \quad r < r_0 \\ -Ze^2/r & \text{when} \quad r > r_0 \end{cases} \tag{2.3.1}$$

2) Applications of this method have been described e.g. by Cohen and Heine (1970).

Here $\hat{P}_l$ is the projection operator that extracts components with the azimuthal quantum number l out of the wave function it is affecting, and E is the eigenvalue of the ion's energy corresponding to this quantum number.

Assuming $A_l \equiv A_2$ for $l \geq 2$, Animalu and Heine (1966) proposed a simpler model pseudopotential:

$$w_{\mathrm{HAA}}(\mathbf{r}) = \begin{cases} -A_2 - (A_0 - A_2)\hat{P}_2 - (A_1 - A_2)\hat{P}_1 & \text{when} \quad r < r_0 \\ -Ze^2/r & \text{when} \quad r > r_0 \end{cases} \tag{2.3.2}$$

and calculated it for many metals.

Among other simple but widely used types of pseudopotentials we should mention the *pointwise-ion pseudopotential* proposed by Harrison (1966), and the *empty-core pseudopotential* proposed by Ashcroft (1968).

(i) The pointwise-ion pseudopotential of Harrison (1966)

$$w_{\mathrm{H}}(\mathbf{r}) = -Ze^2 l|r| + \beta\delta(\mathbf{r}) \tag{2.3.3}$$

consists of the potential of Coulomb attraction and that of electron repulsion by interior shells of the ion which is described by Dirac's δ-function. The only parameter β to be determined in this pseudopotential can be found from the condition of obtaining the correct values of the 'form-factors' at $q = 2k_F$ (see below).

(ii) Ashcroft's pseudopotential is one of the most simplified versions of the Heine-Abarenkov pseudopotential, namely

$$w_{\mathrm{A}}(\mathbf{r}) = \begin{cases} 0 & \text{when} \quad r < r_0 \\ -Ze^2/r & \text{when} \quad r > r_0 \end{cases} \tag{2.3.4}$$

The adjustment parameter r_0 of this pseudopotential is found experimentally from data on the resistance or on the Fermi surface. Ashcroft's pseudopotential is mainly used in the study of the kinetic properties of metals.

Let us touch upon another widely used pseudopotential type, the Cohen-Abarenkov-Heine pseudopotential (see Cohen, 1962):

$$w_{\mathrm{CAH}}(\mathbf{r}) = \begin{cases} Ze^2 u\, l\, r_0 & \text{when} \quad r < r_0 \\ -Ze^2\, l\, r & \text{when} \quad r > r_0 \end{cases} \tag{2.3.5}$$

This pseudopotential turns into Ashcroft's pseudopotential at $u = 0$. If we assume $u = 1$, it turns into Shaw's pseudopotential (Heine *et al.*, 1970; Cohen and Heine, 1970), which is also frequently applied in calculating the thermodynamic properties of a metal. The Fourier transform (or the 'form-factor') of the Cohen-Abarenkov-Heine pseudopotential has the form

$$w_{\mathrm{CAH}}(q) = \int w_{\mathrm{CAH}}(\mathbf{r}) \exp(-i\mathbf{q} \cdot \mathbf{r}) \mathrm{d}^3 r$$
$$= -\frac{4\pi Z e^2}{q^2} \left[(1+u)\cos(qr_0) - u \frac{\sin(qr_0)}{qr_0} \right] \tag{2.3.6}$$

The adjustment parameters u and r_0, as a rule, are found from the comparison of theoretical and experimental data on the metal's static properties. For instance, Brovman *et al.* (1970, 1974) do it with the help of the condition of 'zero pressure' ($p = 0$) at $T = 0$, for the lattice at equilibrium and also the equality of theoretical and experimental values of the shear modulus C_{44}. Such adjustment of the parameters u and r_0 in the pseudopotential (2.3.5) has proved to be quite efficient: one set of u and r_0 is used to give an accurate description of the binding energy, the equation of state at zero isotherm, all the modules of elasticity, the phonon spectrum, etc.

All the above-mentioned pseudopotential types, however, share the same drawback. Their form-factors at $q \to \infty$ do not provide sufficiently quick convergence of series and integrals in perturbation theory. For this reason, Animalu and Heine (1966) have modified the Cohen-Abarenkov-Heine pseudopotential by introducing an exponential 'cut-off function' at $q \to \infty$ (Vaks *et al.*, 1977):

$$w_{\mathrm{AH}}(q) = -\frac{4\pi Z e^2}{q^2} \left[(1+u)\cos(qr_0) - u \frac{\sin(qr_0)}{qr_0} \right] \exp\left(-0.03 \frac{q^4}{16 k_{\mathrm{F}}^4} \right) \tag{2.3.7}$$

Such a pseudopotential was successfully used in calculations of thermodynamic properties of simple metals by Vaks *et al.* (1977). A similar attempt was made by Gurski and Krasko (1971): proceeding from considerations about a 'smooth' distribution of the electronic density in the atom, they proposed the following model pseudopotential:

$$w_{\mathrm{GK}}(\mathbf{r}) = Z e^2 \left\{ \frac{\exp(-r/r_0) - 1}{r} + \frac{a}{r_0} \exp(r/r_0) \right\} \tag{2.3.8}$$

The form-factor of this 'screened pseudopotential' tends to zero at $q \to \infty$ as $1/q^4$. The parameters a and r_0 in Eq. (2.3.8) can be found with the help of experimental values of the true first ionization potential of the ion and the values of the form-factor of the 'screened pseudopotential' in the first sites of the reciprocal lattice. If such measurements are not available, the second condition is replaced by the constraint that $p = 0$ at $T = 0$ in the equilibrium lattice.

We may point out in conclusion that the best consistence of theory and experiment in the present context is achieved by the Animalu-Heine and the Cohen-Abarenkov-Heine pseudopotentials, Eqs (2.3.7) and (2.3.6) respectively. Here we concentrate on 'simple metals' that are sp-bonded. Of course, for very ambitious computer calculations, e.g. density-functional and/or molecular dynamics calculations, for systems including 3d-metals or 4f-metals, there are more recent developments, e.g. based on 'ultrasoft pseudopotentials' first proposed by Vanderbilt

(1990) and applied to magnetic Fe, Co and Ni in the previously cited paper by Moroni *et al.* (1997). Here we only refer briefly to a recent review (Payne *et al.*, 1992), and we also mention the recent technical advancements contained in, for example, the papers of Louie *et al.* (1982) and Troullier and Martin (1991).

2.4 Microscopic Theory of the Homogeneous Electron Gas

If the electron-ion interaction is considered to be small, then the electrons in a metal, in zeroth approximation, form a gas placed in a positive compensating charge evenly spread over the whole volume (see Section 2.1). Let us consider the influence of the electron-electron interaction on the properties of such a gas. (Because this 'electron gas' is rather dense, i.e. the average distance between the electrons is smaller than the average distance between the atoms in the metal, we should refer to the system as an 'electron liquid', although the electron-electron interaction in this system, as assumed, is negligible to zeroth approximation.)

The Hamiltonian of the system is obtained from Eq. (2.1.9), if the electron-ion interaction is neglected:

$$\hat{H}_l = \hat{H}_0 + \hat{H}_j' = \sum_{\mathbf{k}\sigma} \frac{\hbar^2 k^2}{2m} a_{\mathbf{k}\sigma}^+ a_{\mathbf{k}\sigma} + \frac{1}{2} \sum_{\substack{\mathbf{k},\mathbf{k}',\mathbf{q} \\ \sigma,\sigma' (q \neq 0)}} \frac{4\pi e^2}{Vq^2} a_{\mathbf{k}+\mathbf{q},\sigma}^+ a_{\mathbf{k}'-\mathbf{q},\sigma'}^+ a_{\mathbf{k}',\sigma'} a_{\mathbf{k}\sigma} \qquad (2.4.1)$$

Let us define the energy of the electron liquid in its ground state. We will use perturbation theory for this purpose, assuming that the electron-electron interaction is small compared to their kinetic energy. It is easy to show the consistency of this approach for systems with sufficiently high density of electrons: indeed, the mean potential energies and kinetic energies of the electrons are proportional to e^2/a_0 and $\hbar^2/ma_0^2$, respectively. Their ratio will be small at high electron densities, i.e. when $4\pi a_0^3/3$ represents the specific volume corresponding to one electron, then the above-mentioned ratio of the potential energy to the kinetic energy is small in the high-density limit:

$$e^2 ma_0^2/(a_0\hbar^2) = a_0/a =: r_s \ll 1$$
$$\left(a = \hbar^2/(me)^2 = 0.529\ \text{Å is the Bohr radius}, \quad \frac{1}{Z}v_0 = \frac{4\pi}{3}a_0^3 \right)$$

If the potential energy is therefore neglected, the wave function of the ground state $|0\rangle$ coincides with the corresponding wave function for the non-interacting electron gas. In configurational space, this function $|0\rangle$ is a *Slater determinant* composed of plane waves, i.e. it corresponds to an antisymmetrized product of plane-wave states $\psi_{\mathbf{k}\sigma}(\mathbf{r}) = \chi_\sigma \exp(i\mathbf{kr})/V^{1/2}$, with different pairs $(\mathbf{k}\sigma)$, where $\hbar\mathbf{k}$ is the momentum and $\mathbf{k}$ the corresponding wavenumber, which in magnitude must be smaller than the upper limit, the 'Fermi wavenumber' k_F; V is the volume of the sample, and $\sigma = \pm\hbar/2$ is the quantum number corresponding to the z-component of the spin, i.e. the 'spinors' χ_σ correspond formally to column vectors $\binom{1}{0}$ and $\binom{0}{1}$, respectively. In the second quantization representation, we have the usual requirements

$$a_{\mathbf{k}\sigma}|0\rangle = 0, \ |\mathbf{k}| > k_F$$
$$a^+_{\mathbf{k}\sigma}|0\rangle = 0, \ |\mathbf{k}| < k_F$$

For the ground state, the mean value of the *occupation number operator* $n_{\mathbf{k}\sigma}$ is determined by the Fermi distribution:

$$n_{\mathbf{k}\sigma} = \langle 0|a^+_{\mathbf{k}\sigma}a_{\mathbf{k}\sigma}|0\rangle = \begin{cases} 0, & |\mathbf{k}| > k_F \\ 1, & |\mathbf{k}| < k_F \end{cases} \tag{2.4.2}$$

If we take into account the potential energy of interaction between the electrons, the energy of the ground state per electron, according to perturbation theory, can be expressed as

$$\varepsilon = \frac{1}{ZN}\langle 0|\hat{H}_0|0\rangle + \frac{1}{ZN}\langle 0|\hat{H}'_j|0\rangle - \frac{1}{ZN} \times \sum_n \frac{\langle 0|\hat{H}'_j|n\rangle\langle n|\hat{H}'_j|0\rangle}{E_n - E_0} + \frac{1}{ZN}\left\{\sum_{n,m} \frac{\langle 0|\hat{H}'_j|n\rangle\langle n|\hat{H}'_j|m\rangle\langle m|\hat{H}'_j|0\rangle}{(E_n - E_0)(E_m - E_0)} + \langle 0|\hat{H}'_j|0\rangle \sum_n \frac{\langle 0|\hat{H}'_j|n\rangle\langle n|\hat{H}'_j|0\rangle}{(E_n - E_0)^2} - \right\} + \cdots \tag{2.4.3}$$

Confining this expression to the first two summands, we obtain the *Hartree-Fock approximation*:

$$\varepsilon_{HF} = \frac{1}{ZN}\langle 0|\hat{H}_0|0\rangle + \frac{1}{ZN}\langle 0|\hat{H}'_j|0\rangle \tag{2.4.4}$$

The difference between the precise value of ε and the energy value obtained in the Hartree-Fock approximation is called the *correlation energy*:

$$\varepsilon_{cor} = \varepsilon - \varepsilon_{HF} = -\frac{1}{ZN}\sum_n{}' \frac{\langle 0|\hat{H}'_j|n\rangle\langle n|\hat{H}'_j|0\rangle}{E_n - E_0} + \frac{1}{ZN}\left\{\sum_{n,m}{}' \frac{\langle 0|\hat{H}'_j|n\rangle\langle n|\hat{H}'_j|m\rangle\langle m|\hat{H}'_j|0\rangle}{(E_n - E_0)(E_M - E_0)} + \langle 0|\hat{H}'_j|0\rangle \sum_n \frac{|\langle 0|\hat{H}'_j|n\rangle|^2}{(E_n - E_0)^2} - \right\} + \cdots \tag{2.4.5}$$

Let us first calculate the energy in the Hartree-Fock approximation. For this purpose, we will go over from summation to integration in Eq. (2.4.4), and also take into account the relation (2.4.2). Then the first summand is

$$\frac{1}{ZN}\langle 0|\hat{H}_0|0\rangle = \frac{1}{ZN}\sum_{\mathbf{k}\sigma}\frac{\hbar^2 k^2}{2m}n_{\mathbf{k}\sigma} = \frac{8\pi v_0}{Z(2\pi)^3}\int_0^{k_F}\frac{\hbar^2 k^2}{2m}k^2 \mathrm{d}k = \frac{2.21}{r_s^2}Ry \tag{2.4.6}$$

where r_s has been defined above. Here $Ry = e^2/(2a_B) = 13.5\,\mathrm{eV}$ is the ionization energy of a hydrogen atom. In the second summand we will take into account that

$$\begin{aligned}\langle 0|a^+_{\mathbf{k}+\mathbf{q},\sigma}a^+_{\mathbf{k}'-\mathbf{q},\sigma'}a_{\mathbf{k}',\sigma'}a_{\mathbf{k}\sigma}|0\rangle &= \delta_{0\mathbf{q}}\langle 0|a^+_{\mathbf{k}\sigma}a_{\mathbf{k}\sigma}|0\rangle - \langle 0|a^+_{\mathbf{k}+\mathbf{q},\sigma}a_{\mathbf{k}',\sigma'}a^+_{\mathbf{k}'-\mathbf{q},\sigma'}a_{\mathbf{k}\sigma}|0\rangle \\ &= \delta_{0\mathbf{q}}n_{\mathbf{k}\sigma} - \delta_{\mathbf{k}+\mathbf{q},\mathbf{k}'}\delta_{\sigma\sigma'}n_{\mathbf{k}+\mathbf{q},\sigma}n_{\mathbf{k}\sigma}\end{aligned} \tag{2.4.7}$$

Then it will take the form

$$\frac{1}{ZN}\langle 0|\hat{H}'_j|0\rangle = -\frac{1}{ZN}\frac{1}{V}\sum_{\substack{\mathbf{q},\mathbf{k},\sigma\\(\mathbf{q}\neq 0)}}\frac{2\pi e^2}{q^2}n_{\mathbf{k}+\mathbf{q},\sigma}n_{\mathbf{k}\sigma} \tag{2.4.8}$$

This expression corresponds to the process in which two particles inside the Fermi sphere are destroyed, while two others reappear there, so that the whole system returns to its basic state. This process will be different from zero if the creation and annihilation of the particle is determined by the pairs of operators $(a^+_{\mathbf{k}+\mathbf{q},\sigma};\ a_{\mathbf{k}'\sigma'})$ and $(a^+_{\mathbf{k}'-\mathbf{q},\sigma'};\ a_{\mathbf{k}\sigma})$. It is called an *exchange process* and is represented diagrammatically in Fig. 2.4.1. The solid lines in the diagram represent the electrons, and the wavy line represents their interaction.

By substituting the variables $\mathbf{k}' = \mathbf{k} + \mathbf{q}$, Eq. (2.4.8) becomes of the form

$$\frac{1}{ZN}\langle 0|\hat{H}'_j|0\rangle = -\frac{v_0}{Z(2\pi)^6}2\int_{|\mathbf{k}|<k_F}\mathrm{d}^3k\int_{|\mathbf{k}'|<k_F}\mathrm{d}^3k'\frac{2\pi e^2}{|\mathbf{k}-\mathbf{k}'|^2} = -\frac{0.916}{r_s}Ry \tag{2.4.9}$$

Thus, the energy of the electron gas in the Hartree-Fock approximation is

$$\varepsilon_{\mathrm{HF}} = \left(\frac{2.21}{r_s^2} - \frac{0.916}{r_s}\right)Ry \tag{2.4.10}$$

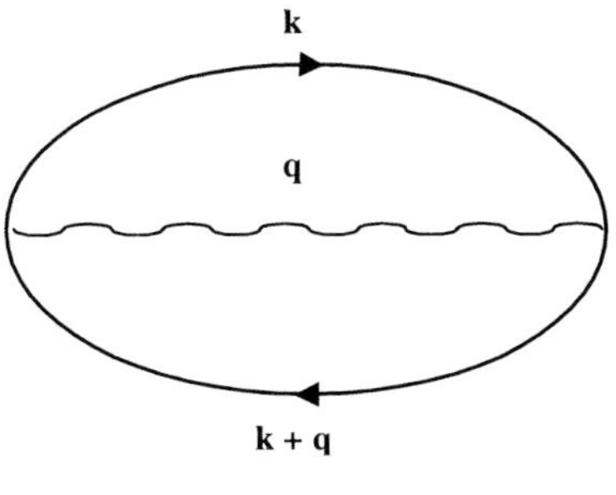

Fig. 2.4.1 An exchange process

Let us now calculate the correlation energy. The first thing to be done here is to calculate the first summand in Eq. (2.4.5):

$$\varepsilon_{2\text{cor}} = -\frac{1}{ZN}\sum_{n}\frac{\langle 0|\hat{H}_j'|n\rangle\langle n|\hat{H}_j'|0\rangle}{E_n - E_0} \tag{2.4.11}$$

We should find out which states contribute to this sum, and how the contribution can be calculated.

A typical scattering process described by the Hamiltonian $\hat{H}_i'$ consists of the transition of electrons from the states $\mathbf{k}$ and $-\mathbf{k}'$ into the states $\mathbf{k}+\mathbf{q}$ and $-\mathbf{k}'-\mathbf{q}$. According to the Pauli principle, the states $\mathbf{k}$ and $-\mathbf{k}'$ should exist *inside* the Fermi sphere, but the states $\mathbf{k}+\mathbf{q}$ and $-\mathbf{k}'-\mathbf{q}$ should exist *outside*. The matrix element of such a transition, if it is allowed, equals $4\pi e^2/(Vq^2)$. The excitation energy is

$$\varepsilon_{\mathbf{k}\mathbf{k}'}(\mathbf{q}) = \frac{\hbar^2(\mathbf{k}+\mathbf{q})^2}{2m} + \frac{\hbar^2(-\mathbf{k}'-\mathbf{q})^2}{2m} - \frac{\hbar^2(k)^2}{2m} - \frac{h^2(k')^2}{2m} = \frac{\hbar^2\mathbf{q}\cdot(\mathbf{k}+\mathbf{k}'+\mathbf{q})}{m} \tag{2.4.12}$$

Now we have to leave the excited state n, characterized by the presence of electrons with the momenta $\mathbf{k}+\mathbf{q}$ and $-\mathbf{k}'-\mathbf{q}$ and holes $\mathbf{k}$ and $-\mathbf{k}'$, to go back to the ground state. Here we should distinguish the following two processes that yield different contributions to the ground state energy.

1. The first is the so-called 'direct' process – when the transition to the ground state occurs in the same way as to the excited state; this process corresponds to the matrix element $4\pi e^2/(Vq^2)$. The transition diagram has the form of Fig. 2.4.2, and the contribution ('Hartree contribution') to $\varepsilon_{2\text{cor}}$ is

$$\varepsilon_{2\text{cor}}^{(a)} = -\frac{4}{ZNV^2}\cdot\frac{1}{4}\sum_{\substack{\mathbf{k},\mathbf{k}',\mathbf{q}\\(q\neq 0)}}\left(\frac{4\pi e^2}{q^2}\right)^2\frac{m}{\hbar^2\mathbf{q}(\mathbf{k}+\mathbf{k}'+\mathbf{q})}n_{\mathbf{k}\sigma}(1-n_{\mathbf{k}+\mathbf{q},\sigma})n_{\mathbf{k}'\sigma'}(1-n_{\mathbf{k}'+\mathbf{q},\sigma'}) \tag{2.4.13}$$

The factor 4 results from the sum over spins.

2. The second process is the so-called 'exchange' process when the electron with momentum $\mathbf{k}+\mathbf{q}$ returns to $-\mathbf{k}$ and the electron with $-\mathbf{k}'-\mathbf{q}$ returns to the state

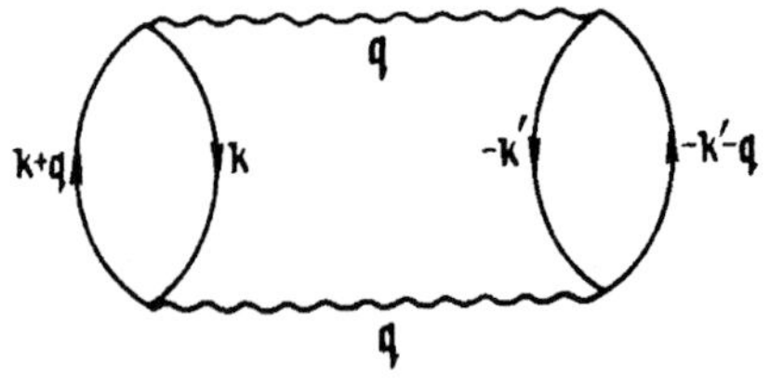

Fig. 2.4.2 The 'direct contribution' ('Hartree contribution') to the correlation energy. The arrows on the **q** lines are omitted as in Fig. 2.4.1, because their mutual direction is obvious

k. The matrix element for this transition is $-4\pi e^2/(V|\mathbf{k}+\mathbf{k}'+\mathbf{q}|^2)$. The minus sign reflects the fact that now the creation and annihilation operators in Eq. (2.4.7) stand in a different order. We should point out that the exchange process can only take place if the electrons **k** and $-\mathbf{k}'$ have parallel spins, while they can have both parallel and antiparallel spins in the 'direct process'. A typical exchange diagram is shown in Fig. 2.4.3, and its contribution to $\varepsilon_{2\text{cor}}$ is

$$\varepsilon_{2\text{cor}}^{(b)} = \frac{2}{ZNV^2}\cdot\frac{1}{4}\sum_{\substack{\mathbf{k},\mathbf{k}',\mathbf{q}\\(q\neq 0)}}\frac{4\pi e^2}{q^2}\,\frac{4\pi e^2}{(\mathbf{k}+\mathbf{k}'+\mathbf{q})^2}\,\frac{m}{\hbar^2\mathbf{q}(\mathbf{k}+\mathbf{k}'+\mathbf{q})}$$
$$\times\, n_{\mathbf{k}\sigma}\left(1-n_{\mathbf{k}+\mathbf{q},\sigma}\right)n_{\mathbf{k}'\sigma}\left(1-n_{\mathbf{k}+\mathbf{q},\sigma}\right) \qquad (2.4.14)$$

The contribution $\varepsilon_{2\text{cor}}^{(a)}$ of direct processes diverges logarithmically at small transferred momenta; in order to be convinced of this, we may observe that, due to the requirement $k<k_F$ and $|\mathbf{k}+\mathbf{q}|>k_F$, the values of k can only be near the Fermi surface in a narrow layer of the order q.

In this case

$$n_{\mathbf{k}\sigma}(1-n_{\mathbf{k}+\mathbf{q},\sigma}) \approx n_{\mathbf{k}\sigma}(1-n_{\mathbf{k}\sigma}-\mathbf{q}\cdot\nabla_{\mathbf{k}}n_{\mathbf{k}\sigma}+\ldots)$$
$$\approx n_{\mathbf{k}\sigma}\,\frac{\mathbf{q}\cdot\mathbf{k}}{k}\,\delta(k-k_F)\,,\quad \text{with}\quad (\mathbf{q}\cdot\mathbf{k})\geq 0$$

We have used here the equality $\nabla_{\mathbf{k}}n_{\mathbf{k}\sigma} = -\frac{\mathbf{k}}{k}\delta(k-k_F)$ and also the condition $k<k_F$; $|\mathbf{k}+\mathbf{q}|>k_F$ which has led to the inequality $(\mathbf{q}\cdot\mathbf{k})\geq 0$. Thus, at small values of q

$$\frac{1}{V}\sum_{\mathbf{k}} n_{\mathbf{k}\sigma}(1-n_{\mathbf{k}+\mathbf{q},\sigma}) = \frac{1}{V}\sum_{\mathbf{k}'} n_{\mathbf{k}'\sigma}\left(1-n_{\mathbf{k}'+\mathbf{q},\sigma}\right) \simeq \frac{q}{k_F}\,\frac{Z}{v_0}$$

Going over from summation to integration in Eq. (2.4.13), we obtain:

$$\frac{1}{V}\sum_{\mathbf{k}} n_{\mathbf{k}\sigma}(1-n_{\mathbf{k}+\mathbf{q},\sigma}) = \frac{1}{V}\sum_{\mathbf{k}'} n_{\mathbf{k}'\sigma}\left(1-n_{\mathbf{k}'+\mathbf{q},\sigma}\right) \simeq \frac{q}{k_F}\,\frac{Z}{v_0} \qquad (2.4.15)$$

which leads to a logarithmic divergence on integration (see below).

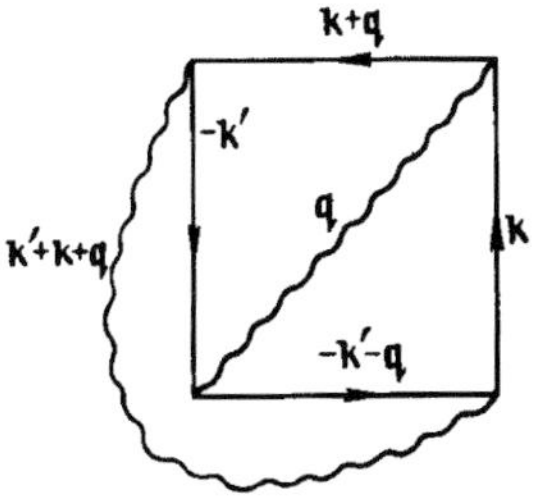

Fig. 2.4.3 The 'exchange contribution' to the correlation energy

This logarithmic peculiarity is a direct consequence of the long-range Coulomb interaction. If the interaction were decreasing quickly with distance, its Fourier transform at small values of q would tend to a constant, and there would be no divergence. We might point out that the existence of this divergence has delayed a systematic study of the electron gas in the framework of perturbation theory for at least 20 years.

Another big step in studying the interacting electron gas was made by Gell-Mann and Brueckner (1957) and Nozieres and Pines (1958) (see also Pines and Nozieres, 1966). They showed that a modified perturbation method is quite applicable to the electron gas, if the analysis is not confined to the first encountered term of divergence, but, instead, the most diverging summands of each term continue being added. In such a way a finite result is obtained.

As we have already seen, the calculation of the ground state energy in the second order of the perturbation theory leads to a logarithmic divergence. Higher-order terms contain even greater divergences connected with the long-range character of the Coulomb forces. The structure of those terms could be found as follows. It was mentioned earlier that the second-order contribution is given by the 'direct' and 'exchange' processes. The divergence in the term corresponding to the direct process (when an electron perturbed from the Fermi sphere falls back with the same momentum transfer) appears to be due to the accumulation of the factors $1/q^2$. On the other hand, the term corresponding to the exchange process has no divergence as it has only one factor $1/q^2$, while the second factor $(\mathbf{k}+\mathbf{k}'+\mathbf{q})^{-2}$ remains finite at $q \to 0$.

The most diverging term in the third order corresponds to electron transitions with the same momentum transfer again (Fig. 2.4.4).

This term describes the following process. First, the excitation of two electron-hole pairs takes place with a transfer of the momentum q. Then one of the pairs annihilates and is created again, due to the interaction $4\pi e^2/q^2$. Finally, both pairs annihilate, and the system returns to the initial state with no excitations. The value of this term is easily estimated at $q \to 0$. In comparison with $\varepsilon^{(a)}_{2\text{cor}}$, this term has an additional factor $4\pi e^2/q^2$ and an additional energy denominator that behaves like $\hbar q k_F$ at $q \to 0$. Besides, there is a factor proportional to q/k_F, related to a decrease in the allowed phase space region due to the Pauli principle. Consequently, the contribution to the ground-state energy from the process described by Fig. 2.4.4 will contain a quadratic divergence:

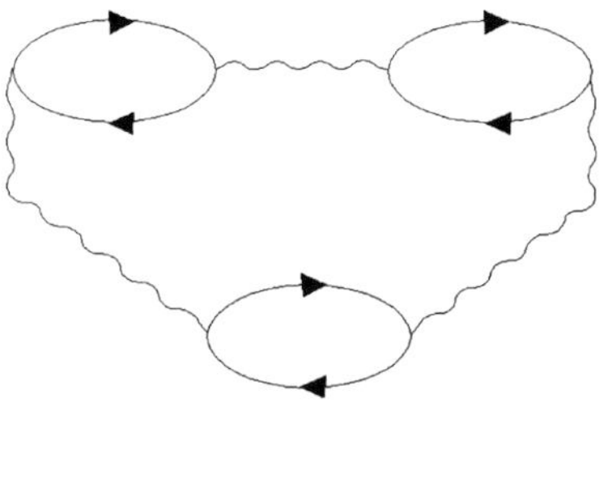

Fig. 2.4.4 The most diverging term of third order

$$\varepsilon_{3\text{cor}}^{(a)} \sim \int \frac{dq}{q} \frac{4\pi e^2}{q^2} \frac{q}{k_\text{F}} \frac{1}{qk_\text{F}} \sim \frac{r_s}{\beta^2}\, Ry$$

A certain minimal wavevector β is introduced here.

On the other hand, if one of these pair scattering processes is an exchange process, we obtain for the corresponding contribution

$$\varepsilon_{3\text{cor}}^{(b)} \sim \int \frac{dq}{q} \frac{4\pi e^2}{(\mathbf{k}+\mathbf{k}'+\mathbf{q})^2} \frac{q}{k_\text{F}} \frac{1}{qk_\text{F}} \sim r_s \ln \beta Ry$$

This expression diverges, too, but more weakly than $\varepsilon_{3\text{cor}}^{(a)}$.

Similarly, the terms including two exchange scattering processes will give the contribution

$$\varepsilon_{3\text{cor}}^{(c)} \sim r_s R_y$$

Further, we may consider the terms of fourth order. The contribution from the terms that include only electron transitions with the same momentum transfer (electron loops) is

$$\varepsilon_{4\text{cor}}^{(a)} \sim \int \frac{dq}{q} \left(\frac{4\pi e^2}{q^2}\right)^2 \left(\frac{q}{k_\text{F}}\right)^2 \frac{1}{(qk_\text{F})^2} \sim \frac{r_s^2}{\beta^4}\, Ry$$

On the other hand, each term including one exchange process gives a contribution of the order $\varepsilon_{4\text{cor}}^{(b)} \sim r_s^2/\beta^2$; the contribution from the terms with two exchange processes is of the order $r_s^2 \ln \beta$; from the terms with three exchange processes, of the order r_s^2, etc.

Thus, the structure of the perturbation theory series is now clear. As we take into account the behaviour of the matrix elements at small momentums of transfer, we obtain the series

$$\begin{aligned}\varepsilon_\text{cor} &= b \ln \beta + c\,\frac{r_s^2}{\beta^4} + \ldots + a_1 + b_1 r_s \ln \beta + \frac{c_1 r_s^2}{\beta^2} + d_1 \frac{r_s^3}{\beta^4} + \ldots \\ &\quad + a_2 r_s + b_2 r_s^2 \ln \beta + c_2 \frac{r_s^3}{\beta^2} + d_2 \frac{r_s^4}{\beta^4} + \ldots\end{aligned}$$

The minimum wavevector β should probably be proportional to the inverse screening length of the Coulomb potential by electrons. The order of this value can be shown to be equal to the ratio of the Fermi velocity to the plasma frequency, i.e.

$$\beta \sim r_s^{1/2}$$

Then the sum of the first set of terms in the series that represents the contribution of scattering processes with the same momentum transfer q gives a constant term and a term proportional to $\ln r_s$. Accordingly, the second set of the series' terms gives a constant term and terms of the order $r_s \ln r_s$ and r_s; the third set gives terms of the order r_s, $r_s^2 \ln r_s$ and r_s^2, etc. Thus, if we add the most diverging terms in the perturbation theory series, we can obtain both a term proportional to $\ln r_s$ and a constant term in the correlation energy. The next terms of the perturbation theory series will lead to terms of order $r_s \ln r_s$, r_s, etc.

Calculations of this kind were carried out by Gell-Mann and Brueckner (1957). They developed a comparatively simple way of summing up the series, which is based on the diagrammatic technique of Feynman propagators. From summing the most diverging terms and taking account of the constant term in the second order of the perturbation theory, Gell-Mann and Brueckner obtained:

$$\varepsilon_{\text{cor}} = \{0.062 \ln r_s - 0.096 + ar_s + br_s \ln r_s + \ldots\} \, Ry \tag{2.4.16}$$

This expression, however, is correct only in the limiting case of high electron densities $r_s \ll 1$, whereas the electron concentration in real metals corresponds to r_s-values in the interval

$$1.8 \leq r_s \leq 5.5 \tag{2.4.17}$$

In order to get an approximate theory that would be correct for electron densities corresponding to typical metals it is necessary to carry out a more detailed analysis of the contributions with different momentum transfers in the correlation energy.

It has been shown that the contributions from small momentum transfers play a significant role at small electron densities. At intermediate densities, it is necessary to take into account the contributions from large momentum transfers as well. There, however, the 'exchange' part of the perturbation theory series (which is only present for electrons with equal spins) is compensated together with the half of the 'direct' interaction which is to be attributed to electrons with parallel spins. Such compensation is simply explained: any direct transition process ('Hartree' process) for electrons with parallel spins described by the matrix element $4\pi e^2/q^2$ is always accompanied by the 'conjugate exchange' process with the matrix element $-4\pi e^2/|\mathbf{q}+\mathbf{k}+\mathbf{k}'|^2$. We have seen that these exchange terms are negligible at small momentum transfers. At large momentum transfers, however, they are mutually compensated by those 'Hartree' terms, which describe the 'direct' interaction between electrons with parallel spins.

These considerations form the basis of the approximate analysis proposed by Hubbard (1957) and Nozieres and Pines (1958) (see also Pines and Nozieres, 1966). Hubbard proposed avoiding the difficulty connected with large momentum transfers by modifying the dielectric function formula. By using the well-known correlation between the dielectric function and the ground state energy, he makes a numerical estimate of the correlation energy values for different intermediate concentrations.

Nozieres and Pines (1958) first determine the contributions to the correlation energy in the domains of large and small momentum transfers, respectively, and then perform a smooth interpolation between large and small q values to obtain the contributions from the intermediate values of the momentum q. The correlation energy they obtain is

$$\varepsilon_{\mathrm{cor}} = (-0.1156 + 0.0315 \ln r_s)\, Ry \tag{2.4.18}$$

Hubbard's interpolation formula and that of Nozieres and Pines give very close results for the correlation energy. This is not simply a coincidence, because both schemes are based on the same physical picture. We will use the formula of Nozieres and Pines in Eq. (2.4.18).

In the Nozieres-Pines approximation, the ground state total energy of the electron gas is then

$$\varepsilon_{\mathrm{NP}} = \left(\frac{2.21}{r_s^2} - \frac{0.916}{r_s} - 0.1156 + 0.0315 \ln r_s\right) Ry \tag{2.4.19}$$

2.5
The Effective Interaction Between Ions in Liquid or Amorphous Metals

If we are only interested in the equilibrium (rsp. quasi-equilibrium) properties of liquid (rsp. amorphous) metals, it is sufficient to calculate the grand canonical thermodynamic potential Ω of the system (the thermodynamic potential in the variables V, T, μ), and then to determine other thermodynamic parameters through commonly known relations. For liquid metals, the thermodynamic potential Ω is determined (see, for example, Landau and Lifshitz, 1976) by the relation

$$\Omega = -k_B T \cdot \ln\left\{\mathrm{Tr}\left[\exp\left(\frac{-\hat{H} + \mu_e \hat{N}_e + \mu_i N_i}{k_B T}\right)\right]\right\} \tag{2.5.1}$$

in which μ_e and μ_i are the chemical potentials of the electrons and ions of the system, $\hat{H}$ is the effective Hamiltonian of the system determined by Eqs (2.1.8) to (2.1.10), and $\hat{N}_e = \sum_{\vec{k}\sigma} a^+_{\vec{k}\sigma} a_{\vec{k}\sigma}$ is the number operator of the electrons. The 'tracing' operation Tr means taking the sum of all the diagonal elements of the matrix, evaluated in any orthonormal basis. The summation is taken with respect both to the number of particles in the system and to all the possible states, e.g. of the liquid metal with given number of particles.

For amorphous metals, however, as seen below, for a given ionic configuration i, *only the trace over the electronic states* is performed, and then the result Ω_i is averaged over the essentially fixed typical configuration of the ions. This final average over Ω_i for amorphous systems, which may be unnecessary for an extremely large, i.e. 'self-averaging', sample, is called a 'quenched average', in contrast to the 'annealed averaging' performed by taking the trace over all electronic plus ionic states.

It is also necessary to point out here that all the consequences following from Eq. (2.5.1) will only be correct for a choice of μ_i and μ_e that fulfills the requirement of electrical neutrality of the whole system, i.e. the total electronic charge should be equal to the total ionic charge (in the absolute value):

$$\partial\Omega/\partial\mu_e = Z\partial\Omega/\partial\mu_i \tag{2.5.2}$$

In the adiabatic approximation (often also called the 'Born-Oppenheimer approximation'), it is convenient to calculate Ω in two stages. First, in Eq. (2.5.1), we should perform the trace over the states of the electronic subsystem at fixed positions of the ions, and then the trace over the states of the ionic subsystem:

$$\Omega = -k_B T \cdot \ln \mathrm{Tr}^{(i)} \left\{ \exp\left(\frac{-\hat{H}_i + \mu_i N_i}{k_B T}\right) \mathrm{Tr}^{(e)} \left[\exp\left(\frac{-\hat{H}_{ie} + \mu_e \hat{N}_e}{k_B T}\right)\right]\right\} \tag{2.5.3}$$

This is for *liquid* metals, where the ionic state is 'ergodic', i.e. thermodynamic equilibrium, with respect to the position of the *ions,* is obtained in times which are short compared with the timescales of the measurement; in contrast, for *amorphous* metals, the ionic subsystem is 'non-ergodic', i.e. it is essentially frozen into a metastable state, and remains far from thermodynamic equilibrium during the experiments considered. In this case the usual 'annealed averaging', i.e. with the trace, is only performed with respect to the electrons, whereas the ions are considered as 'quenched', i.e. fixed into a metastable quasi-equilibrium configuration. Only the 'fast' vibrational oscillations around this fixed configuration are taken into account in the 'trace', in contrast to transitions between different metastable states, which *do* take place, although these are very rare events. In fact, the transitions between different metastable states give rise to the *'aging phenomena'* already mentioned, but here they will be neglected unless explicitly mentioned. More details are given below.

For averaging over the electronic states we can use the thermodynamic perturbation theory, assuming the electron-ion and electron-electron interactions to be small compared to the kinetic energy of the electrons. Then, the expression denoted by $\mathrm{Tr}^{(e)}$ may be represented as follows (Abrikosov *et al.*, 1965):

$$\exp\left(\frac{-\hat{H}_{ie} + \mu_e \hat{N}_e}{k_B T}\right) = \exp\left(\frac{-\hat{H}_0 + \mu_e \hat{N}_e}{k_B T}\right) \hat{S} \tag{2.5.4}$$

in which

$$\hat{S} = \sum_n \frac{(-1)^n}{n!} \int_0^{1/k_B T} \mathrm{d}\tau_1 \ldots \int_0^{1/k_B T} \mathrm{d}\tau_n T_\tau [\hat{H}_j(\tau_1) \ldots \hat{H}_j(\tau_n)] = T_\tau \exp\left[- \int_0^{1/k_B T} \hat{H}_j(\tau)\, \mathrm{d}\tau \right] \tag{2.5.5}$$

$$\hat{H}_j(\tau) = \exp[\tau(\hat{H}_0 - \mu_e\hat{N}_e)]\hat{H}_j \exp[-\tau(\hat{H}_0 - \mu_e\hat{N}_e)] \tag{2.5.6}$$

Here Dyson's 'time-ordering symbol' T_τ indicates that all the operators $\hat{H}_j(\tau_i)$ in the operator product on the first line of Eq. (2.5.5) should be arranged from left to right in the same order as τ_l decreases. It follows from Eq. (2.5.4) that

$$\mathrm{Tr}^{(e)}\left[\exp\left(\frac{-\hat{H}_{ie} + \mu_e\hat{N}_e}{k_B T}\right)\right] = \langle\hat{S}\rangle \exp\left(-\frac{\Omega_0}{k_B T}\right) \tag{2.5.7}$$

in which

$$\Omega_0 = -k_B T \cdot \ln\left\{\mathrm{Tr}^{(e)}\left[\exp\left(\frac{-\hat{H}_0 + \mu_e\hat{N}_e}{k_B T}\right)\right]\right\} \tag{2.5.8}$$

is the thermodynamic potential of the ideal electron gas, and

$$\langle\hat{S}\rangle = \mathrm{Tr}^{(e)}\left[\hat{S}\exp\left(\frac{\Omega_0 - \hat{H}_0 + \mu_e\hat{N}_e}{k_B T}\right)\right] \tag{2.5.9}$$

By using *Wick's theorem* in the previous equation (see, for example, Abrikosov *et al.*, 1965), we obtain a perturbation series for $\langle\hat{S}\rangle$. This series can be represented very visually, if each term of the series is associated with the corresponding Feynman diagram; details can be found in the well-known textbooks of Abrikosov *et al.* (1965) or Fetter and Walecka (1971). The diagrams are of two types: connected and disconnected. *Disconnected* diagrams consist of at least two closed loops, which are *not* connected by any lines. *Connected* diagrams, in contrast, are obtained when an arbitrary term of the $\langle\hat{S}\rangle$ series is rewritten according to Wick's theorem as

$$\frac{(-1)^n}{n!}\int_0^{1/k_B T} d\tau_1 \ldots \int_0^{1/k_B T} d\tau_n \langle T_\tau\{\hat{H}_j(\tau_1)\ldots\hat{H}_j(\tau_n)\}\rangle$$

so that the usual contraction starts with some operator within this expression and then returns to it *without omitting any of the* $\hat{H}_j$. A disconnected diagram is obtained in all other cases.

Let us use $\langle S\rangle_C$ to denote the sum of all connected diagrams of all orders in $\langle S\rangle$. We can show then that

$$\langle\hat{S}\rangle = \exp\{\langle S\rangle_C - 1\} \tag{2.5.10}$$

By substituting this expression into Eq. (2.5.7), we obtain the thermodynamic potential of the electron gas in a metal, for a fixed ionic subsystem:

$$\Omega_{\text{ie}} = -k_{\text{B}}T \cdot \ln\left\{\text{Tr}^{(\text{e})}\left[\exp\left(\frac{-\hat{H}_{\text{ie}} + \mu_{\text{e}}\hat{N}_{\text{e}}}{k_{\text{B}}T}\right)\right]\right\} = \Omega_0 - k_{\text{B}}T\left\{\langle\hat{S}\rangle_{\text{C}} - 1\right\} \quad (2.5.11)$$

Now we can expand the matrix $\hat{S}$ into a power series with respect to the pseudopotential (i.e. into a $\hat{H}_i''$ power series):

$$\hat{S} = \hat{S}' + \sum_{n=1}^{\infty} \frac{(-1)^n}{n!} \int_0^{1/k_{\text{B}}T} \mathrm{d}\tau_1 \ldots \int_0^{1/k_{\text{B}}T} \mathrm{d}\tau_n T_\tau \left\{\hat{H}_j''(\tau_1)\ldots\hat{H}_j''(\tau_n)\,\hat{S}'\right\} \quad (2.5.12)$$

in which

$$\hat{S}' = T_\tau \exp\left\{-\int_0^{1/k_{\text{B}}T} \mathrm{d}\tau \hat{H}_j'(\tau)\right\} \quad (2.5.13)$$

Then Ω_{ie} can be expressed as

$$\Omega_{\text{ie}} = \Omega_0 - k_{\text{B}}T\left\{\langle\hat{S}'\rangle_{\text{C}} - 1\right\} - \sum_{n=1}^{\infty} \Omega_n \quad (2.5.14)$$

Here, if we use the explicit form of $\hat{H}_j''$, we can see that Ω_n has the following structure:

$$\begin{aligned}\Omega_n = &- k_{\text{B}}T\frac{(-1)^n}{n!}\int \mathrm{d}\tau_1\,\mathrm{d}^3r_1 \ldots \int \mathrm{d}\tau_n\,\mathrm{d}^3r_n \sum_{\nu_1}\Big[w(\vec{r}_1 - R_{\nu_1}) \\ &- \frac{1}{V}\int w(\vec{r}\,')\mathrm{d}^3r'\Big] \ldots \sum_{\nu_n}\Big[w(\vec{r}_n - R_{\nu_n}) - \frac{1}{V}\int w(\vec{r}\,'_n)\mathrm{d}^3r'_n\Big] \\ &\times \sum_{\sigma_1\ldots\sigma_n} \langle T_\tau\{\hat{\Psi}^+_{\sigma_1}(\vec{r}_1\tau_1)\hat{\Psi}_{\sigma_1}(\vec{r}_1\tau_1)\ldots\hat{\Psi}^+_{\sigma_n}(\vec{r}_n\tau_n)\Psi_{\sigma_n}(\vec{r}_n\tau_n)\hat{S}'\}\rangle_C\end{aligned} \quad (2.5.15)$$

As we said above, Ω_0 is the thermodynamic potential of the *ideal* electron gas, while the second summand in Eq. (2.5.14) shows the changes in the thermodynamic potential of the electron gas on account of the electron-electron interaction. The sum of these two summands,

$$\Omega_{\text{e}} = \Omega_0 - k_{\text{B}}T\{\langle S'\rangle_{\text{C}} - 1\} \quad (2.5.16)$$

is the thermodynamic potential of the interacting electron gas in the neutralizing background.

The third term in Eq. (2.5.14), $\sum_{n=1}^{\infty}\Omega_n$, describes changes in the electronic thermodynamic potential caused by the electron-ion interaction.

Let us have a closer look at the structure of the summands in the third term. The function $\sum_{\sigma}\langle T_\tau\{\hat{\Psi}^+_\sigma(\vec{r},\tau)\Psi_\sigma(\vec{r}\,',\tau')S'\}\rangle_C$ is the one-particle thermodynamic Green's function $G(\vec{r}-\vec{r}\,',\tau-\tau')$. Due to the homogeneity of space-time, it depends only on the difference in the arguments. This is why the first summand, Ω_1, is equal to zero:

$$\Omega_1 = -k_B T \int_0^{1/k_B T} d\tau \int_V d^3r \sum_\nu \left[w(\vec{r}-R_\nu) - \frac{1}{V}\int w(\vec{r}\,')d^3r'\right] G(0)$$
$$= -G(0)\sum_\nu \int_V \left[w(\vec{r}-R_\nu) - \frac{1}{V}\int w(\vec{r}\,')d^3r'\right] d^3r = 0\,. \tag{2.5.17}$$

The second summand Ω_2 contains the so-called *Polarization Green's function*:

$$\Gamma_2(\vec{r}_1-\vec{r}_2;\,\tau_1-\tau_2) = -\sum_{\sigma_1\sigma_2}\langle T_\tau\{\hat{\Psi}^+_{\sigma_1}(\vec{r}_1\tau_1)\hat{\Psi}_{\sigma_1}(\vec{r}_1\tau_1)\hat{\Psi}^+_{\sigma_2}(\vec{r}_2\tau_2)\hat{\Psi}_{\sigma_2}(\vec{r}_2\tau_2)S'\}\rangle_C \tag{2.5.18}$$

The Fourier transform of this function is defined by the formula

$$\Gamma_2(\vec{r},\tau) = k_B T \sum_{\omega_n} \exp[i\omega_n\tau]\frac{1}{V}\sum_{\vec{k}} \exp(i\vec{k}\vec{r})\tilde{\Gamma}_2(\vec{k},\omega_n) \tag{2.5.19}$$

Here the $\omega_n = (2n+1)\pi k_B T$ are the so-called *Matsubara frequencies* (energies), and $\tilde{\Gamma}_2(\vec{k},\omega_n)$ can be expressed by means of the diagrammatic expansion in Fig. 2.5.1.

Let us now formulate the rules for the expression corresponding to a certain diagram on the r.h.s. of Fig. 2.5.1 (Abrikosov *et al.*, 1965).

1. Each diagram line and each vertex (i.e. a vertex in which only two fermion lines meet) is juxtaposed with the momentum $\hbar\vec{k}$ and the frequency ω_n. The momenta and frequencies of all the lines should fulfill the 'conservation laws'

$$\sum \vec{k} = 0,\ \sum \omega_n = 0$$

$\tilde{\Gamma}_2(k,\omega_n) \equiv$ 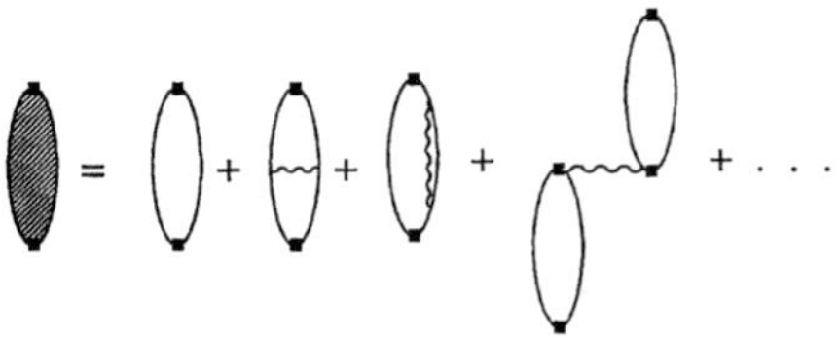

Fig. 2.5.1 Diagrammatic representation of $\tilde{\Gamma}_2(\vec{k},\omega_n)$

2. Each solid interior line $\xrightarrow{\vec{k}}$ $\xrightarrow{\omega_n}$ is juxtaposed with the quantity $(i\omega_n - \varepsilon_0(\vec{k}) + \mu_e)^{-1}$ in which $\varepsilon_0(\vec{k}) = \pi^2 k^2/(2m)$. Each wavy line is juxtaposed with the quantity $-4\pi e^2/q^2$.
3. All independent *internal* momenta are to be either summed up as $V^{-1}\sum_{\vec{k}}\ldots$, or integrated as $(2\pi)^{-3}\int d^3k\ldots$; all independent *internal* frequencies are to be summed up as $k_B T\sum_{\omega_n}\ldots$, whereas the momenta and frequencies of *external* vertices are kept fixed.
4. The expression obtained with rules 1, 2 and 3 for a diagram should be multiplied by a factor $(2s+1)^F(-1)^F$, where F is the number of closed loops formed by the fermion lines in the diagram, and s is the spin of the particle.

The functional expansion for the function $\tilde{\Gamma}_2(q,\omega_n)$ discussed above contains two types of diagrams: *strongly connected* and *weakly connected*. The latter group is usually made up of those diagrams that can be split into two disconnected ones by removing one wavy line. A diagram is called 'strongly connected' if this cannot be done.

Let us assume that we can sum up all the strongly connected diagrams in $\tilde{\Gamma}_2(q,\omega_n)$. This sum is called the *polarization operator* ($F_0 := \Pi(q,\omega_n)$) and is shown in Fig. 2.5.2.

Then, the function $\tilde{\Gamma}_2(q,\omega_n)$ can be expressed as the series shown in Fig. 2.5.3.

By using Eqs (2.5.18) and (2.5.19) and the series shown in Fig. 2.5.3, we can express Ω_2 as

$$\Omega_2 = \frac{V}{2v_0^2}\sum_{\vec{q}\neq 0}\tilde{\Gamma}_2(\vec{q},0)|\rho(\vec{q})w(\vec{q})|^2 = -\frac{V}{2v_0^2}\sum_{\vec{q}\neq 0}\frac{\pi(\vec{q})}{1+(4\pi e^2/q^2)\pi(\vec{q})}|\rho(\vec{q})w(\vec{q})|^2 \tag{2.5.20}$$

in which $\pi(\vec{q}) \equiv \pi(\vec{q},0)$. Because the dielectric function of the electron gas is related to the polarization operator through the relationship

$$\varepsilon(\vec{q},\omega_n) = 1 + \frac{4\pi e^2}{q^2}\pi(\vec{q},\omega_n) \tag{2.5.21}$$

the expression for Ω_2 may be rewritten as:

?tpt=3.5mm>

$$\Omega_2 = -\frac{V}{2v_0^2}\sum_{\vec{q}\neq 0}\pi(\vec{q})\frac{1}{\varepsilon(\vec{q})}|\rho(\vec{q})w(\vec{q})|^2 \tag{2.5.22}$$

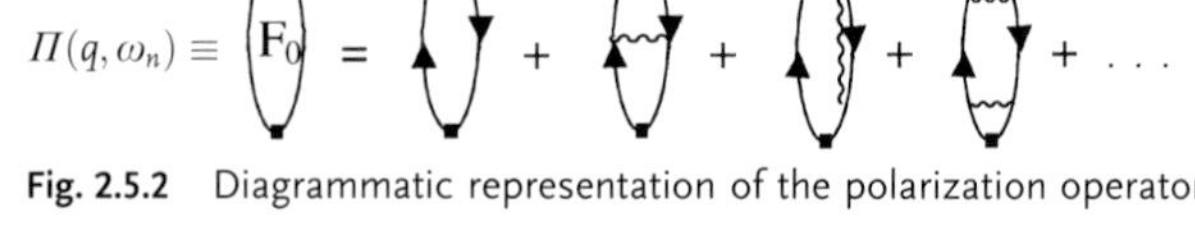

Fig. 2.5.2 Diagrammatic representation of the polarization operator

$-\tilde{\Gamma}_2(q,\omega_n) =$

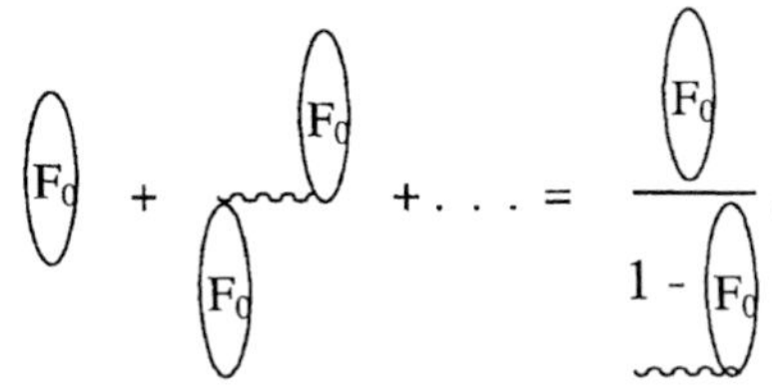

$$= \frac{\Pi(q,\omega_n)}{1 + \frac{4\pi e^2}{q^2}\cdot \Pi(q,\omega_n)}$$

Fig. 2.5.3 Diagrammatic series for $\tilde{\Gamma}_2(q,\omega_n)$

When the expansion of the polarization operator is confined to the first terms on the r.h.s. of Fig. 2.5.2, the approximation is called the *'random phase approximation'* (RPA), which is famous and treated in every textbook on solid-state theory. The polarization operator in this approximation has the form first derived by Lindhard (1954), whose name is given to the final result (the Lindhard function), namely

$$\pi_0(\vec{q}) = -2k_B T \sum_{\omega_n} \int \frac{d^3k}{(2\pi)^3} \cdot \frac{1}{[i\omega_n - \varepsilon_0(\vec{k}) + \mu_e][i\omega_n - \varepsilon_0(\vec{k}+\vec{q}) + \mu_e]}$$
$$\approx 8\pi \int_0^\infty \frac{kdk}{(2\pi)^3} \cdot \frac{n_{\vec{k}+\vec{q},\sigma} - n_{\vec{k}}}{-\varepsilon_0(\vec{k}+\vec{q}) + \varepsilon_0(\vec{k})} \approx \frac{mk_F}{2\pi^2\hbar^2}\left[\frac{4k_F^2 - q^2}{4k_F q}\ln\left|\frac{2k_F + q}{2k_F - q}\right| + 1\right] \tag{2.5.23}$$

However, the RPA cannot be used for electron liquids with densities corresponding to typical metals. It is necessary to go beyond the limits of the random phase approximation by performing a selective summation of an infinite number of certain diagrams in the expansion in Fig. 2.5.2. Such a generalization was first made by Hubbard (1957), who showed that the structure of the polarization operator is

$$\pi(\vec{q}) = \frac{\pi_0(\vec{q})}{1 - G(\vec{q})\pi_0(\vec{q})} \tag{2.5.24}$$

For $G(q) = 0$ we obtain the random phase approximation. If the exchange interaction is taken into account, the structure of $G(q)$ turns out to be

$$G(q) = \frac{1}{2}\frac{4\pi e^2}{q^2 + \lambda k_F^2} \tag{2.5.25}$$

When calculating the ground state energy of the electron gas, Hubbard assumed $\lambda = 1$.

If we use other approximations to find the ground state energy, the parameter λ, for internal consistency, will be defined from the correlation between $\pi(0)$ and the compressibility κ_e of the electron gas (Pines and Nozieres, 1966):

$$\pi(0) = Z^2 \frac{\kappa_e}{v_0^2} = Z\left(v_0^3 \frac{\partial^2 \varepsilon}{\partial v_0^2}\right)^{-1} \tag{2.5.26}$$

For example, if we use the approximation of Nozieres and Pines, λ is expressed by the so-called *Geldart-Vosko formula* (Geldart and Vosko, 1965, 1966):

$$\lambda = 2 \bigg/ \left(1 + \frac{0.0155\pi}{k_F a_B}\right) \tag{2.5.27}$$

Vashista and Singwi (1972) used another simple expression to approximate the function:

$$G(\vec{q}) = \frac{4\pi e^2}{q^2} A\left[1 - \exp(-Bq^2/k_F^2)\right] \tag{2.5.28}$$

The parameters A and B were determined from the sum rule in Eq. (2.5.26) and on the basis of the pair correlation function of the electron gas (Vashista and Singwi, 1972).

In Eq. (2.5.14), an arbitrary term Ω_n with $n>2$ can be expressed as follows:

$$\Omega_n = k_B T \frac{(-1)^n}{n!} \int d^3 r_1 d\tau_1 \ldots \int d^3 r_n d\tau_n \Gamma_n(\vec{r}_1\tau_1 \ldots \vec{r}_n\tau_n)$$
$$\times \sum_{\nu_1}\left[w(\vec{r}_1 - \vec{R}_{\nu_1}) - \frac{1}{V}\int w(\vec{r}_1')d^3\vec{r}_1'\right] \ldots \sum_{\nu_n}\left[w(\vec{r}_n - \vec{R}_{\nu_n}) - \frac{1}{V}\int w(\vec{r}_n')d^3\vec{r}_n'\right]$$

Here

$$\Gamma_n(\vec{r}_1\tau_1 \ldots \vec{r}_n\tau_n) = -\frac{1}{(n-1)!}$$
$$\times \sum_{\sigma_1 \ldots \sigma_n} \langle T_\tau\{\hat{\Psi}_{\sigma_1}^+(\vec{r}_1\tau_1)\hat{\Psi}_{\sigma_1}(\vec{r}_1\tau_1) \ldots \hat{\Psi}_{\sigma_n}^+(\vec{r}_n\tau_n)\Psi_{\sigma_n}(\vec{r}_n\tau_n)\}\rangle_C$$

is the *n-particle polarization Green's function,* which can be represented as the sum of connected and topologically non-equivalent diagrams with n fixed external vertices. If we look at the Fourier representation,

$$\Gamma_n(\vec{r}_1\tau_1 \ldots \vec{r}_n\tau_n) = \frac{(k_B T)^n}{V} \sum_{\omega_1 \ldots \omega_n} \sum_{k_1 \ldots k_n} \exp[-i(\omega_1\tau_1 + \ldots + \omega_n\tau_n)]$$
$$\times \exp[i(\vec{k}_1\vec{r}_1 + \ldots + \vec{k}_n\vec{r}_n)]\tilde{\Gamma}_n(\vec{k}_1 \ldots \vec{k}_n; \omega_1 \ldots \omega_n)$$
$$\times \frac{V}{k_B T}\delta(\omega_1 + \ldots + \omega_n)\delta(\vec{k}_1 + \ldots + \vec{k}_n)$$

where $\delta(x_1 + \ldots + x_n)$ is the Kronecker symbol, and if we then denote

$$\tilde{\Gamma}_n(\vec{k}_1 \dots \vec{k}_n) \equiv \tilde{\Gamma}_n(\vec{k}_1 \dots \vec{k}_n;\, 0 \dots 0)$$

we will obtain another expression for Ω_n:

$$\Omega_n = V\frac{(-1)^n}{nv_0^n}\sum_{\vec{k}_1\neq 0}\dots\sum_{\vec{k}_n\neq 0}\tilde{\Gamma}_n(\vec{k}_1\dots\vec{k}_n)\rho(\vec{k}_1)w(\vec{k}_1)\dots\rho(\vec{k}_n)w(\vec{k}_n)\delta(\vec{k}_1+\dots+\vec{k}_n) \tag{2.5.29}$$

Unfortunately, the expressions for $\tilde{\Gamma}_n$ with $n>2$ cannot be obtained in a form as simple as with $\tilde{\Gamma}_2$; however, a partial summation can be made by introducing the polarization operator $\pi(\vec{k}, \omega_n)$ again (Brovman and Kagan, 1974).

Then, for $(-1)^n\tilde{\Gamma}_n(\vec{k}_1 \dots \vec{k}_n;\, \omega_1 \dots \omega_n)$ we get the diagrammatic representation of Fig. 2.5.4, where the 'bubble' denotes the polarization propagator (we have omitted the symbol $\mathbf{F_0}$ inside the bubble for simplicity) and $f = 1$.

Here, Λ_n is determined by a combination of diagrams with n entrances and without parts that could be referred to one of the polarization parts. For instance, at $n = 3$, $(-1)^3\Lambda_3$ is given by the diagram in Fig. 2.5.5.

Therefore, the expression for Ω_n will be

$$\Omega_n = \frac{V}{nv_0^n}\sum_{\vec{k}_1\neq 0}\dots\sum_{\vec{k}_n\neq 0}\Lambda_n(\vec{k}_1\dots\vec{k}_n)\frac{\rho(\vec{k}_1)w(\vec{k}_1)}{\varepsilon(\vec{k}_1)}\dots\frac{\rho(\vec{k}_n)w(\vec{k}_n)}{\varepsilon(\vec{k}_n)}\delta(\vec{k}_1+\dots+\vec{k}_n) \tag{2.5.30}$$

After we have found the structure of the thermodynamic potential of the electron gas in the presence of ions, let us single out the terms without ion coordinates in the sum $H_\mathrm{i} + \Omega_\mathrm{ie}$:

$$H_\mathrm{i} + \Omega_\mathrm{ie} = \Omega_\mathrm{e} + \frac{NZb}{v_0} + N\Delta E + H^*(\vec{R}_1 \dots \vec{R}_N;\, \vec{P}_1 \dots \vec{P}_N) \tag{2.5.31}$$

Here,

$$\Delta E = \sum_{n\geq 2}\Delta E^{(n)} \tag{2.5.32}$$

$$\Delta E^{(2)} = -\frac{1}{2(2\pi)^3}\int\frac{\pi(q)}{1+(4\pi e^2/q^2)\pi(q)}\mathrm{d}^3q - \frac{bZ}{v_0} - \frac{Z}{2}v_0^2\frac{\partial^2\varepsilon}{\partial v_0^2} \tag{2.5.33}$$

ε is the energy per electron of a homogeneous electron gas, and $\Delta E^{(n)}$, with $n>2$, is the structureless contribution to Ω_n. Then, according to Eqs (2.5.3) and (2.5.7), the thermodynamic potential of the whole system will be expressed for liquid or amorphous metals as

$(-1)^n \cdot \tilde{\Gamma}_n(k_1, \ldots, k_n; \omega_1, \ldots, \omega_n)$

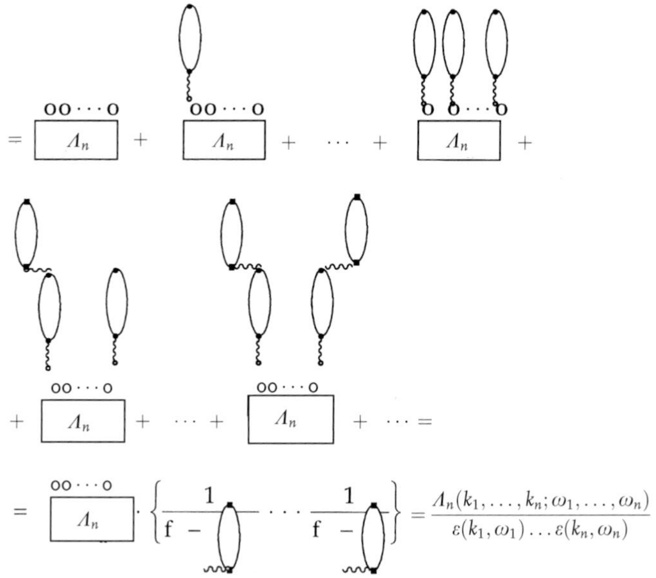

Fig. 2.5.4 Diagrammatic representation of $\Gamma_n(k_1, \ldots, k_n, \omega_1, \ldots, \omega_n)$. Here the 'bubble' represents the polarization propagator $\mathbf{F_0}$, and $f=1$

$$\Omega = \Omega_e + \frac{NZb}{v_0} + N\Delta E + \Omega_i \tag{2.5.34}$$

in which

$$\Omega_i = -k_B T \cdot \ln\left\{ \mathrm{Tr}^{(i)}\left[\exp\left(\frac{-H^* + \mu_i N}{k_B T} \right) \right] \right\} \tag{2.5.35}$$

The operation $\mathrm{Tr}^{(i)}[\ldots]$ is understood as summation by the number of ions and integration over the whole phase space of the ionic subsystem with a fixed number of ions. Equation (2.5.35) expresses the thermodynamic potential of a system of classical particles (ions), whose interaction may be described by a certain effective many-particle potential, namely:

$$H^* = \sum_{\nu=1}^{N} \frac{\vec{P}_\nu^2}{2M} + U^*(\vec{R}_1 \ldots \vec{R}_N) \tag{2.5.36}$$

Fig. 2.5.5 Diagrammatic representation of $\Lambda_3(k_1, k_2, k_3, \omega_1, \omega_2, \omega_3)$

in which

$$U^*(\vec{R}_1 \ldots \vec{R}_N) = U_2(\vec{R}_1 \ldots \vec{R}_N) + U_3(\vec{R}_1 \ldots \vec{R}_N) + \ldots \tag{2.5.37}$$

$$U_m(\vec{R}_1 \ldots \vec{R}_N) = \sum_{n \geq m} U_m^{(n)}(\vec{R}_1 \ldots \vec{R}_N) \tag{2.5.38}$$

$U_m^{(n)}(\vec{R}_1 \ldots \vec{R}_N)$ is the potential energy of the effective ion-ion interactions as far as they are of the nth order in the pseudopotential and describe non-reducible interactions between m particles. In particular, the pairwise interactions of particles in the second order in the pseudopotential are expressed as

$$\begin{aligned} U_2^{(2)}(\vec{R}_1 \ldots \vec{R}_N) &= \frac{1}{2} \sum_{\substack{\mu,\nu \\ \mu \neq \nu}} \Phi_2^{(2)}(\vec{R}_\mu, \vec{R}_\nu) \\ &= \frac{N}{2v_0} \sum_{\vec{q}} \left(|\rho(\vec{q})|^2 - \frac{1}{N} \right) \left\{ \frac{Z^2 4\pi e^2}{q^2} - \frac{\pi(\vec{q})|w(\vec{q})|^2}{1 + (4\pi e^2/q^2)\pi(\vec{q})} \right\} \end{aligned} \tag{2.5.39}$$

Equations (2.5.36) to (2.5.39) allow us to see quite clearly how the electrons influence the indirect interaction between ions: The term U_2 describes a central interaction between pairs of ions that consists (i) of a *direct* Coulomb repulsion and (ii) the *indirect* ion-electron-ion interaction. The terms U_m with $m>2$ describe indirect interactions among m ions. Here successive terms U_m contain at least one additional factor of the small parameter $\tilde{w}(\vec{q})/E_F$. For this reason, the terms corresponding to non-pair interactions have a higher order of smallness in the pseudopotential than U_2 and so they can be neglected in a quadratic approximation. The analysis of various properties of metals, however, shows that this approximation is sufficient for univalent (alkaline) metals only. With polyvalent metals, the expression for U^* should take into account the terms that describe many-particle interactions of ions.

2.6 The Effective Two-Particle Ion-Ion Interaction (to Second Order in the Pseudopotential)

We cannot study thermodynamic properties without a detailed analysis of the ion-ion interaction. Let us start this analysis with alkaline metals in the liquid or crystalline state because the many-particle forces can be neglected in this case to a good accuracy.

According to Eq. (2.5.39) the effective potential of the pairwise ion-ion interaction is expressed by

$$\Phi(\vec{R}) = \frac{1}{2\pi^2 R} \int_0^\infty \sin(qR) \left\{ \frac{Z^2 4\pi e^2}{q^2} - \frac{\pi(\vec{q})|w(\vec{q})|^2}{1 + (4\pi e^2/q^2)\pi(\vec{q})} \right\} q\mathrm{d}q \tag{2.6.1}$$

We can find the typical properties of the interion potential by using for instance the Cohen-Abarenkov-Heine two-parameter pseudopotential in the Fourier representation (see Eq. (2.3.5) above)

$$\Omega_0 = -k_B T \cdot \ln\left\{ \mathrm{Tr}^{(e)} \left[\exp\left(\frac{-\hat{H}_0 + \mu_e \hat{N}_e}{k_B T} \right) \right] \right\} \tag{2.6.2}$$

and also the polarization operator of Geldart and Vosko, see Eqs (2.5.24)–(2.5.27).

The u and r_0 parameters of this pseudopotential, taken from Brovman *et al.* (1970) for Na and K, and from Cohen (1962) for Na, K, Rb and Cs, are given in Table 2.6.1.

Numerical calculations have been performed for the temperatures and densities corresponding to eight points on the coexistence curve (Krasny and Onishchenko, 1972). Figure 2.6.1 shows the calculated results for all the alkaline metals.

We can see from Fig. 2.6.1 (A to C) that the ion-ion interaction hardly depends on the parameters of the pseudopotential: the difference between the potential curves calculated at different parameters, i.e. the dashed and solid lines, does not exceed 10%. The curves for *sodium* begin to differ noticeably from the results for 0 K close to the critical temperature, and for *potassium* at 1000 K and higher. The interion potential proves to be highly dependent on the polarization properties of the electron gas.

This follows directly from Figs 2.5.2 and 2.5.3 above, which demonstrate how the depth ε of the potential well and the effective diameter σ of the ion (at $R = \sigma$ and $\Phi(\sigma) = 0$) depend on the electron concentration. The depth ε of the potential well increases with decreasing particle density: it is 3 to 4 times higher near the critical point. The effective diameter σ of the ion also decreases, but only a little. The shape of the potential curve, and the position of its first minimum R_e (see Fig. 2.6.3), remain practically unchanged. There are some Friedel oscillations at long distances, but they are very small.

The potential curves in the *reduced coordinates* $\Phi(R)/\varepsilon$ and R/R_e, which correspond to points on the coexistence curve with equal reduced values $T^* := k_0 T/\varepsilon$ and $v_0^* := v_0/R_e^3$, agree within 1–3% for all alkaline metals (Fig. 2.6.4).

The curves corresponding to different points on the coexistence curve up to the boiling point also agree within 2–5% (curve 1). A distinction will only be made for the curves at temperatures higher than the boiling point (curve 2 in Fig. 2.6.3 corresponds to the critical point of the metals). This distinction however is insignificant.

Tab. 2.6.1 Parameters of the model pseudopotential (atomic units)

Parameter	*Na*	*K*	*Rb*	*Cs*
r_0	2.0731	2.9726	–	–
r_0	2.1	3.2	4.6	5.5
u	–0.3632	–0.5399	–	–
u	–0.4452	–0.6960	–1.0373	–1.1000

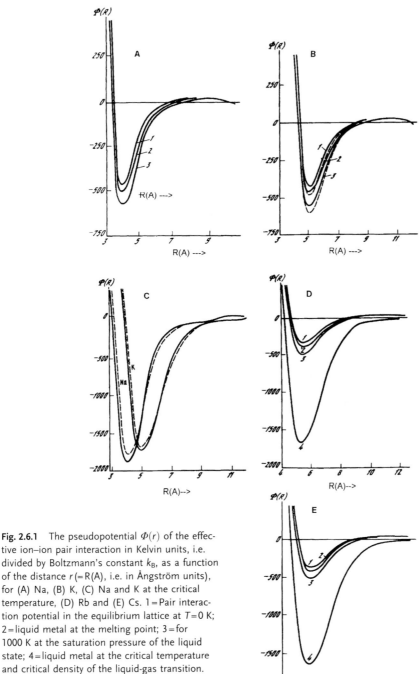

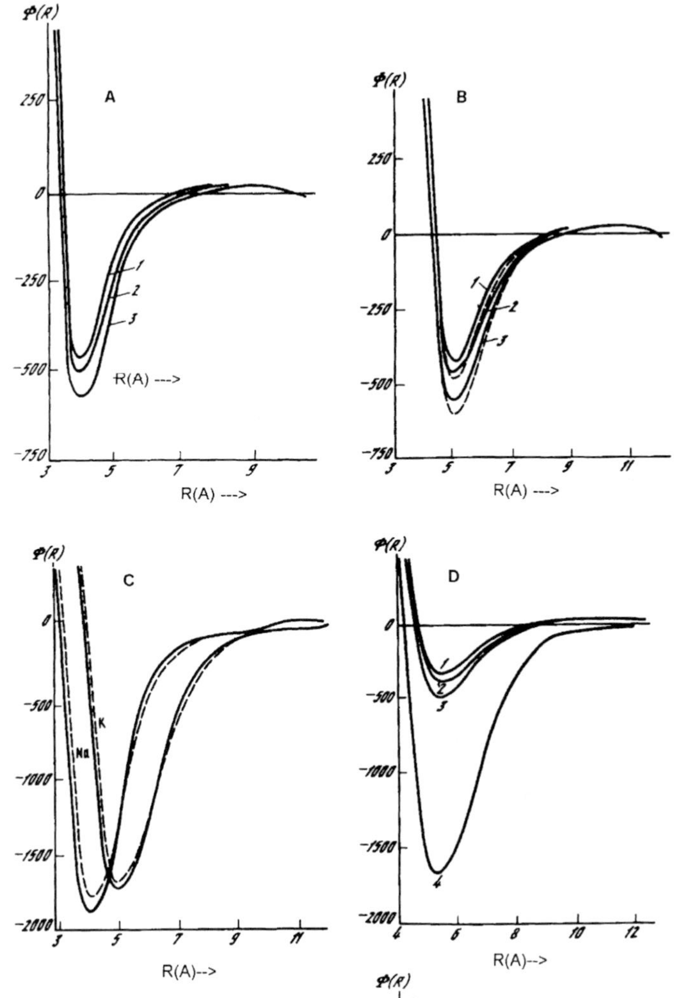

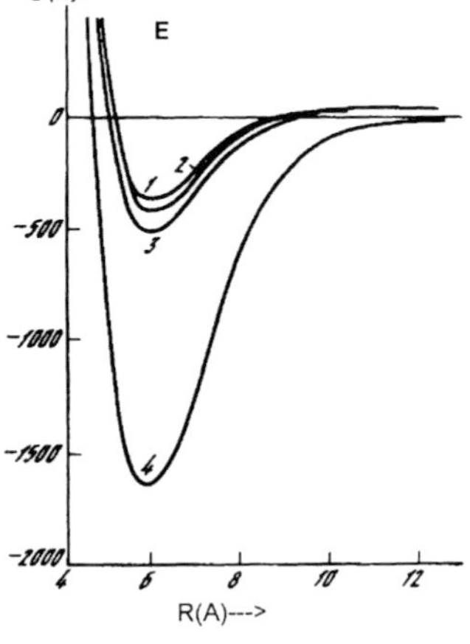

Fig. 2.6.1 The pseudopotential $\Phi(r)$ of the effective ion–ion pair interaction in Kelvin units, i.e. divided by Boltzmann's constant k_B, as a function of the distance r (= R(A), i.e. in Ångström units), for (A) Na, (B) K, (C) Na and K at the critical temperature, (D) Rb and (E) Cs. 1 = Pair interaction potential in the equilibrium lattice at $T = 0$ K; 2 = liquid metal at the melting point; 3 = for 1000 K at the saturation pressure of the liquid state; 4 = liquid metal at the critical temperature and critical density of the liquid-gas transition. Dashed lines: calculated with parameters from Brovman *et al.* (1970). Solid lines: calculated with Cohen's parameters (Cohen, 1962)

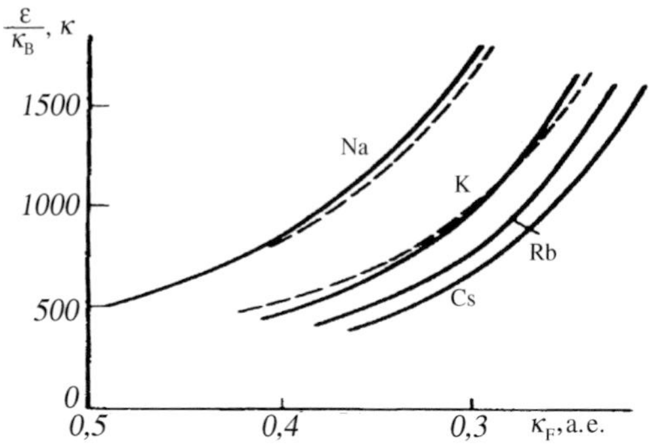

Fig. 2.6.2 Dependence of the potential well depths ε/k_B (see Fig. 2.6.1) on the Fermi wavenumber k_F in atomic units

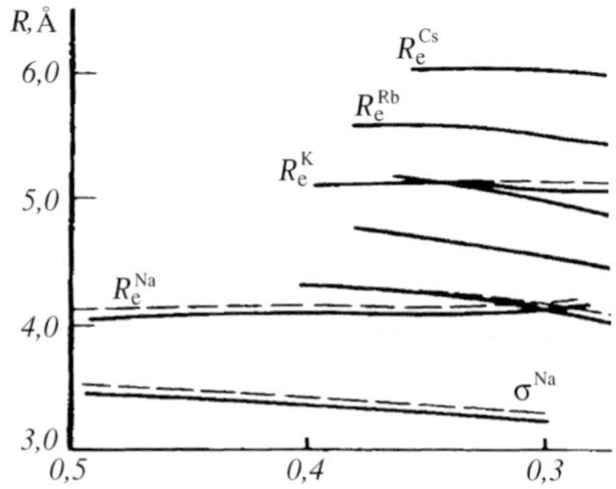

Fig. 2.6.3 Effective collision diameter σ and equilibrium diameter R_e in the liquid alkaline metals Cs, Rb, K and Na (from top); see Fig. 2.6.1

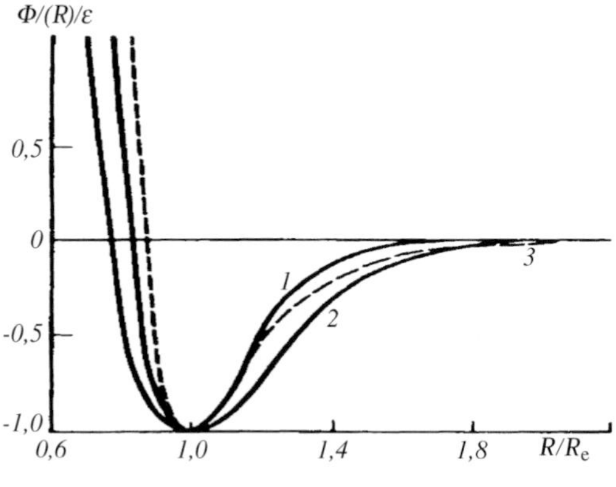

Fig. 2.6.4 Dimensionless pair-interaction pseudopotential $\Phi(R)/\varepsilon$ as a function of R/R_e for liquid alkaline metals. Curve 3 shows the Lennard-Jones potential for comparison. Lines 1 and 2 are on the coexistence curve and for the critical point, respectively

Therefore the interion potential of all the alkaline metals can be represented with good accuracy as

$$\Phi(R) = \varepsilon\varphi(R/R_e) \tag{2.6.3}$$

in which ε (the depth of the potential well) and R_e (the position of the minimum of the potential curve) are functions of the density. The function $\varphi(x)$ is common ('*universal*') for all the alkaline metals up to the boiling point. However, above the boiling point this function depends both on x and on the density. This dependence however is so small that it can be neglected in a qualitative analysis of thermodynamic properties. The *Lennard-Jones potential* (LJP) is shown for comparison in the same coordinates in Fig. 2.6.4 (curve 3). We can see that the function $\varphi(x)$ is approximated by the LJP curve quite well, however the repulsive core in the metals is much softer than in the Lennard-Jones liquid.

Due to the presence of the function $\varphi(x)$, which is common for all liquid alkaline metals, the thermodynamic potential of the ion subsystem Ω_i can also be expressed with the help of a universal function Ω_i^*:

$$\Omega_\mathrm{i}/\varepsilon = \Omega_\mathrm{i}^*(T^*, \mu^*, V^*) \qquad (2.6.4)$$

in which $T^* = k_\mathrm{B}T/\varepsilon$; $V^* = V/R_\mathrm{e}^3$; $\mu^* = \mu/\varepsilon$ are the reduced temperature, volume and chemical potential, respectively. Due to the presence of the universal function Ω_i^* and a due to a rather weak temperature dependence of the thermodynamic potential Ω_e of the electron subsystem, the reduced parameters of the triple point, and also for the critical point, should be the same for all the liquid alkaline metals. This is confirmed by the data in Table 2.6.2, which shows results derived from experiment.

It was shown by Krasny and Onishchenko (1972), Krasny and Kovalenko (1981) and Bratkowskii *et al.* (1984) that the ion-ion interaction in liquid alkaline metals hardly depends on the already chosen pseudopotential parameters; moreover it is barely sensitive to the type of pseudopotential chosen. For instance we can obtain similar results by means of the modified pseudopotential of Animalu and Heine, which has already been mentioned, see Eq. (2.3.7) above:

$$w(q) = -\frac{4\pi Ze^2}{q^2}\left[(1+u)\cos(qr_0) - u\frac{\sin(qr_0)}{qr_0}\right]\exp(-\delta q^4) \qquad (2.6.5)$$

Tab. 2.6.2 Reduced parameters of the triple and critical points for alkaline metals

Parameter	*Na*	*K*	*Rb*	*Cs*	*Average*
T_melt, K [8]	371.00	336.66	312.65	301.55	
$(\varepsilon/k_\mathrm{B})_{T=T_\mathrm{melt}}$, K	501.9	469.5	396.0	418.5	
T^*_melt, K	0.739	0.717	0.790	0.721	0.742
V_melt, Å/particle	41.15	78.37	96.42	120.1	
$(R_\mathrm{e})_{T=T_\mathrm{melt}}$, Å	4.09	5.12	5.57	6.06	
V^*_melt	0.601	0.584	0.558	0.532	0.569
T_cr, K	2573	2223	2093	2057	
$(\varepsilon/k_\mathrm{B})_{T=T_\mathrm{cr}}$	1864	1708	1658	1644	
T^*_cr, K	1.380	1.301	1.263	1.251	1.299
V_cr, Å/particle	185.3	334.7	410.2	515.6	
$(R_\mathrm{e})_{T=T_\mathrm{cr}}$, Å	4.16	5.10	5.44	5.91	
V^*_cr, Å/particle	2.57	2.52	2.55	2.50	2.53

2.7
Effective Ion-Ion Pair Interaction to Cubic Order in the Pseudopotential

It was mentioned earlier that the electron liquid in a metal causes non-pair forces of the covalent type. *Pair forces* are naturally caused by the direct and indirect ion-ion interaction determined by the first and second summands in Eq. (2.6.1), respectively. In an amorphous system, where microscopically the translation invariance is broken, the pair potentials should also depend, at least in principle, on the true positions of the considered atoms and their surroundings, i.e. these pair potentials should be non-central and can in principle only be determined – if at all – by extensive computer simulations, separately for a given realization of the amorphous system. Here the recent numerical study of Moroni *et al.* (1997) is mentioned again; it is particularly relevant for the pair forces concerning the *metalloid* atoms such as *B, P, Si,* in metal-metalloid amorphous alloys. However, as in the preceding paragraph, in the following we concentrate on translational invariant systems as liquid metals, rsp. on (s-p-dominated) metal-metal amorphous metals such as the $Mg_{1-x}Zn_x$ system (von Heimendahl, 1979), where usually only translational invariant averages are considered and the pair potentials are also assumed to be rotationally invariant. In fact, however, for an amorphous metal with 3d electrons this may be not correct, if the contribution from the metallic d electrons to the binding forces is non-negligible.

Along with the pair potentials, there are also non-reducible interactions binding n ions, with $n \geq 3$. We can see from Eq. (2.5.31) that Ω_n contains contributions from n interacting particles as well as from a smaller number of ions, because scattering may occur a number of times on one and the same ion (multiple scattering). Therefore the effective pair interaction is described in fact by a *series in powers of the pseudopotential* (PP):

$$\Phi_2(R) = \Phi_2^{(2)}(R) + \Phi_2^{(3)}(R) + \ldots \tag{2.7.1}$$

in which the upper index shows the order in PP, and $\Phi_2^{(2)} \equiv \Phi$. Similarly, if we introduce the potential $\Phi_n(\vec{R}_1 \ldots \vec{R}_N)$ of a non-reducible n-particle interaction with $n \geq 3$, it will also be described by a series similar to Eq. (2.7.1) and its first term will be of nth order in the PP.

Brovman and Kagan (1970) carried out a qualitative analysis of this series. It is important for our further analysis that the terms of different orders in the PP differ in their roles. According to calculations made for crystal structures it is the third-order terms in the PP that are marked out for many metals, especially the *polyvalent* ones. This is due to a significant compensation of the contributions from the ion lattice and from the electron energy in the second order in PP, which gives the terms of third order a notable weight. Higher-order terms with $n \geq 4$, as a rule, play an insignificant role.

The effective Hamiltonian of the ionic subsystem, correct to third order in PP, is

$$H^* = \sum_{\nu=1} \frac{\vec{P}_\nu^2}{2M} + \frac{1}{2} \sum_{\nu_1 \nu_2} \Phi_2(\vec{R}_{\nu_1} - \vec{R}_{\nu_2}) + \frac{1}{3!} \sum_{\nu_1 \nu_2 \nu_3}{}' \Phi_3(\vec{R}_{\nu_1}, \vec{R}_{\nu_2}, \vec{R}_{\nu_3}) \tag{2.7.2}$$

in which Φ_2 is determined by the terms of Eq. (2.7.1) and by

$$\Phi_3(\vec{R}_1, \vec{R}_2, \vec{R}_3) = \Phi_3^{(3)}(\vec{R}_1, \vec{R}_2, \vec{R}_3)$$

Using Eq. (2.5.31) we obtain:

$$\Phi_3^{(3)}(\vec{R}_1, \vec{R}_2, \vec{R}_3) = \frac{3!}{3} \frac{v_0^3}{V^2} \sum_{\substack{\vec{k}_1 \vec{k}_2 \vec{k}_3 \\ (k \neq 0)}} \tilde{\Lambda}_3(\vec{k}_1, \vec{k}_2, \vec{k}_3)$$
$$\times \exp[-i\vec{k}_1 \cdot \vec{R}_1 - i\vec{k}_2 \cdot \vec{R}_2 + i(\vec{k}_1 + \vec{k}_2) \cdot \vec{R}_3] \Delta(\vec{k}_1 + \vec{k}_2 + \vec{k}_3) \tag{2.7.3}$$

in which

$$\tilde{\Lambda}_3 = \frac{1}{v_0^3} \Lambda_3 \frac{w(\vec{k}_1)}{\varepsilon(\vec{k}_1)} \frac{w(\vec{k}_2)}{\varepsilon(\vec{k}_2)} \frac{w(\vec{k}_3)}{\varepsilon(\vec{k}_3)}$$

It follows from Eq. (2.7.3) that, if the third-order terms are taken into account, then

(i) the two-particle interaction is renormalized, and
(ii) a non-reducible three-particle interaction appears, both being of the same order of magnitude and both must be equally taken into account in any successive calculation of the free energy of the metal.

Let us now consider the $\Phi_2^{(3)}$ contribution to the effective ion-ion interaction. Figures 2.7.1 and 2.7.2 show the numerical results for the effective pair potential of liquid K and Al with different choices of the model PP and of the screening function.

In Fig. 2.7.1, with $\Phi_2^{(3)}$ taken into account, the potential well of the lines corresponding to line 2 and 2′, i.e. the 'renormalized potential' Φ_2, is much deeper and the position of the potential minimum is slightly shifted towards smaller distances, compared with the non-renormalized potential $\Phi_2^{(2)}$ i.e. with the lines 1 and 1′. Similarly to the unrenormalized potential, in square approximation in the PP the renormalized $\Phi(R)$ also has a typical behavior for alkaline metals: the depth of the potential well increases significantly with increasing temperature (along the coexistence curve).

Figure 2.7.2 presents the results of a calculation of $\Phi(R)$ for a non-alkaline metal (liquid aluminum) at the melting temperature. The effect of the renormalization of the pair potential is most significant here: the depth of the well grows by an order of magnitude, which happens irrespective of the chosen dielectric function.

Our conclusion about the influence of $\Phi_2^{(3)}$ on Φ_2 for the alkaline metals is qualitatively consistent with data from other authors (Hasegawa, 1976; Evans and Kumaravadivel, 1976). Hasegawa (1976) attempted to define the influence of the

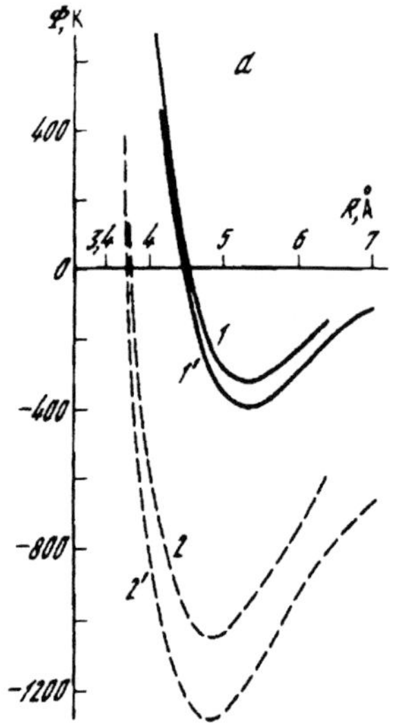

Fig. 2.7.1 The interion interaction potential in liquid K, in Kelvin units. Dielectric function: Geldart-Vosko; pseudopotential: Krasko-Gurski (a) and Ashcroft (b); 1, 1′, $\Phi_2^{(2)}(R)$; 2, 2′, $\Phi_2(R)$; 1, 2, at T_{melt}; 1′, 2′, at $T = 1000$ K

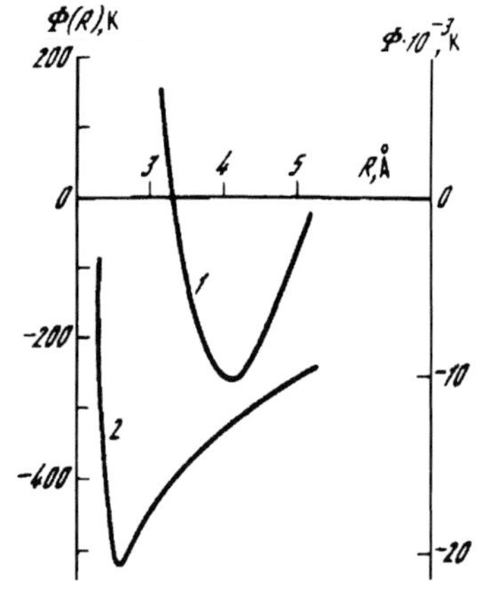

Fig. 2.7.2 The same for liquid Al at T_{melt}. Pseudopotential: Krasko-Gurski; 1, $\Phi_2^{(2)}(R)$ (left axis); 2, $\Phi_2(R)$ (right axis); $\Phi_2(R)$ is the 'renormalized' interaction, see text

third-order terms on the structure of liquid Na and K. He calculates the sum of the first two summands in the series in Eq. (2.7.1). The role played by the summand $\Phi_2^{(3)}(R)$ is firstly a significant (40–60%) increase on the potential well determined by $\Phi_2^{(2)}$, and secondly a slight reduction of the potential's repulsive 'core'. Such a big increase in the depth of the potential well makes a significant impact on the system's structure.

However, the agreement with experiment reached in the second order becomes much poorer when the structure factor $S(k)$ is calculated on the basis of the Gibbs-Bogolyubov variational principle (the real metal was modeled through a system of hard spheres with a diameter that minimizes the free energy of the system) and with the addition of $\Phi_2^{(3)}$ to the pair potential taken into account.

Kumaravadivel and Evans (1976) came to the same conclusion. They calculated the entropy and the heat capacity of liquid sodium and potassium with the help of the Weeks-Chandler-Anderson perturbation theory (Weeks *et al.*, 1971), in which the system in zero order is substituted by hard spheres; the hard sphere diameter is determined by the pair interaction potential; the calculations show satisfactory agreement with experiments for entropy and heat capacity with the diameter value obtained in the second order, i.e. *without* the renormalization. When $\Phi_2^{(3)}$ is included, the agreement is much worse, and the authors conclude that there is no need to take the third order into account for sodium and potassium (although it may perhaps be necessary in calculating the properties of polyvalent metals).

Let us now take into account the irreducible three-particle ion-ion interaction described by the potential $\Phi_3^{(3)}(\vec{R}_1, \vec{R}_2, \vec{R}_3)$. If higher-order terms in the PP are taken into account, as suggested by Kovalenko and Kuz'mina (1981), the potential energy

$$u = \frac{1}{2} \sum_{\nu_1 \nu_2}{}' \tilde{\Phi}_2(\vec{R}_{\nu_1}, \vec{R}_{\nu_2})$$

in which

$$\tilde{\Phi}_2(\vec{R}_{\nu_1}, \vec{R}_{\nu_2}) = \Phi_2^{(2)}(\vec{R}_{\nu_1}, \vec{R}_{\nu_2}) + \Phi_2^{(3)}(\vec{R}_{\nu_1}, \vec{R}_{\nu_2}) + \frac{2!}{3!} \sum_{\nu_3} \Phi_3^{(3)}(\vec{R}_{\nu_1}, \vec{R}_{\nu_2}, \vec{R}_{\nu_3})$$

This is in fact a formal way of expressing u as a sum of 'pair' interactions because the last summand contains the coordinates of three ions. Let us average $\Phi_3^{(3)}(\vec{R}_1, \vec{R}_2, \vec{R}_3)$ over the position of one particle in order to change over to the effective pair potential. Because two particles remain in a fixed position in this averaging, we have to use the conditional distribution functions $F_1(R_1/(\vec{R}_2, \vec{R}_3))$. Thus we obtain finally

$$\left\langle \sum_{\nu_3} \Phi_3^{(3)}(\vec{R}_{\nu_1}, \vec{R}_{\nu_2}, \vec{R}_{\nu_3}) \right\rangle = \frac{N-2}{V} \int \Phi_3^{(3)}(\vec{R}_{\nu_1}, \vec{R}_{\nu_2}, \vec{R}_{\nu_3}) F_1(\vec{R}_{\nu_3}/(\vec{R}_{\nu_1}, \vec{R}_{\nu_2})) \mathrm{d}^3 \vec{R}_{\nu_3}$$

As we go over to ordinary distribution functions the effective pair potential in the thermodynamic limit can be expressed as follows:

$$\begin{aligned} \Phi_2^{\mathrm{eff}}(\vec{R}_1 - \vec{R}_2) &= \Phi_2^{(2)}(\vec{R}_1 - \vec{R}_2) + \Phi_2^{(3)}(\vec{R}_1 - \vec{R}_2) \\ &+ \frac{n}{3} \int \Phi_3^{(3)}(\vec{R}_1, \vec{R}_2, \vec{R}_3) \frac{F_3(\vec{R}_1, \vec{R}_2, \vec{R}_3)}{g(\vec{R}_1, \vec{R}_2)} \mathrm{d}^3 R_3 \end{aligned} \qquad (2.7.4)$$

The effective pair potential introduced in this way allows us to express the total free energy of the metal correctly.

For numerical calculations let us use Kirkwood's superposition approximation:

$$F_3(\vec{R}_1, \vec{R}_2, \vec{R}_3) \cong g(\vec{R}_1 - \vec{R}_2) g(\vec{R}_2 - \vec{R}_3) g(\vec{R}_3 - \vec{R}_1) \qquad (2.7.5)$$

The accuracy of this approximation for calculations of the effective interaction (in the case of the Lennard-Jones pair potential) was estimated by Ram and Singh (1977). They showed that Eq. (2.7.5) leads to an effective pair interaction, which fits accurate results in a better way than the pretty successful Percus-Yevick theory, which includes three-particle interactions (Rushbrooke and Silbert, 1967). In the superposition approximation we get

$$\Phi_2^{\text{eff}}(R) = \Phi_2^{(2)}(R) + \Phi_2^{(3)}(R) + \Delta\Phi(R)$$

in which the decisive quantity is

$$\begin{aligned}\Delta\Phi(R) = \frac{2n}{3(2\pi)^6}\Bigg\{&\frac{1}{(2\pi)^3}\iint \mathrm{d}^3 k_1\, \mathrm{d}^3 k_2 \exp(i\vec{k}_2\cdot\vec{R})\Lambda_3(\vec{k}_1,\vec{k}_2,-\vec{k}_1-\vec{k}_2)\\ &\times\int \mathrm{d}^3k_3\, h(\vec{k}_3)\, h(\vec{k}_1+\vec{k}_2-\vec{k}_3)\exp(-i\vec{k}_3\cdot\vec{R})\\ &+\iint \mathrm{d}^3 k_1\, \mathrm{d}^3 k_2 \exp(i\vec{k}_2\cdot\vec{R})\tilde{\Lambda}_3(\vec{k}_1,\vec{k}_2,-\vec{k}_1-\vec{k}_2)\, h(\vec{k}_1+\vec{k}_2)\\ &\times[\exp(-i\vec{k}_1\cdot\vec{R})+\exp(i\vec{k}_2\cdot\vec{R})]\Bigg\}\end{aligned} \tag{2.7.6}$$

Figure 2.7.3 shows the effective ion-ion interaction calculated for liquid potassium. The Gurski-Krasko pseudopotential was used, and $h(K)$ was taken from the solution of the Percus-Yevick equation for a system of hard spheres. We can see from Fig. 2.1.6 that the potential $\Delta\Phi(R)$ is hardly changed by introducing $\Phi_2^{(2)} + \Phi_2^{(3)}$, although there is partial mutual compensation of the third-order contributions to the effective potential. The obtained potential has the same form as in the second order.

Hasegawa (1976) suggests a different way of taking $\Phi_3^{(3)}$ into account: according to this method, $\Phi_3^{(3)}$ is considered as a perturbation for $\Phi_2^{(3)}$ taken into account ac-

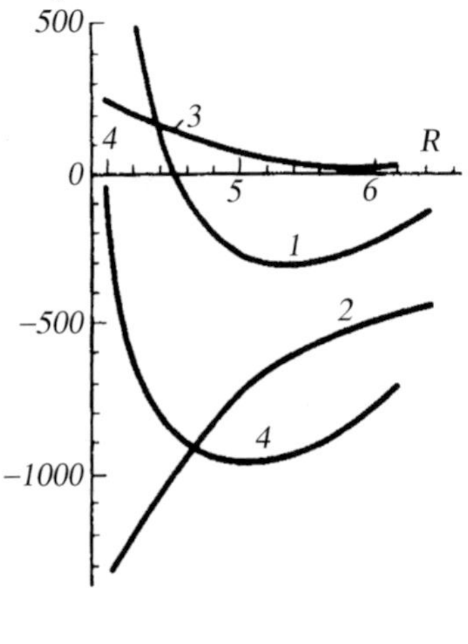

Fig. 2.7.3 The effective interion interaction potential $\Phi(R)$ in Kelvin units at T_{melt} as a function of R (Å). Pseudopotential: Gurski-Krasko (1971); dielectric function: Geldart-Vosko, see above (Eqs. (2.5.24) to (2.5.27)). 1, $\Phi_2^{(2)}(R)$; 2, $\Phi_2^{(3)}(R)$; 3, $\Delta\Phi(R)$; 4, $\Phi_2^{\text{eff}}(R)$

curately. The author uses an effective potential, which leads to the correct radial distribution function (RDF). Calculations with the experimental RDF show the potential (2.7.2) growing *deeper* on account of $\Delta\Phi(R)$. If we use instead the non-perturbative Percus-Yevick equation for $g(R)$ and then reconstruct the inter-particle potential with the help of the experimental structure factor, the 'empirical' potential thus obtained will show better agreement with $\Phi_2^{(2)}$, while again $\Phi_2^{(3)}$ and $\Delta\Phi$ make this agreement worse.

We now discuss the results obtained. It has been shown that the ion-ion interaction calculated in the second order of perturbation theory is significantly renormalized by the terms of the third order; the renormalization effect remains large with any PP type and any form of dielectric function and grows still bigger with the transition from alkaline to polyvalent metals. On the other hand, the calculation of the *compressibility* of these metals in the second order of PP shows that the third-order terms should be quite small for the alkaline metals. These results are apparently contradictory.

However, a significant decrease in the effective scattering diameter due to $\Phi_2^{(3)}$ should have some influence on the compressibility because the compressibility of a dense liquid is determined by the 'hard core' of the inter-particle potential. These results suggest the following interpretation: the third-order terms should largely compensate each other in the calculation of the equilibrium properties of liquid metals. This is also indicated by the calculated effective pair interaction potential (see Fig. 2.1.2) where partial mutual compensation of the third-order terms can be observed. Unfortunately a direct comparison between $\Phi_2^{(3)}$ and $\Phi_3^{(3)}$ is not possible, and therefore $\Phi_3^{(3)}$ should be averaged in order to keep within the scheme of purely two-particle interaction. The expression (2.7.4) is one of the ways of such averaging. It is clear that the averaging has resulted in a smaller amplitude of $\Phi_3^{(3)}$ and thus concealed the effects of actual compensation. Such effects are well known for metals in the crystalline phase.

In fact, Brovman and Kagan (1974) showed that it is not expedient to carry out separate calculations for pair and non-pair interactions of terms of a certain order in PP because the contributions from $\Phi_2^{(3)}$ and $\Phi_3^{(3)}$ to the metal's properties are of the same order of magnitude and can largely compensate each other. The crystalline phase of metals is traditionally described in terms of momentum representation, which does *not* require calculation of the potentials of inter-particle interaction first to obtain the metal's physical properties on their basis. Moreover, one can do totally without explicit use of inter-particle potentials due to crystal symmetry so that all the summation can be done in reciprocal space. In that case the order of smallness of this or that term in the series of the perturbation theory is naturally evaluated from the PP power and not from the rank of pair or non-pair interactions.

In dealing with liquids or amorphous state two-particle interaction, three-particle interaction, etc., have to be treated separately in the potential energy because, when equilibrium properties are calculated, pair and non-pair interactions require using many-particle distribution functions of different orders, which cannot be reduced to one another in principle. For non-conducting liquids such as argon the three-particle interaction is usually taken into account with the help of thermodynamic perturbation theory where the potential of non-reducible three-particle interaction acts as the perturbation (Ram and Singh, 1977; Sinha *et al.*, 1977). Hasegawa (1976) used the same technique for liquid metals (Na and K). Our analysis shows that this approach is unsatisfactory for liquid metals. A consistent approach demands that $\Phi_2^{(3)}$ and $\Phi_3^{(3)}$ should be considered with equal significance because they have the same order in PP.

2.8
Many-Particle Interactions in Metallic Hydrogen

2.8.1
Introduction

This section shows that for ions with small masses the many-body terms of the ion-ion interaction really become important and may give rise, for example, to two-level subsystems that somehow reflect collective quantum oscillatory states of three (or more) ion groups. This problem is reminescent of tunneling states in amorphous systems, which will be one of the main topics of Chapter 3.

The significance of many-particle interactions and their role in forming the structure and properties of various condensed systems are not yet well known (Croxton, 1974). For the calculation of many-particle interaction potentials, knowledge of the linear and non-linear response functions of the electron gas is necessary. At present the explicit form of the three-pole diagram expression in the so-called 'ring approximation' for the homogeneous electron gas is known (Lloyd and Sholl, 1968; Brovman and Kagan, 1974). It was possible to calculate the equilibrium structure and dynamic properties of simple metal crystals and metallic hydrogen to the third order to the electron-ion potential (Brovman and Kagan, 1974; Brovman *et al.*, 1971, 1972; Kagan *et al.*, 1977; Hasegawa, 1976; Chakravarty *et al.*, 1981; Hammerberg and Ashcroft, 1974), as well as to study three-ion interaction potentials in simple metals (Brovman and Kagan, 1974; Hasegawa, 1976).

The study of the possibility of a metastable state of metallic hydrogen at zero pressure has special importance (Brovman *et al.*, 1971). In the case of solid metallic hydrogen, calculations have been performed by Brovman and Kagan (1974). They calculated the structure, elastic properties and phonon spectrum of solid metallic hydrogen at zero pressure and proved the local stability of the metastable metallic hydrogen phase in the framework of the many-particle theory of metals (Brovman *et al.*, 1971). They found that, in third order of the electron-ion potential, an energy minimum exists for two-parameter hexagonal lattices with triangular Bravais lattices (Brovman *et al.*, 1971): For all structures with an energy minimum the elementary cell volume was $\Omega_0 = 20.8a_B^3$ (a_B is the Bohr radius) and the interproton spacing fixed along the z-axis was equal to $d=2.04a_B$. Investigations of higher-order approximations did not change these conclusions essentially. This is connected with the fact that eventually a small parameter in the perturbation expansion for metallic hydrogen energy at zero pressure approximately equals 1/5 (Brovman *et al.*, 1971; Hasegawa, 1976). In such a case the sum of all terms with orders $n=5$ of the perturbation expansion gives an error as the dielectric permittivity of the uniform electron gas does (Hasegawa, 1976; Hammerberg and Ashcroft, 1974).

The evaluation of the lifetime of the metastable *metallic* hydrogen phase relative to the spontaneous quantum tunneling transition to the *insulating* phase is one of the most important problems. This estimate cannot be made without considering the specific microscopic mechanism of the nucleation of the H_2 molecules or the H_2^+ ions in the metallic phase. For metallic hydrogen as a system, in which the tendency to diatomic ordering exists (Chakravarti *et al.*, 1981), the calculations of interionic interaction potentials, and the elucidation of this tendency in the language of interaction potentials of ion groups in configuration space, are of great interest. In the quantum kinetics of the formation of the new phase two characteristic times may be distinguished (Lifshitz and Kagan, 1972). In the first place there is the *characteristic tunneling time* through the potential barrier (in our case the lifetime of the metallic phase relative to nucleation of the H_2 molecule). Secondly there is the *total time of formation of the new phase.* As a consequence of the significant difference between the electron and proton masses, during the process of homogeneous nucleation of the molecular hydrogen phase the characteristic time will be determined by slow ionic motion in the potential relief created by nearest neighboring ion groups. Diatomic ordering in metallic hydrogen corresponds to the possibility of drawing together the ions to a distance corresponding to the interatomic separation in a hydrogen molecule. Inasmuch as the direct interionic interaction corresponds to the Coulomb repulsion, the possibility of drawing ions together may be ensured only by means of an *indirect* interaction of the ions induced by the surrounding electron gas.

The purpose of this section is to consider pair- and three-ion interactions in *metastable metallic hydrogen at zero pressure.* The consideration is based on the many-particle approach of Brovman and Kagan (1974); see also Brovman *et al.* (1971, 1972) and Kagan *et al.* (1977). The peculiarities of the potential relief for ion groups are discovered and interpreted from the point of view of formation of the molecular hydrogen phase by many-particle tunneling.

2.8.2
Many-body Interactions in Metals

In the framework of the many-particle theory of non-transition metals the energy of the electron subsystem in the field of fixed ions may be written in the adiabatic approximation as (Brovman and Kagan, 1974):

$$E_e = \varphi_0 + \sum_n \varphi_1(\mathbf{R}_n) + \frac{1}{2!}\sum_{m\neq n} \varphi_2(\mathbf{R}_n, \mathbf{R}_m) + \frac{1}{3!}\sum_{n\neq m\neq l} \varphi_3(\mathbf{R}_n, \mathbf{R}_m, \mathbf{R}_l) + \ldots \tag{2.8.1}$$

where each term of the series describes interactions of ion groups through the surrounding electron gas and can be represented as a power series in the electron-ion potential, for example (Brovman and Kagan, 1974):

$$\varphi_2(\mathbf{R}_1,\mathbf{R}_2)=\sum_{i=2}^{\infty}\Phi_2^{(i)}(\mathbf{R}_1,\mathbf{R}_2)\,,$$

$$\varphi_3(\mathbf{R}_1,\mathbf{R}_2,\mathbf{R}_3)=\sum_{i=3}^{\infty}\Phi_3^{(i)}(\mathbf{R}_1,\mathbf{R}_2,\mathbf{R}_3) \tag{2.8.2}$$

etc. The indirect interaction of two ions in second-order perturbation theory with respect to the electron-ion potential is well known (e.g. Harrison, 1966) and equals

$$\Phi_2^{(2)}(R)=\frac{1}{\pi^2}\int_0^{\infty}\mathrm{d}q\,q^2\Gamma^{(2)}(q)\,|V(q)|^2\,\frac{\sin qR}{qR} \tag{2.8.3}$$

where $V(q)=-4\pi e^2/q^2$ is the form factor of the ion-electron interaction potential; $\Gamma^{(2)}(q)$ is the sum of two-pole diagrams.

In the third-order perturbation theory in $V(q)$ the indirect ion-pair interaction is defined by the expression

$$\Phi_2^{(3)}(R)=\frac{3}{4\pi^4}\int_0^{\infty}\mathrm{d}q_1\,q_1^2\int_0^{\infty}\mathrm{d}q_2\,q_2^2\int_{-1}^{1}\mathrm{d}x\,V(q_1)V(q_2)V(q_3)\Gamma^{(3)}(q_1,q_2,q_3)\times\frac{\sin q_1R}{q_1R} \tag{2.8.4}$$

where $\Gamma^{(3)}(q_1,q_2,q_3)$ is the sum of three-pole diagrams, and $q_3=(q_1^2+q_2^2+2q_1q_2x)^{1/2}$.

The indirect three-ion interaction potential may be written after double integration as (Hasegawa, 1976):

$$\begin{aligned}\Phi_3^{(3)}(R_{12},R_{23},R_{13})&=\frac{3}{2\pi^4}\int_0^{\infty}\mathrm{d}q_1\,q_1^2\int_0^{\infty}\mathrm{d}q_2\,q_2^2\int_{-1}^{1}\mathrm{d}z V(q_1)V(q_2)V(q_3)\Gamma^{(3)}\\&\times(q_1,q_2,q_3)\int_0^{1}\mathrm{d}x\cos\left[x\left(q_1\,R_{12}\,\frac{R_{12}^2+R_{23}^2-R_{13}^2}{2R_{12}R_{23}}+q_2\,R_{23}\,z\right)\right]\\&\times J_0\left(q_1\,R_{12}(1-x^2)^{1/2}\left(1-\frac{(R_{12}^2+R_{23}^2-R_{13}^2)}{4R_{12}^2R_{23}^2}\right)^{1/2}\right)\\&\times J_0[q_2R_{23}(1-x^2)^{1/2}(1-z^2)^{1/2}]\end{aligned} \tag{2.8.5}$$

where $J_0(x)$ is the Bessel function of zeroth order; $z=\cos(\mathbf{q}_1\cdot\mathbf{q}_2)$; R_{12},R_{23},R_{13} are the distances between the vertices of a triangle formed by the protons.

The potentials were calculated with a Wigner-Seitz radius $r_s=1.65$, which corresponds to zero pressure in the zero-order model of a metal (Brovman *et al.*, 1971).

A permittivity function in the Geldart-Vosko form was employed (see Eq. (2.5.27) and Geldart and Vosko (1965, 1966)). (The potential values were calculated on a set with step width equal to 1 Bohr radius for each dimension. In the drawings, a cubic 'spline interpolation' was used.)

2.8.3
Results

The following results and figures are taken from Kaim *et al.* (1997). The interionic pair-potential in third order has the form

$$\varphi^*(R) = \frac{e^2}{R} + \Phi_2^{(2)}(R) + \Phi_2^{(3)}(R) \tag{2.8.6}$$

where e is the electronic charge.

Figure 2.8.1 shows the computed potential $\varphi^*(R)$ and its components. Curve 1 corresponds to the contribution $e^2/R + \Phi_2^{(2)}(R)$, curve 2 to $\Phi_2^{(3)}(R)$, and finally curve 3 to $\varphi^*(R)$.

The interionic pair-potential $e^2/R + \Phi_2^{(2)}(R)$ has no deep potential well. In fact, Stevenson and Ashcroft (1974) obtained an inter-proton pair-potential $e^2/R + \Phi_2^{(2)}(R)$ for metallic hydrogen at a density corresponding to r_S=1.6 analogous to the one shown in Fig. 2.8.1 (curve 1). The indirect interionic interaction $\Phi_2^{(3)}(R)$, curve 2, is attractive in nature and forms a potential well for the nearest neighboring ion with a minimum at $R \approx 3.3$ a.u. in the repulsive part of the potential $\varphi^*(R)$. The decrease in the density from r_S=1.65 to r_S=1.72 leads to a strong increase in the depth of the minimum in the repulsive part of the potential $\varphi^*(R)$ (=curve 3 in Fig. 2.8.1).

Curve 1 in Fig. 2.8.2 corresponds to the potential $\varphi^*(R)$ at r_S=1.72. At r_S<1.65 the minimum in the repulsive part of $\varphi^*(R)$ turns shallow and its position shifts towards a smaller R, and at r_S=1.55 this minimum disappears (curve 2 in

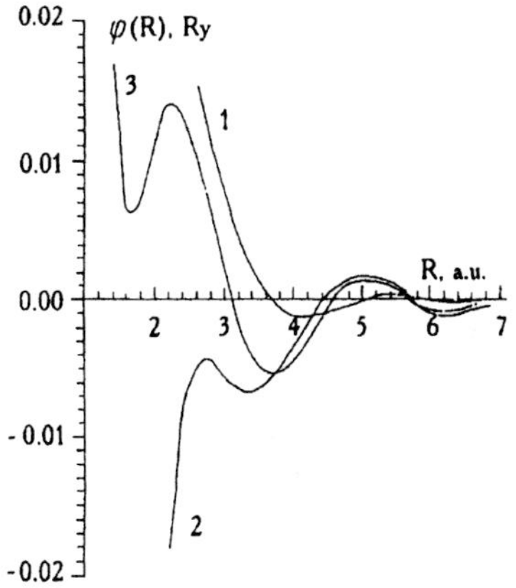

Fig. 2.8.1 The potential $\varphi^*(R)$ in Ry and its components for a density corresponding to r_S=1.65. The radius is given in atomic units

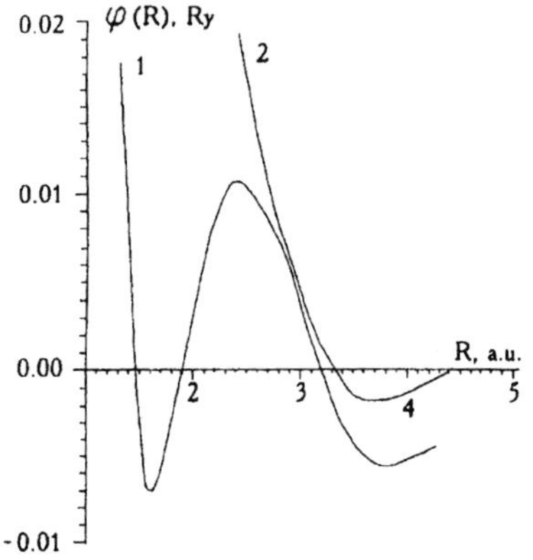

Fig. 2.8.2 The same as curve 3 in Fig. 2.8.1, but for $r_S = 1.72$ (curve 1) and $r_S = 1.55$ (curve 2)

Fig. 2.8.2). The position of the minimum in the repulsive part of $\varphi^*(R)$ in Fig. 2.8.1 corresponds to the separation of $R = 1.6a_B$. It should be mentioned that the separation between the nuclei in the H_2 molecule is equal to $1.4a_B$ and the binding length in H_2^+ ions is equal to $2a_B$ (Murrell *et al.*, 1965). The formation of the minimum in $\varphi^*(R)$ at a distance around $R \approx 1.6$, which is significantly smaller than the average interproton distance, and the strong dependence of the depth of this minimum on the density of metallic hydrogen, may be considered to be a tendency to diatomic ordering in the electron-proton plasma.

The results of the calculation of the irreducible three-ion interaction potential $\Phi_3^{(3)}$ for $r_S = 1.65$ are presented in Fig. 2.8.3 in the form of a *potential landscape*.

The minimal interproton separation in the metallic phase at $P = 0$ equals $2.04a_B$ (Brovman *et al.*, 1971). Two ions are placed on the ordinate axis at a distance $2a_B$ and the third ion is situated in the plane XOY. For small distances the potential $\Phi_3^{(3)}(2, R_{23}, R_{13})$ has an attractive character. The potential $\Phi_3^{(3)}(2, R_{23}, R_{13})$ has an anisotropic form and its most rapid change takes place in the OX direction. For large distances the potential $\Phi_3^{(3)}$ has an oscillatory character. Note that $\Phi_3^{(3)}(0,0,0) \cong -1Ry$.

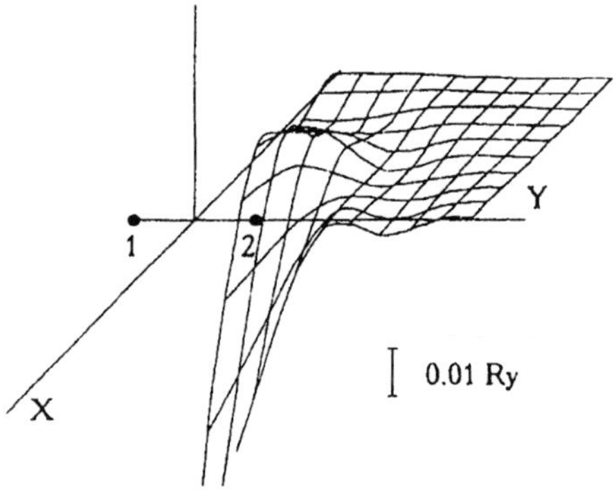

Fig. 2.8.3 Irreducible three-ion interaction potential $\Phi_3^{(3)}(2, R_{13}, R_{23})$

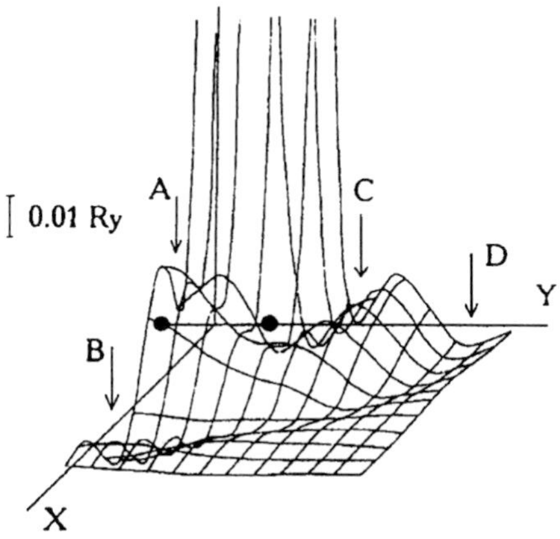

Fig. 2.8.4 Potential relief for the third ion in the field of two ions with pair interactions taken into account

Figure 2.8.4 (the geometry coincides with Fig. 2.8.3) shows the potential landscape for the third ion in the fields of two other ions with pair interactions taken into account, i.e. the potential $\varphi^*(R_{23}) + \varphi^*(R_{13})$. The minima B and D, and the valley that connects them, correspond to the possible positions of the nearest neighbouring ion in the field of two fixed ions (see Fig. 2.8.4). It should be noted that ion transitions from minimum B to minimum D along the valley entails the crossing of a potential barrier (or tunneling through it) with a height of ~ 1200 K. This situation remains unchanged when the irreducible three-proton interaction is taken into account because of the short-range character of the potential $\Phi_3^{(3)}$. Quantum effects of the ion motion in the field of two fixed ions result in the *splitting of the energy level* for the ion that is the nearest neighbor for the pair considered. This is strongly reminiscent of the *'tunneling levels'* discussed in the next chapter. The local minima A and C in Fig. 2.8.4 correspond to distances to ions 1 and 2, which are less than the average interionic distance.

Figure 2.8.5 shows the potential landscape that is created by the ion pairs, taking into account pair- and three-ion interactions. It is described by the function $\varphi^*(R_{23}) + \varphi^*(R_{13}) + \Phi_3^{(3)}(2, R_{23}, R_{13})$. A change in the positions A,B,C,D of the local minima, as compared with Fig. 2.8.4, is unimportant. As a consequence of the short-range and anisotropic character of the potential $\Phi_3^{(3)}$ the depth of the minimum A changes markedly, and the depth of the minimum B increases slightly. The ion wave function in the potential landscape, which is shown in Fig. 2.8.5, has a non-zero value near the local minima A and C. This means that ions have finite probability of being drawn together for separations of the order of internuclei distances in H_2 or H_2^+. By comparison of Fig. 2.8.1 and Fig. 2.8.5 we conclude that the presence of the third ion essentially increases this probability.

Figure 2.8.6 shows the computed distribution of the conditional probability density for the ion positions in the pair-interaction field of the ions, with the pair and three-ion interactions taken into account in Boltzmann approximation at a temperature $T = 1000$ K. In this approximation the distribution of the conditional probability density is

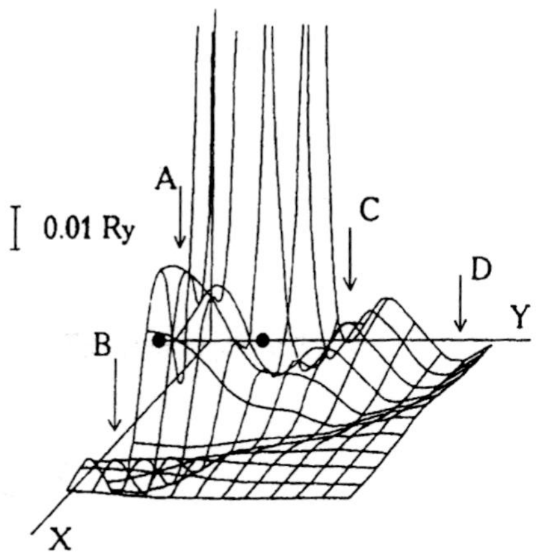

Fig. 2.8.5 The potential relief for the third ion in the field of two ions (the dots) with pair- and three-ion interactions taken into account

$$F_1(\mathbf{R}_1|\mathbf{R}_2, \mathbf{R}_3) \sim \exp[(-(\varphi^*(R_{12}) + \varphi^*(R_{13}) + \Phi_3^{(3)}(R_{12}, 2, R_{13}))/k_B T) \tag{2.8.7}$$

where k_B is Boltzmann's constant. The *local minima* of the potential landscape in Fig. 2.8.5 correspond to the *peaks* of the function $F_1(\mathbf{R}_1|\mathbf{R}_2, \mathbf{R}_3)$ of Eq. (2.8.7), which are presented in Fig. 2.8.6.

It is interesting to consider the potential landscape created by a group of ions. Figure 2.8.7 shows the potential landscape, which three ions forming an equilateral triangle with side $2a_B$ and placed in the XOY plane create in the same plane for the fourth ion. In Fig. 2.8.7 the ion-pair interaction is taken into account and the potential landscape is assigned to the function $\varphi^*(R_{14}) + \varphi^*(R_{24}) + \varphi^*(R_{34})$. It is easy to see two minima A and B and the saddle point C in the XOY plane.

The potential landscape in Fig. 2.8.8 (the geometry is the same as in Fig. 2.8.7) is described by the formula:

$$\varphi^*(R_{14}) + \varphi^*(R_{24}) + \varphi^*(R_{34}) + \Phi_3^{(3)}(2, R_{24}, R_{14}) + \Phi_3^{(3)}(R_{24}, 2, R_{34}) + \Phi_3^{(3)}(R_{14}, R_{34}, 2) \tag{2.8.8}$$

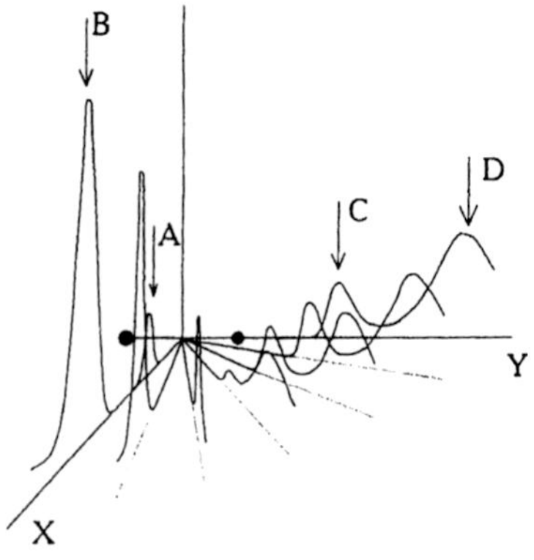

Fig. 2.8.6 Distribution of the conditional probability density for the position of a third ion in the field of a pair of ions (the dots)

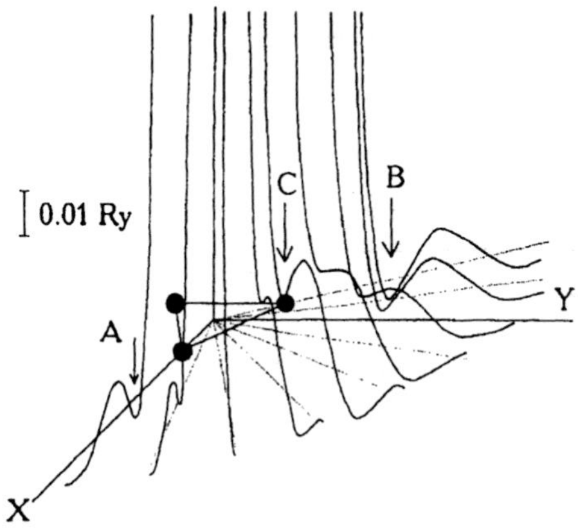

Fig. 2.8.7 Potential relief that three ions (the dots) create to a fourth ion

It takes into account three-ion interactions in a cluster of four ions. Three-ion interactions transform the saddle C into a local minimum, and the minima A and B become considerably deeper. This means that an increase in the particle numbers in the cluster lowers the potential barriers to diatomic ordering.

If the three ions are situated in the way shown in Fig. 2.8.8 and the fourth ion is placed on the z-axis, the dependence of potential $\Phi(z)$, with ion-pair and three-ion interactions taken into account, is shown in Fig. 2.8.9.

The presence of the deep minimum at $z = a_B$ is a characteristic peculiarity of this dependency. The second minimum at $z = 3.5a_B$ corresponds to the equilibrium position of the fourth ion.

The quantum description of the ionic motion in the two-well potential of Fig. 2.8.9 yields the conclusion about the possibility of a tunneling transition from well A to well B. The probability rate of ion transitions per 1 s is equal to the product of the ion oscillation frequency in the well A and of the transmission coefficient through the potential barrier. The transmission coefficient can be estimated by a quasi-classical approach. The ion (ground state) oscillation frequency in the well A is equal to $\upsilon = 3.61 \times 10^{13}$ Hz. The transmission coefficient through the po-

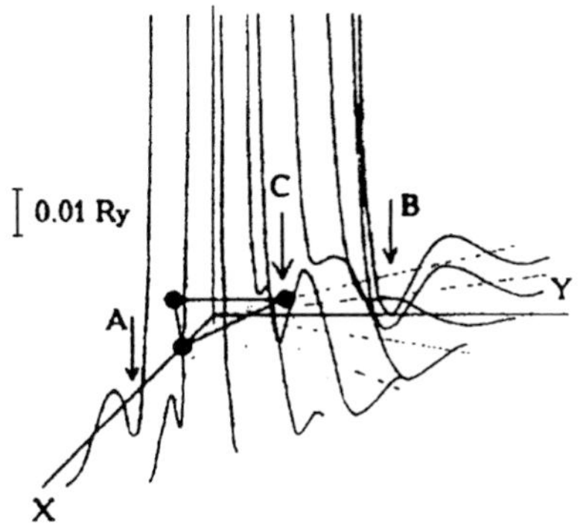

Fig. 2.8.8 As in Fig. 2.8.7, but now three-ion interactions are taken into account as well as the pair interactions

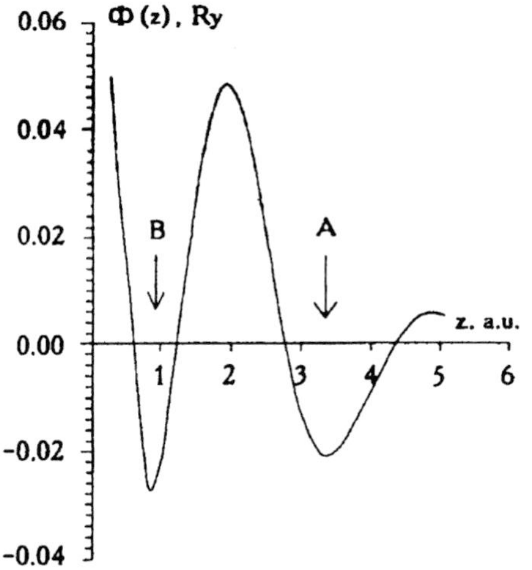

Fig. 2.8.9 The dependence of the potential $\Phi(z)$ (measured in Ry units) on the coordinate z (measured in atomic units)

tential barrier is $D=3.85\times10^{-12}$. Then the 'lifetime' τ of the ion in well A equals $\tau=(D\upsilon)^{-1}=0.0072$ s. If transitions of the ion into well B take place preferentially, this corresponds to preferred *localization of the electrons on ionic pairs.* Taking into consideration also the small characteristic time of an electron subsystem it is possible to consider the obtained time as a characteristic time of the tunneling nucleation of the H_2 molecule in metallic hydrogen at zero pressure. If the identity of the ions is taken into account it is possible to reach a conclusion about a many-ion tunneling mechanism for a *transition from metallic hydrogen into a molecular phase.*

Quantum Monte Carlo simulations of molecular hydrogen (Ceperley and Alder, 1987) and calculations of the energy of the diatomic phase as a function of the interproton separation (Chakravarti *et al.*, 1981) indicate that for given total density

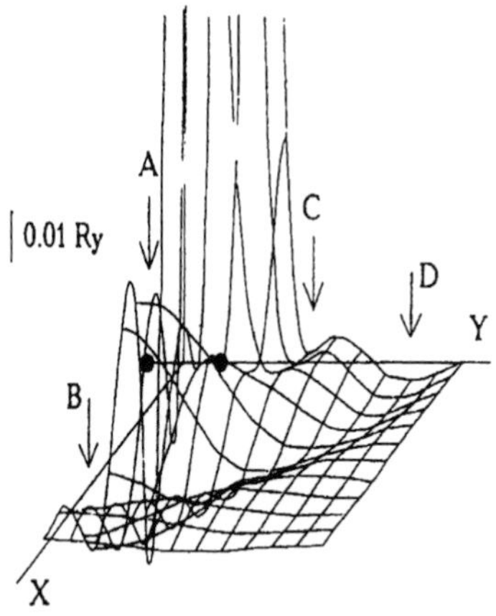

Fig. 2.8.10 Potential relief for the third ion in the field of two ions (interatomic distance 1.5 a.u.) with the pair interaction taken into account in third-order perturbation theory (in contrast to Fig. 2.8.4)

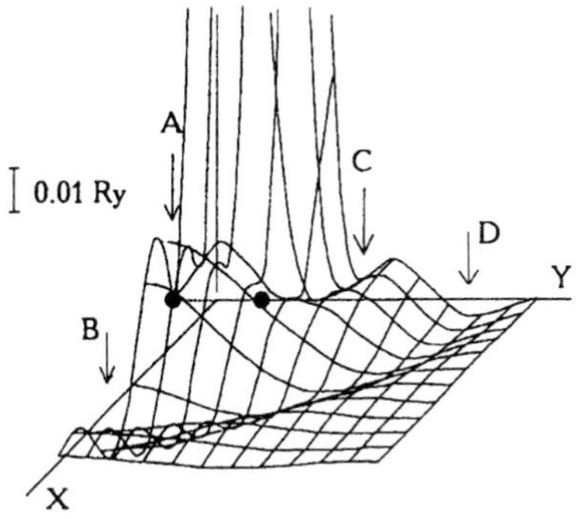

Fig. 2.8.11 As in Fig. 2.8.10, but with pair- and three-ion interactions taken into account in third-order perturbation theory, in contrast to Fig. 2.8.5

the interproton separation in a molecular unit of this phase is somewhat less than the corresponding separation $1.4a_B$ in the neutral molecule. We calculated the *potential landscape created by ion pairs* with interproton separation $1.5a_B$ (Figs 2.8.10, 2.8.11). Figure 2.8.10 shows the potential relief for the third ion in the field of two other ions with pair-ion interaction in the third-order perturbation theory (analogous to Fig. 2.8.4).

Figure 2.8.11 (the geometry coincides with that of Fig. 2.8.10) shows the potential relief for the third ion with pair and irreducible three-ion interactions taken into account, i.e. the potential $\varphi^*(R_{12}) + \varphi^*(R_{23}) + \Phi_3^{(3)}(1.5, R_{13}, R_{23})$.

As is easily seen from Figs 2.8.10 and 2.8.11, the irreducible three-ion interaction essentially deepens the potential well A (Fig. 2.8.11 is analogous to Fig. 2.8.5). Comparison of Figs 2.8.5 and 2.8.11 yields the conclusion about a strong dependence on the interproton separation R_{12} of the probability of triplets of ions being drawn together to interproton separations as in the H_2 molecule. The time of the proton transition from well B into well A (Fig. 2.8.11) cannot be considered as a characteristic time of the tunneling nucleation of H_2 molecules in the metallic phase because the interproton separation $R_{12} = 1.5a_B$ is less than the minimal interproton separation in the metallic hydrogen phase at zero pressure, which equals $2.04a_B$ (Brovman *et al.*, 1971).

2.8.4 Discussion and Conclusions

Our consideration is based on the assumption that metastable metallic hydrogen exists at zero pressure. The calculations show that irreducible three-ion indirect interactions have a decisive influence on the local structure of the ionic subsystem. The calculations of the possible metallic hydrogen structures at zero pressure show the tendency for metallic hydrogen to crystallize in a triangular Bravais lattice with a two-dimensional periodic structure (Brovman *et al.*, 1971). Our calculations of the 'potential landscape' that three ions create towards the fourth ion confirm this conclusion and show that taking into account $\Phi_3^{(3)}$ really leads to the formation of a *hexagonal structure* in a planar arrangement of four ions. Taking into account $\Phi_3^{(2)}$ leads to the transformation of the saddle C (Fig. 2.8.7) into a local minimum (Fig. 2.8.8). The minimum C is formed along the direction that forms an angle of 60° with the OX axis (the curves in Figs 2.8.7 and 2.8.8 are drawn with an angular step of 15°).

It is possible to observe an interesting peculiarity of the potential landscape that a pair of ions creates towards the third ion (Figs 2.8.4–2.8.6): This is the appearance of the local minima B and D, and the valley that connects them and has a saddle point. To make a transition from the minimum B to that of D it is necessary to cross over (or tunnel through) a potential barrier of typically ~1000 K. This means that the energy levels of the third ion in the field of the two fixed ions are split. The quantum effect of the splitting of the energy levels will be felt at temperatures less than the potential barrier height. It should be noted here that as a consequence of a short-range character of the potential $\Phi_3^{(3)}$, it has a weak influence on this effect (Figs 2.8.4 and 2.8.5).

Analogously it can be supposed that the existence of two-level subsystems in amorphous solids at low temperature is somehow a display of collective quantum oscillatory states of groups of (at least) three particles.

In this study the peculiarities of the pair- and three-ion interactions are interpreted as the *vestiges of the molecular phase* in metastable metallic hydrogen. These peculiarities occur in the *metastable* metallic phase and are absent at megabar pressures (Fig. 2.8.2). This means that an electron-proton system possesses a self-organization effect of 'preparing' a transition to the molecular phase. Such an effect increases sharply with decreasing electron gas density.

The probability of the ions being drawn together to interproton separations as in the H_2 molecule (or even somewhat less) depends essentially on the density of the electron gas. With decreasing density the probability of the many-ionic tunneling nucleation of H_2 molecules increases sharply. Therefore of some interest is the study of the lifetime of metastable metallic hydrogen relative to the homogeneous nucleation of the molecular phase in the pressure interval $0 < P < P_t$ (P_t is the pressure of the molecular hydrogen transition in the metallic phase).

The calculation of the lifetime of the *metallic hydrogen* as a macroscopic system calls for consideration of both the mechanism of *homogeneous nucleation of the molecular phase*, and of the kinetics of formation of *droplets of the insulating phase* in the metastable metal. In this section we did not consider the kinetic stage. Such a consideration must include a search of the volume- and surface-free energy minimum of the heterogeneous system consisting of a metastable metal and the insulating droplets. In accordance with the classical and quantum theories of the formation of the new phase (Lifshitz and Kagan, 1972; Folmer, 1986; Iordansky and Finkel'shtein, 1972) calculated in this chapter, the lifetime of the metallic phase relative to tunneling nucleation of the H_2 molecule may differ from the lifetime of metallic hydrogen as a macroscopic system by a factor of 3–5. The new experimental data on the structures and phase diagrams for molecular hydrogen and deuterium at megabar pressure found recently (see e.g. Mao and Hemley, 1994; Cui *et al.*, 1994) may be very useful for consideration of the kinetic stage of the formation and growth of droplets of the insulating phase in metastable metallic hydrogen.

2.9
Computer Calculations of the Electronic Structure in Metallic Amorphous Alloys

In Sections 2.1–2.7 we have concentrated on liquid alkali metals and have seen that there the indirect pair-interaction between the ions dominates and can be described by radially symmetric pseudopotentials, which were universal for the alkali metals at corresponding densities, temperatures, etc. These pseudopotentials were in fact similar to the phenomenological Lennard-Jones potentials (see Eq. (1.3.1)), which serve for a rough calculation of the structural properties of various crystalline, liquid or amorphous compounds throughout the periodic system of elements (e.g., Kittel, 1971). Of course, for non-alkaline systems, the calculation of interaction pseudopotentials is a more ambitious task, which has been described for sp-bonded metals in the book by Hafner (1987). For amorphous $Mg_{1-x}Zn_x$ alloys, the results, which are unspectacular, are presented in Fig. 2.9.1 (Hafner 1976, 1977 a, b).

For transition metal (TM)-metalloid (M) amorphous alloys the situation is more subtle; in this case we expect non-spherical contributions from the TM-M pair interactions. For example, it is well known that Si or C favor tetrahedral hybridized sp^3 bonds, e.g. the corresponding linear independent one-electron states are

$$\psi \sim \psi_s + a(\pm\psi_x \pm \psi_y + \psi_z) \tag{2.9.1}$$

Here the states with the four different $(\pm, \pm)$ combinations correspond to the four different body diagonals in a cube, and ψ_x, ψ_y and ψ_z correspond to atomic orbitals of the corresponding p-symmetry on the metalloid atoms, e.g. $\psi_x \sim xf(r)$. Finally the factor $a = \pm 1$ determines whether we get *bonding* or *antibonding* states.

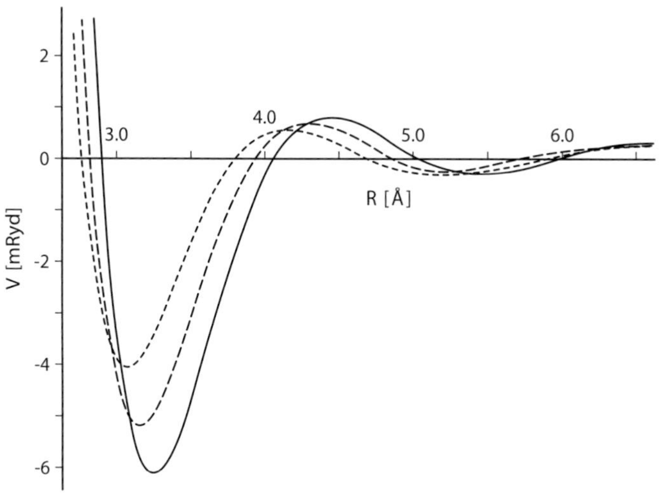

Fig. 2.9.1 The three effective pair-potentials as calculated by J. Hafner from the general non-local pseudopotential theory. Lowest line: Mg-Mg, broken line: Mg-Zn; upper line: Zn-Zn

Of course in an amorphous system, where the surroundings of an Si metalloid atom generally strongly deviate from the ideal tetrahedral symmetry, there will be variations of the prefactors, which may lead to the preference of *distorted* tetrahedral surroundings of an Si metalloid, where the distortion, i.e. the variational parameters, should depend on the configuration of all atoms, and the bonding (rsp. antibonding) states will no longer be degenerate. Then, of course, all local interaction potentials between the ions depend on the global positional configuration, at least in some neighborhood of the atoms considered. This means that the pseudopotentials should *not* be fixed *before* the computer simulation, but should be determined by an '*online calculation*' of the *electronic* state for every positional configuration of the ions used in the 'molecular dynamics' simulation' of the system.

This is in fact an ambitious program, an '*ab initio molecular dynamics*', but it is being realized, although in *approximate* form, by recent calculations of Hausleitner and Hafner (1993). In these calculations, the s-bands of all atoms are treated 'as before', i.e. in a 'nearly-free electron approximation', whereas the d-states of the metals rsp. the p-states of the metalloids are treated in a 'one-band tight-binding approximation', which gives rise to an additional contribution to the electronic energy, namely

$$E_{\mathrm{bond}} := (2\mathrm{N})^{-1} \sum_{\mathrm{l,m}} \mathrm{t}_{\mathrm{l,m}}(\mathrm{R}_{\mathrm{l,m}})\theta_{\mathrm{l,m}} \tag{2.9.2}$$

Here the sum is over all atoms l and m ($l \neq m$), the $t_{l,m}$ are the hopping integrals and the $\theta_{l,m}$ are the differences in the occupation numbers in the bonding rsp. antibonding eigenfunctions formed by the tight-binding states on the sites l and m. This is again a pair approximation, but radial-symmetric effective pair potentials are this time *only obtained after an averaging procedure*. Moreover, the results for the pair potentials ϕ are no longer 'additive', i.e. in contrast to the Lennard-Jones potentials of Eq. (1.3.1), the condition $\phi_{AB} = (\phi_{AA} + \phi_{BB})/2$ no longer applies as we can see, for example, from Figs 2.9.2–2.9.4, where the 'pair potentials' obtained in this way by Hausleitner and Hafner (1993) for $Fe_{80}B_{20}$, $Ni_{80}P_{20}$ and $Ni_{64}B_{36}$ are plotted.

Note that, in the last two cases, the 'effective pair potentials' are different! Of course, we should also stress that the results for the partial distribution functions compare sufficiently with experiments, although it is clear that future calculations of the same sort will still bring improvement in this respect.

If we are less ambitious, we generate structural models of amorphous systems from phenomenological model potentials, as for example the Lennard-Jones potentials (1.3.1), which have already been mentioned several times. Having generated reasonable models in this way, we then calculate the electronic structure of the systems by an adequate semi-empirical method. Such *semi-empirical calculations* have been performed by one of the authors (Krey *et al.*, 1987; Ostermeier and Krey, 1988) for amorphous $Fe_{1-x}B_x$ alloys in the non-magnetic state, and also for amorphous $Cu_{1-x}Zr_x$ alloys. In these calculations not only the electronic struc-

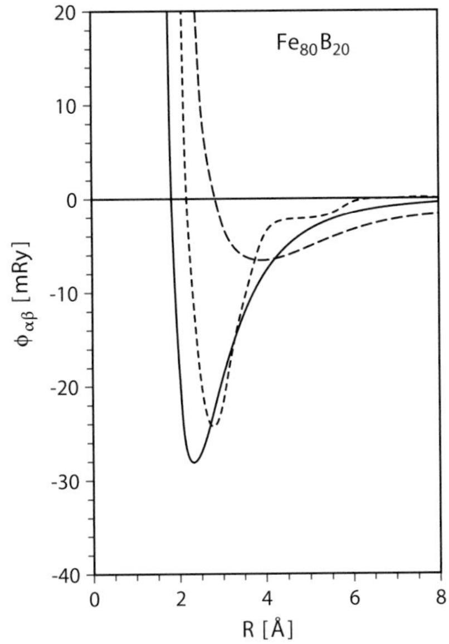

Fig. 2.9.2 The effective pair potentials for $Fe_{80}B_{20}$. Solid line: metal-metalloid; dotted line: metal-metal; dashed line: metalloid-metalloid (from Hausleitner and Hafner, 1993)

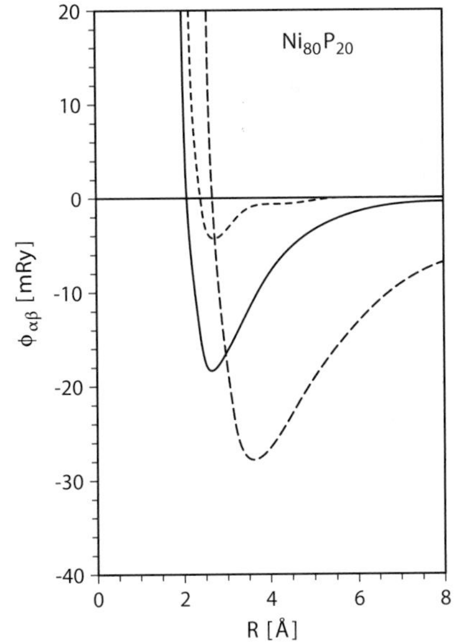

Fig. 2.9.3 The effective pair potentials for $Ni_{80}P_{20}$. Solid line: metal-metalloid; dotted line: metal-metal; dashed line: metalloid-metalloid (from Hausleitner and Hafner, 1993)

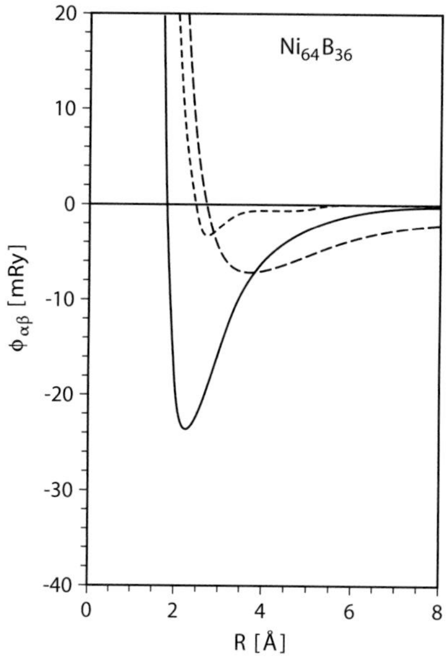

Fig. 2.9.4 The effective pair potentials for $Ni_{64}B_{36}$. Solid line: metal-metalloid; dotted line: metal-metal; dashed line: metalloid-metalloid (from Hausleitner and Hafner, 1998)

ture of the system has been studied, but also the electric resistivity, by a so-called *Kubo formalism*, which is not described here.

In these studies, the computer models of the amorphous systems were generated by distributing 2000 atoms at first randomly in a cubic box with the correct volume and then relaxing the positions of the atoms until convergence to a metastable amorphous state in the appropriate Lennard-Jones potentials. The subsequent calculation of the electronic structure proceeded in a tight-binding formalism of the Slater-Koster type, taking into account nine valence electrons (e.g. five 3d, one 4s and three 4p orbitals) per metal atom and the appropriate valence states for the metalloids (e.g. the 2s and 2p states for B). Interactions ranging up to second-nearest neighbors were taken into account in these calculations, where the eigenvalue equation is

$$H_{l,m;\alpha,\beta} c^{\nu}_{m,\beta} = \varepsilon_{\nu} c^{\nu}_{l,\alpha} \tag{2.9.3}$$

Here the ε_{ν} are the eigenvalues of the electronic Hamiltonian, which in the usual one-particle approximation has the Hamiltonian matrix $H_{l,m;\alpha,\beta}$, where l and m denumerate the sites of the atoms and α and β the orbitals considered and, as usual, on the left-hand side of Eq. (2.9.3), a summation over the indices m and β, which appear twice, is understood (Einstein's summing convention). Furthermore, as usual, $|c^{\nu}_{m,\beta}|^2$ is the probability of an electron in the eigenstate indexed by ν to be found in the orbital β at the atom m.

The matrix elements $H_{l,m;\alpha,\beta}$ appearing in this equation are calculated for the given positions of the atoms in the model as usual by formulae of the type defined by Slater and Koster (1954):

$$H_{l,m;\alpha,\beta} = \sum_{w}(\mathrm{dir.cos}(l,m))_{w} I_{\alpha,\beta;w}(r_{l,m}) \tag{2.9.4}$$

where $(\mathrm{dir.cos}(l,m))_w$ denotes certain functions of the direction cosines of the vector $(\mathbf{r}_l-\mathbf{r}_m)$ as given in the table by Slater and Koster (1954), and the $I_{\alpha,\beta;w}(r_{l,m})$ are the corresponding 'two-center integrals', where for example for d electrons the index w takes the three values σ, π or δ, according to the possible rotational symmetries around the axis between l and m. The distance dependence of the two-center integrals has been parametrized according to the semi-empirical rules of Harrison (1980), namely

$$I_{\alpha,\beta;w}(r_{l,m}) = \eta_{\alpha,\beta,w}(\hbar^2/m_e)r_d^{-2}(r_d/r_{l,m})^k \tag{2.9.5}$$

where the dimensionless parameters $\eta_{\alpha,\beta,w}$, which are almost 'universal' for different compounds, and the 'd radii' r_d are tabulated in the book by Harrison (1980); m_e is the electron mass, and $\hbar$ is Planck's constant divided by 2π. Finally in Eq. (2.9.5) we have $k=5$ for $(\alpha,\beta)=(d,d)$; $k=3.5$ for $(\alpha,\beta)\in(s,d)$ or (p,d); $k=2$ for $(\alpha,\beta)\in(s,s)$, (s,p) or (p,p).

To the Hamiltonian matrix H there corresponds the '*resolvent matrix*' $G(z) = (z-H)^{-1}$, where $z=E+i\varepsilon$ is a complex number with a positive 'infinitesimal' ε. Diagonal matrix elements of the 'resolvent' are of the type $\langle\Phi|G|\Phi\rangle$, where Φ is an arbitrary vector describing the electronic system; they can be calculated by a *continued fraction algorithm,* which is used extensively by Krey *et al.* (1988), Ostermeier and Krey (1988) and Ostermeier (1988). In this way we can calculate various *local densities of states,* e.g. the 'local DOS'

$$g_{m\alpha}(E) = -\lim_{\varepsilon\to 0}\pi^{-1}\mathrm{Im}\langle m\alpha|G(E+i\varepsilon)m\alpha\rangle \tag{2.9.6}$$

where Im means 'imaginary part'. Averages of this quantity, averaged over different sites or samples, can of course also be calculated. In fact, Eq. (2.9.6) represents the *product* of the total density of states (i.e. the total number ΔZ of different solutions of Eq. (2.9.3) with energies between E and $E+\Delta E$, divided by the product of ΔE times the total number of orbitals in the sample, e.g. ΔZ divided by $\Delta E\times 2000\times 9$ for amorphous 3d alloys) *multiplied* by the conditional probability of the electron considered, to be found at the atom m in the orbital α. In this way, by averaging over all electrons of a small energy interval, we get the so-called *partial densities of states* ('partials'), which are especially interesting in the present context.

A particular result concerns the B-s and B-2p 'partials', i.e. the total density of states for the valence electrons of amorphous $Fe_{1-x}B_x$ at a given energy, weighted with the probability of finding the electron at a B site in a 2s or 2p state, as presented in Fig. 2.9.5.

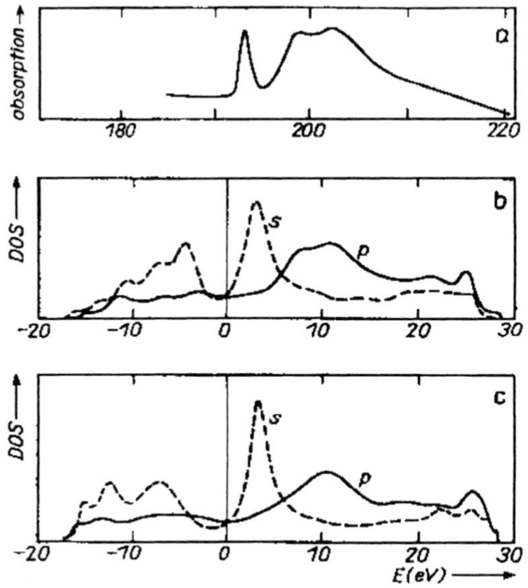

Fig. 2.9.5 Results for amorphous $Fe_{80}B_{20}$. The upper curve (a) shows the X-ray absorption from 1s boron states. Here we measure transitions to B-2s or B-2p states in the amorphous system (see text, Kizler *et al.*, 1986); the curve (b) denotes the partial densities of states for B-2s and B-2p states calculated numerically with the model of Krey *et al.* (1987), whereas (c) presents similar results obtained with the model of Fujiwara *et al.* (1982)

The X-ray absorption from the B-1s shell, which is shown in part (a) of Fig. 2.9.5, is taken from Kizler *et al.* (1986), and compared with partial densities of states for B-2s and B-2p orbitals calculated numerically with the model of Krey *et al.* (1987), part (b), and Fujiwara *et al.* (1982), part (c). In an atom, transitions from B-1s to B-2s would be absolutely forbidden by 'parity conservation', but in an amorphous system the surrounding of a B atom may be not at all invariant against inversion $\mathbf{r} \rightarrow (-\mathbf{r})$. Thus in an amorphous system, such transitions should be allowed. In fact, two typical configurations are plotted in Fig. 2.9.6, which appear in the model of Krey *et al.* (1987), see also Ostermeier (1988), taken from the

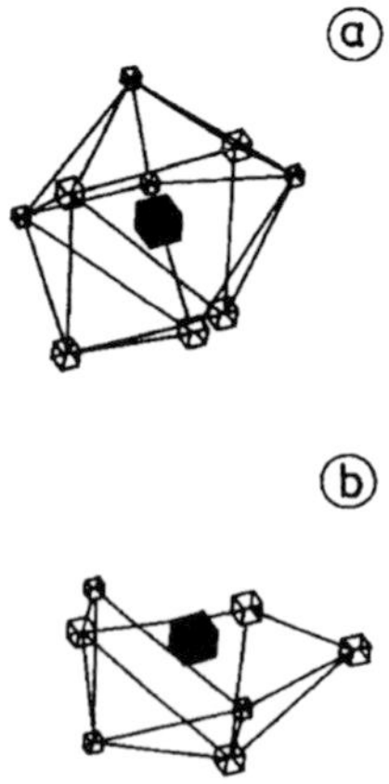

Fig. 2.9.6 Two typical B atoms and their surroundings in the model for amorphous $Fe_{80}B_{20}$ used by Krey *et al.* (1987). All atoms are plotted that are at a distance from the B atom (the black diamond) less than the distance to the first minimum of the radial distribution functions. These 'neighbors' of a B atom always turn out to be Fe atoms

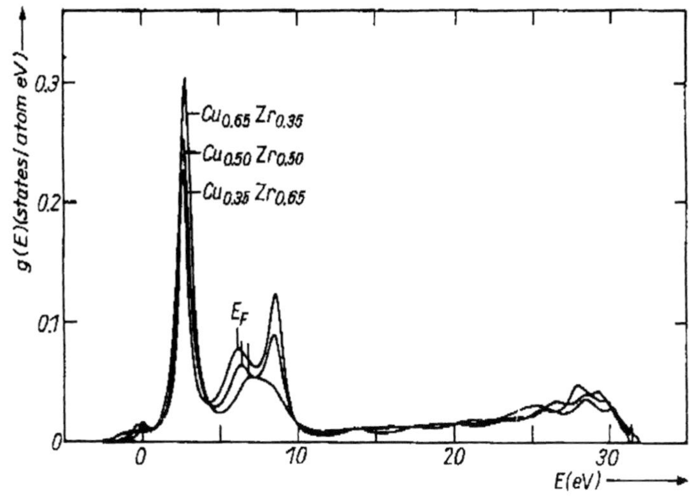

Fig. 2.9.7 The total density of states per atom and per ΔE [eV] for amorphous $Cu_{1-x}Zr_x$ alloys, $x=0.25$, 0.5 and 0.65, taken from Krey *et al.* (1987)

detailed comparisons performed by Kizler (1988). In Fig. 2.9.6, the surroundings of the B atom, the black diamond, are very asymmetric for the lower example.

Furthermore, due to the energy relaxation (i.e. *not* by construction), there are no B-B neighbor pairs. Moreover, we find that the 'splitting' of the B-2p absorption in part (a) of Fig. 2.9.5 is reproduced by the B-2p 'partial' in part (b), but not in part (c), so that we may conclude from this discrepancy that model (c), which was constructed by certain rules, may be somehow still 'too symmetric'. The third observation is the pronounced minimum around the Fermi energy E_F in the B-2s 'partial'. This minimum may be relevant for the *stability* of the amorphous alloy (see below).

In fact, the well-known stability of the amorphous $Cu_{1-x}Zr_x$ alloys over a wide concentration region may be related to the fact that there the Fermi energy E_F is situated in a pronounced minimum of the total density of states, as shown in Fig. 2.9.7, where the pronounced peak below E_F results from the Cu-3d states, whereas the peak above E_F is related to the Zr-4d states.

Such strong scattering effects – almost 'anti-resonance effects' – leading to deep minima in the density of states as observed in the B-2s 'partial' in Fig. 2.9.6(b), happen typically at wavenumbers **k** with magnitude k_p, i.e. the particular magnitude, where the structure function $a(k)$ has its pronounced first maximum. This means that the wavelength of the incoming electronic plane wave state should correspond to typical 'nearest-neighbor distances' in the radial distribution function. Then multiple scattering should be especially effective, if these wavelengths at the same time correspond to the wavelengths that are characteristic of the spatial charge density oscillations (so-called 'Friedel oscillations') around impurities in metals (see every textbook on solid state theory, e.g. Kittel, 1963). In a simple metal, where the s-electrons dominate, these oscillations have characteristic wavenumbers of magnitude $2k_F$, where k_F is the wavenumber corresponding to the Fermi energy. This leads immediately to the famous '*Nagel-Tauc criterion*' (Nagel and Tauc, 1975). These authors argued that a metallic glass whose electronic structure at the Fermi energy is dominated by s electrons should be particularly stable, if the condition $2k_F \approx k_p$ is fulfilled.

However with p and d electrons the situation is more complicated, and the 'criterion' that the Fermi energy should be in a valley of the density of states has probably not much more than a 'heuristic value'.

3
Atomic Properties of Amorphous Metals: Low-energy Excitations

3.1
Experiments on the Atomic Dynamics in Glasses

This chapter summarizes modern concepts on the atomic properties of metallic glasses at sufficiently low temperatures ($T<100\,K$). The properties of metallic glasses in this temperature range resemble in many ways the behavior of other amorphous solids such as dielectric or semiconductor glasses. Experimental investigations of various atomic properties of glasses have revealed the fundamental result that these materials demonstrate a number of universal properties independent of their chemical constitution. Non-characteristic of other crystalline and non-crystalline materials, such properties are often called 'anomalous' and include the low-temperature behavior of the heat capacity, heat conductivity and thermal expansion, as well as ultrasonic propagation and other properties involving the atomic dynamics.

Today it is firmly established that for all amorphous solids – insulators and conductors – the universal temperature behavior of vibrational contributions to the *heat capacity* C_L and *heat conductivity* κ_L differs essentially from that of crystals. This point is illustrated in Fig. 3.1.1, where on a log-log scale the typical behavior of C_L/T_3 and κ_L versus temperature ($0.1<T<100\,K$) is shown for amorphous and crystalline materials.

Two different temperature ranges are observed for amorphous materials: $T<2\,K$ and $T>2\,K$. At very low temperatures ($T<2\,K$) we have

$$\begin{cases} C_L \sim T^{1+n};\ 0 \le n \le 0.3 \\ \kappa_L \sim T^{2-m};\ 0 \le m \le 0.25 \end{cases} \qquad (3.1.1)$$

In the moderate temperature range ($2<T<100\,K$) the C_L/T^3 curve shows a peak whose position and relative magnitude differs slightly for various glasses. The heat conductivity κ_L demonstrates a small plateau at $T\sim 10\,K$, then it increases, but not so steeply as T^2. There is a correlation between the positions of the peak in the curve C_L/T^3 and the plateau in $\kappa(T)$ (Güntherodt and Beck, 1981; Beck and Güntherodt, 1983; Galperin *et al.*, 1989; Phillips, 1980). The *thermal expansion coefficient* a in the low-temperature region ($T<2\,K$) is a linear function of temperature, whereas for crystals it is proportional to T^3. The *Grüneisen coefficient* $\Lambda = aB/$

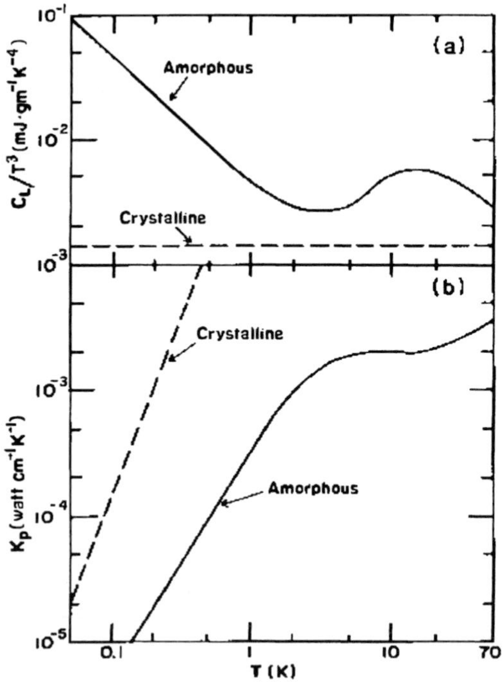

Fig. 3.1.1 (a) Dependence of C_L/T^3, where C_L is the heat capacity and T the Kelvin temperature, for amorphous systems (solid line) and crystalline systems (dashed line); (b) phonon contribution to the thermal conductance κ_p versus T for typical amorphous rsp. crystalline systems (from D. G. Onn, 1983, Fig. 23.1). Note the logarithmic scales.

C_L is temperature-independent for glasses just as for crystals. But at the same time, in glasses Λ can be large and negative (Galperin *et al.*, 1989; Phillips 1980).

The *acoustic absorption* as a function of temperature is shown in Figs 3.1.2 and 3.1.3.

Figure 3.1.2 shows that the absorption coefficient at low temperatures is proportional to T^3 and frequency independent. At very low temperatures ($T < 1$ K) the absorption becomes dependent on the ultrasound intensity: it decreases with growing intensity I, and at some threshold value I_H a saturation occurs (see Fig. 3.1.3, Güntherodt and Beck, 1981). At moderately low temperatures ($10 < T < 100$ K) the ultrasound absorption has a maximum, whose magnitude decreases with increasing frequency (Galperin *et al.*, 1989; Phillips, 1980).

A variety of experiments have investigated the temperature dependence of the sound velocity. Much more pronounced than in crystals, this dependence is logarithmic (Phillips, 1980). Neutron scattering experiments supplied considerable information about low-energy vibrational excitations in glasses. The density of states $n(E)$ of elementary excitations derived from these experiments is shown in Fig. 3.1.4 as a function of energy (expressed in frequency units) (Galperin *et al.*, 1989; Buchenau *et al.*, 1986). For comparison, the Debye density of states is also depicted in Fig. 3.1.4.

We can see that $n(E)$ initially outgrows the Debye analog, then it increases more slowly. Direct inelastic neutron scattering experiments in $Mg_{70}Zn$ revealed a mini-

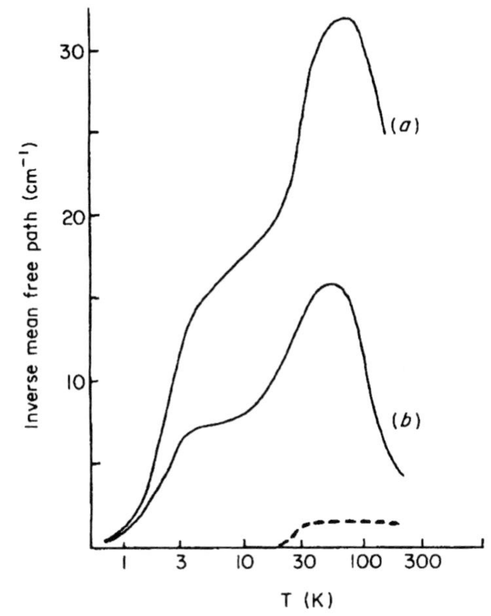

Fig. 3.1.2 Temperature dependence of the absorption of longitudinal acoustic sound waves in glassy SiO_2 (a) ω=930 Mhz, (b) ω=1000 MHz. The dashed line is for crystalline SiO_2 (quartz) (taken from Galperin *et al.*, 1989, Fig. 2)

mum in the dispersion curve of low-energy vibrational excitations (Suck *et al.*, 1983). This minimum was also deduced from calculations of the normal modes for the Lennard–Jones glass by Grest *et al.* (1982), who used molecular dynamics methods. It had been predicted earlier by L. D. Landau for liquids and, for historical reasons, it is called the '*roton minimum*' below.

All the above-mentioned experiments demonstrate that in glasses there are *additional low-energy vibrational excitations* that are non-existent in crystals. These excitations are responsible for the 'anomalous' thermodynamic and kinetic properties of glasses at low temperatures. The generality of these properties, which are common both to conventional silicate glasses, and to metallic glasses, suggests that such elementary excitations are universal and exist only in amorphous materials.

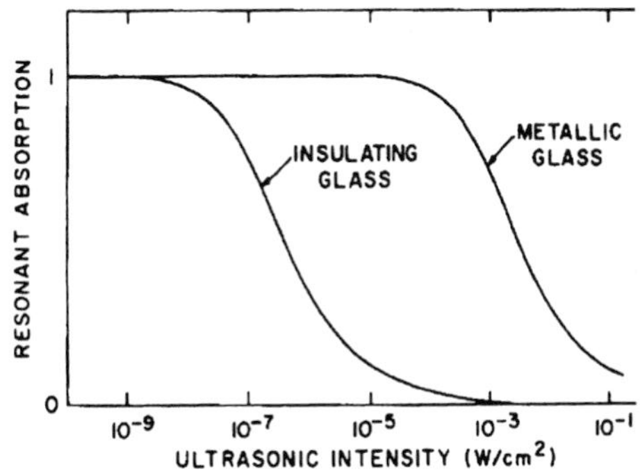

Fig. 3.1.3 Typical metallic versus insulating behavior in the saturation absorption (normalized to its low-power value) at temperatures near 10 mK and frequencies near 1 GHz (taken from Güntherodt and Beck, 1981, Fig. 8.2)

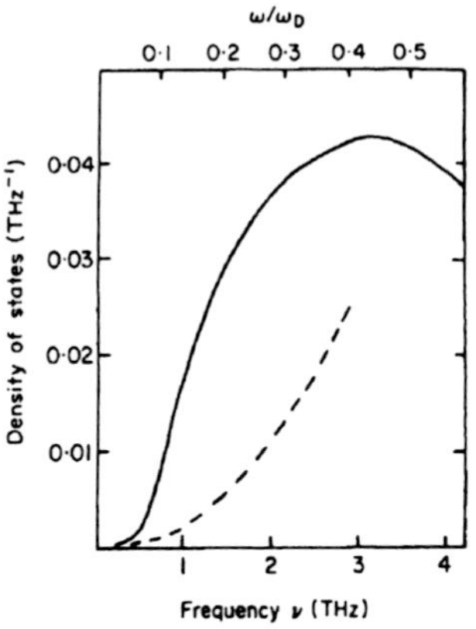

Fig. 3.1.4 The excess density of states for vibrational modes determined by inelastic neutron scattering (solid line) compared with the Debye density for vibrational modes (taken from Galperin *et al.*, 1989, Fig. 3, according to Buchenau *et al.*, 1986)

Additionally, we would also like to stress the following: although here in the present chapter we concentrate on vibrational excitations, we should mention that in magnetic metallic glasses spin excitations with a 'roton-like anomaly' also exist, at the same wavenumbers ($\approx 3.1\,\text{Å}^{-1}$) where the vibrational anomalies usually happen. In such metallic glasses we may also have magnetic analogs of the 'tunneling states' treated in the following subsections. We will come back to these 'universal aspects' at the appropriate parts of this book.

3.2 A Tunneling Model

It should be mentioned here that the following chapter applies both to metallic and non-metallic glasses. However, although we are concentrating on the first of these, the 'tunneling levels', which are treated below, are more easily studied in non-metallic systems, although most of their properties are 'universal' (see below).

A material is usually called *amorphous* or *glassy*, if the positions of the atoms or ions are *fixed* (apart from thermal or quantum fluctuations on a short timescale or changes on a very long 'aging' timescale) and if these fixed positions cannot be represented (not even approximately on a 10 nm scale, say) by a *regular* crystalline or quasi-crystalline lattice. In particular these systems are neither polycrystalline nor micro- or nanocrystalline, which can, however, be stated only after thorough diffraction experiments, or by electron microscopy, atomic force microscopy, or similar advanced methods. In fact, as already mentioned in the introductory chapters, amorphous or glassy materials can be produced typically by *rapid quenching* of the corresponding melt: the *kinetics* of the production is of the kind as to *trap* the system into a region of configuration space corresponding to a *metastable* state, i.e. to a local minimum of the potential energy. Because this usually is a sufficiently high-energy region of configurational space, there has to be a variety of other energetically equivalent 'glassy regions' where the system can be trapped. That is, in the system there exists usually a multitude of different atomic configurations $\{\mathbf{R}_j\}^{(m)}$ describing metastable states (where m is the configurational index and $\mathbf{R}_j$ is the radius vector of the jth atom). However, the fixed positions $\mathbf{R}_j$ of the atoms in amorphous solids are not in true thermodynamic equilibrium, and by thermal or quantum fluctuations the system may drift slowly from one metastable state to another, which leads to the already mentioned 'aging' of the physical properties, or the system may even relax very slowly to the basic configuration of a crystalline lattice. However *most* of the metastable states are widely apart in configurational space, and the relaxation times are therefore large – significantly larger than the characteristic times of a typical experiment. So on the timescales of most experiments amorphous solids may be considered as thermodynamically stable systems with quasi-equilibrium atomic coordinates. Consequently, the measurable characteristics of a quasi-equilibrium configuration of atoms may be calculated with the methods of equilibrium statistical mechanics, namely by Gibbs averaging over the degrees of freedom of quickly relaxing subsystems (the phonons, electrons, spins, etc.) (Mazo, 1963; Juchnovsky and Gurski, 1991). However the observables (such as free energy, heat capacity, magnetic moment, etc.) should not depend on the concrete positions of atoms, in view of the macroscopic number of equivalent configurations. Because of this *self-averaging property* of these variables in amorphous systems, the observables are usually averaged over all typical realizations of the random configurations of atoms in the system. (There are also certain correlation functions that are not self-averaging, but these will not be discussed here, unless explicitly stated.)

To illustrate what has been said, the free energy of such a 'quenched system' is, according to Mazo (1963) or Juchnovsky and Gurski (1991):

$$F_{am} = -k_{\mathrm{B}}T \cdot \overline{\ln Z} \tag{3.2.1}$$

where

$$Z = \mathrm{Tr}\left[\exp\left(-\frac{\hat{H}}{k_{\mathrm{B}}T}\right)\right] \tag{3.2.2}$$

The overbar in Eq. (3.2.1) stands for the configurational averaging with respect to a normalized probability function describing the random configuration of the glass. Depending on the structure concept of the particular disordered system the choice of this function is a problem in the amorphous state theory. For instance, for glassy metals the model of 'densely packed frozen liquids' (Juchnovsky and Gurski, 1991; Ziman, 1979) is preferable among existing models of topological disorder, whereas for random tetrahedrally coordinated silicate glasses, so-called 'Polk networks' are more appropriate (Polk, 1972), as already mentioned in the introduction.

In terms of the 'frozen liquid model' the structure of an amorphous system is described, as in the theory of liquids, by correlation functions characterizing the relationships between the atomic positions in the system. For, N and $V \to \infty\left(v_0 = \frac{V}{N} = \mathrm{const}\right)$, these functions are defined for $s = 1, 2, \ldots$ as follows

$$F_s(\mathbf{r}_1\ldots\mathbf{r}_s) = v_0^s \overline{\sum_{\substack{l_1=1 \\ (l_1\neq l_2\neq\ldots l_s)}}^{N} \cdots \sum_{l_s=1}^{N} \delta(\mathbf{r}_1 - \mathbf{R}_{l_1})\ldots\delta(\mathbf{r}_s - \mathbf{R}_{l_s})} \tag{3.2.3}$$

There is yet no microscopic theory to provide an exact calculation of the functions $F_s(\mathbf{r}_1\ldots\mathbf{r}_s)$. Therefore, we will consider them as phenomenological functions or use models well known in the theory of liquids (Fisher, 1961). For instance, in numerical calculations, when $F_2(\mathbf{r}_1,\mathbf{r}_2) \equiv g(\mathbf{r}_1 - \mathbf{r}_2)$, a hard-sphere model is frequently used, or an improved version of it, where a dense random packing of hard spheres is relaxed in Lennard-Jones potentials or realistic pseudopotentials (see above).

In any case, it should again be stressed here that the operation Tr{...} in Eq. (3.2.2) refers to the Trace with respect to the set of variables *only* of the quickly relaxing subsystems corresponding to the mth configuration of atoms, and the Hamiltonian $\hat{\mathrm{H}}(=\hat{\mathrm{H}}^{(m)})$ refers to these fast degrees of freedom for this particular configuration. The Gibbs averaging over these variables will then be denoted in the following by angular brackets

$$\langle\ldots\rangle = \frac{1}{Z}\mathrm{Tr}\left\{\ldots\exp\left(-\frac{\hat{\mathrm{H}}}{k_\mathrm{B}T}\right)\right\} \tag{3.2.4}$$

A specific model with a considerable amount of 'universality', explaining the anomalous low-temperature ($T < 2$ K) behavior of metallic and non-metallic glasses mentioned previously, is the *tunneling model* (Anderson *et al.*, 1972; Phillips, 1972), which is treated in the following. Not all metastable states with approximately equal energies are widely separated in configurational space; some of them, even if very small, can be so close to each other that the system passes from one state to another *within* the characteristic times of experiment. Because such transitions are predominantly realized by 'tunneling' through a potential barrier separating the metastable states, these states are called the 'tunneling states'. Clearly these 'tunneling processes' make a contribution to the general heat capacity of the system, and it should be stressed here that tunneling of an atom through a barrier is so strongly 'anharmonic' (i.e. so far from being a simple vibrational motion in a parabolic potential) that the 'tunneling excitations' should be treated as a separate class of excitations in addition to phonons.

It is obvious in the context of this general concept that the 'tunneling states' are related to a small group of atoms undergoing a local rearrangement. The number of atoms involved should be sufficiently small as to minimize the distance between the states in configurational space. On the other hand, the requirement of the approximate equivalence of energies can be met more easily as the number of involved atoms increases. It is the *competition* between the degree of accessibility and the degree of degeneracy that determines the size of a 'tunneling state' [1]. The

1) The same competition arises in connection with a famous formula of N. Mott for the 'hopping conductivity' in amorphous semiconductors (see Mott and Davis, 1979).

question of what the tunneling states specifically are has no simple answer. In amorphous SiO_2, for instance, the tunneling states can be represented by passages of an oxygen atom along the Si-Si axis between two quasi-equilibrium states, or by small rotations of the Si-O-Si molecule about the Si–Si axis (Galperin *et al.*, 1989); see Fig. 3.2.1 depicting the structures of crystalline and amorphous SiO_2 (the dotted curves show different interconverting configurations).

In glasses with laminated or cluster structures the fluctuating distances between layers or clusters cause variations in bonding between them, which in turn could result in tunneling states. Many different microscopic models of the tunneling states have been presented (Güntherodt and Beck, 1981; Beck and Güntherodt, 1983). Analysis of all these models suggests that the considered excitations are a fundamental property (e.g. unrelated to any specific impurities) of glasses due to their structural disorder. It should be noted that more and more data are indicative of 'tunneling' states in systems traditionally not regarded as glasses. Among these are the superionic conductors (Anthony and Anderson, 1976; Laermans, 1979), neutron-exposed quartz (Laermans, 1979), and some crystals with 'disordered' inclusions (Lou, 1976). Also in 'spin glasses', i.e. disordered spin systems with competing exchange interactions, tunneling excitations of the spin system appear to exist. At the same time there are amorphous materials (for in-

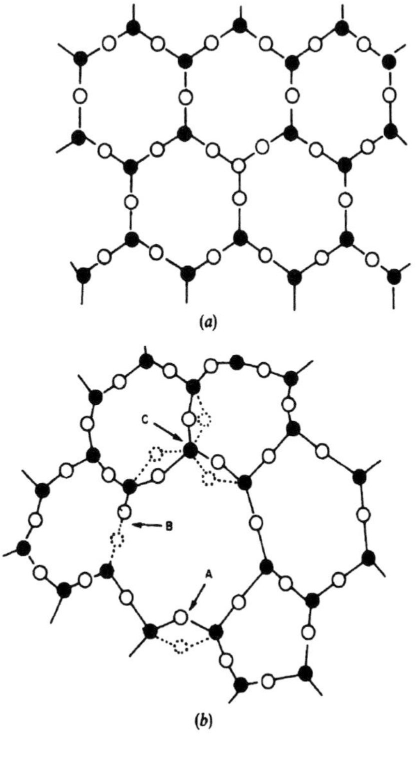

Fig. 3.2.1 Two-dimensional fragment of the structure of SiO_2 in the crystalline (upper) and amorphous (lower) part. The full rsp. empty circles represent Si and O; the dashed line represents a metastable 'tunneling state' (from Galperin *et al.*, 1989, Fig. 5)

stance amorphous Ge, Si, As, etc.) where 'tunneling' states have not been observed. All the above suggests that the 'glassiness' – as we understand it – is neither fully sufficient nor absolutely necessary for realization of the tunneling states. It was pointed out by Phillips (1978) that this realization is likely to require only a degree of irregularity in a small space along with 'not too restrictive' bonding.

Disregarding the microscopic structure, let us consider a local minimum on the potential energy surface of the glass. Our concern here is to find the closely-spaced minima with approximately equal energy. The anomalous contribution $\propto T$ of the heat capacity of the glass (see Stephens, 1973) is so small that the probability (per atom) of finding only one such close state is very low – of the order of 10^{-6}. Thus, to an excellent approximation, the probability of finding a third minimum in the vicinity may be neglected.

A section of the potential energy surface in configuration space of the glass is shown in Fig. 3.2.2. The section plane goes along 'the least resistance' direction (in terms of distance and potential barrier) between the two chosen minima.

When the generalized coordinate corresponding to the chosen direction is denoted by x, the part of the glass potential energy surface along x is a double-well potential $U(x)$ with a barrier height V, asymmetry energy $\Delta(u)$, and generalized distance d – all these parameters are random variables. Then the rearrangement of a group of atoms, corresponding to the transition from one metastable state to another, will be equivalent to the motion of a particle with effective mass M in the x direction under the influence of the potential $U(x)$. The particle mass, M, is of the order of the mass of the rearranged atoms. Because $U(x)$ is a double-well potential, the considered 'tunneling states' are a *two-level system*. The system correlates with its surroundings through the dependence of the potential $U(x)$ on the parameters on the deformation field $u_{ik} = \dfrac{\partial u_i}{\partial x_k} + \dfrac{\partial u_k}{\partial x_i}$ in the vicinity of the 'tunneling' states (u_i is the local displacement vector of the medium). Subsequently neglecting a weak dependence of d and V on u_{ik} we approximate the dependence of the energy difference $\varDelta$ of the two minima of the potential energy $U(x)$ (see Fig. 3.2.2) on u_{ik} by a linear function (Güntherodt and Beck, 1981; Anderson *et al.*, 1972; Phillips, 1972)

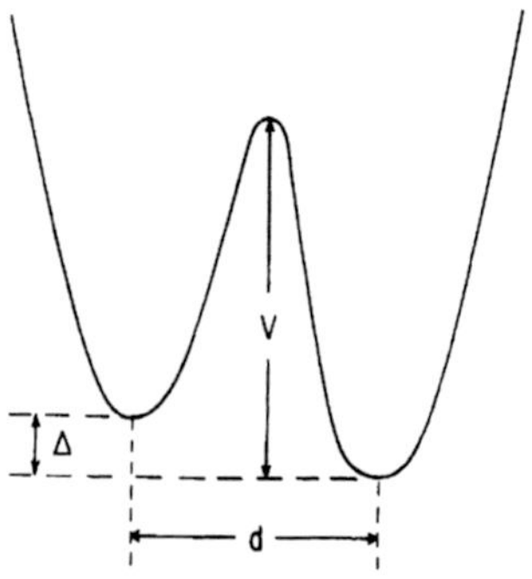

Fig. 3.2.2 Part of the potential energy surface producing a tunneling state characterized by a barrier height V, asymmetry energy $\varDelta$, and generalized distance d (taken from Güntherodt and Beck, 1981, p. 178, Fig. 8.1)

$$\Delta(u) = \varDelta + \gamma_{ik}u_{ik} \tag{3.2.5}$$

The so-called *deformation potential* $\gamma_{ik} = \dfrac{d\Delta}{du_{ik}}$ is determined by the relative positions of the nearby atoms. Because amorphous solids are *isotropic* on average we can put on average $\gamma_{ik} = \gamma\delta_{ik}$ thus assuming that the two-level system interacts only with *longitudinal* phonons, because in the limit of long wavelengths transversal phonons are energetically much less effective, as they do not lead to local changes of the specific volume.

When $U(x)$ is time independent, the Schrödinger equation describing the motion of a particle with mass M in the field $U(x)$ is as follows:

$$\hat{\mathbf{H}}\varphi(x) = E \cdot \varphi(x) \tag{3.2.6}$$

where

$$\hat{\mathbf{H}} = -\frac{\hbar^2}{2M}\frac{d^2}{dx^2} + U(x) \tag{3.2.7}$$

Now assume for simplicity that the origin of the coordinate system is placed at the minimum of the 'left' potential well (see Fig. 3.2.2) and assume that both wells have the same characteristic frequency ω_0, i.e.

$$\begin{cases} \left.\dfrac{d^2U}{dx^2}\right|_{x=0} = \left.\dfrac{d^2U}{dx^2}\right|_{x=d} = M\omega^2 \\ U(0) = \dfrac{\Delta(u)}{2}; \quad U(d) = -\dfrac{\Delta(u)}{2}; \quad U_{\max} = V \end{cases} \tag{3.2.8}$$

Then either well has its own set of vibrational states beginning with the ground state. The wave function of the ground state of the 'left' well may be written as

$$\varphi_L(x) = \left[\sqrt{\pi} \cdot x_0\right]^{-1/2} \exp\left[-\frac{x^2}{2x_0^2}\right] \tag{3.2.9}$$

and for the 'right' well as

$$\varphi_R(x) = \left[\sqrt{\pi} \cdot x_0\right]^{-1/2} \exp\left[-\frac{(x-d)^2}{2x_0^2}\right] \tag{3.2.10}$$

With our choice for the coordinates the energies of these states are respectively:

$$\begin{cases} E_L = \dfrac{1}{2}(\hbar\omega_0 + \Delta(u)) \\ E_R = \dfrac{1}{2}(\hbar\omega_0 - \Delta(u)) \end{cases} \tag{3.2.11}$$

Here

$$x_0 = \sqrt{\frac{\hbar}{M\omega_0}} \tag{3.2.12}$$

The distance between energy levels of subsequent excited states in a single well is the vibrational energy $\hbar\omega_0$. Because $\hbar\omega_0$ is of the order of the Debye energy, which is of the order of 500 K times k_B, the excitation of vibrational states within the wells is insignificant in the temperature range we are interested in ($T<2$ K). Therefore we are left with the opportunity to take into account transitions between E_L and E_R states. Neglecting a doubtful possibility of classical transitions over the barrier, which for $T<2$ K would only be important for those few barriers with extremely low barrier heights $<2\times10^{-4}$ eV, we can see that only in the case of *quantum tunneling* do such transitions appear to be frequent and the states E_L and E_R become observable.

To solve the problem under consideration we employ the variational method with a trial function in the form

$$\varphi(x) = C_L\,\varphi_L(x) + C_R\varphi_R(x) \tag{3.2.13}$$

From the normalization condition

$$\int_{-\infty}^{-\infty} |\varphi(x)|^2 dx = 1$$

it follows that

$$C_L^2 + C_R^2 + 2C_LC_R \cdot S = 1 \tag{3.2.14}$$

where

$$S = \int_{-\infty}^{+\infty} \varphi_L(x) \cdot \varphi_R(x) dx = e^{-\lambda} \tag{3.2.15}$$

is the overlap integral. Here

$$\lambda = \frac{d^2}{4x_0^2} = \frac{1}{4}\frac{d^2 M\omega_0}{\hbar} \tag{3.2.16}$$

The average energy of the investigated system is

$$E = \int \varphi^*(x)\hat{\mathbf{H}}\varphi(x)dx = C_L^2 \int \varphi_L(x)\hat{\mathbf{H}}\varphi_L(x)dx$$
$$+ C_R^2 \int \varphi_R(x)\hat{\mathbf{H}}\varphi_R(x)dx + 2C_LC_R \int \varphi_R(x)\hat{\mathbf{H}}\varphi_L(x)dx$$

Equating the variations of E with respect to the parameters C_L and C_R to zero under the constraint (3.2.15) we have the following set of equations for C_L and C_R:

$$\begin{cases} C_L \int \varphi_L \hat{\mathbf{H}} \varphi_L \mathrm{d}x + C_R \int \varphi_R \hat{\mathbf{H}} \varphi_L \mathrm{d}x = E(C_L + C_R S) \\ C_R \int \varphi_R \hat{\mathbf{H}} \varphi_R \mathrm{d}x + C_L \int \varphi_R \hat{\mathbf{H}} \varphi_L \mathrm{d}x = E(C_R + C_L S) \end{cases} \tag{3.2.17}$$

It is easy to estimate the integrals multiplied by C_L and C_R in Eq. (3.2.17) remembering that the major contribution comes from the region of maximal exponent value. Hence under the integral it is sufficient to expand the pre-exponent factor into a Taylor series in the neighborhood of the point where the exponent is maximal, and retain the first three terms of the expansion:

$$\int \varphi_L(x) \hat{\mathbf{H}} \varphi_L(x) \mathrm{d}x \approx \frac{1}{2}\hbar\omega_0 + \frac{\Delta + \gamma \cdot u_{ii}}{2}$$

$$\int \varphi_R(x) \hat{\mathbf{H}} \varphi_R(x) \mathrm{d}x \approx \frac{1}{2}\hbar\omega_0 + \frac{\Delta + \gamma \cdot u_{ii}}{2}$$

$$\int \varphi_R(x) \hat{\mathbf{H}} \varphi_L(x) \mathrm{d}x \approx \mathrm{e}^{-\lambda} \left\{ \frac{1}{4}\hbar\omega_0 - \frac{\mathrm{d}^2}{8x_0^2} \cdot \hbar\omega_0 + U\left(\frac{d}{2}\right) + \frac{1}{4}\frac{\mathrm{d}^2 U}{\mathrm{d}x^2}\bigg|_{x=\frac{d}{2}} \cdot \frac{\hbar}{M\omega_0} \right\}$$

For simplification the function $U(x)$ is chosen so that

$$\begin{cases} V = U_{\max} \approx U\left(\frac{d}{2}\right) \approx \frac{\mathrm{d}^2}{8x_0^2}\hbar\omega_0 = \frac{2\hbar^2}{M\mathrm{d}^2} \\ \frac{\mathrm{d}^2 U}{\mathrm{d}x^2}\bigg|_{x=\frac{d}{2}} = -M\omega_0^2 \end{cases} \tag{3.2.18}$$

Then

$$\int \varphi_R \hat{\mathbf{H}} \varphi_L \mathrm{d}x \approx 0$$

Converting Eq. (3.2.18) to the new unknowns C_1 and C_2 we have

$$\begin{cases} C_1 = C_L + C_R \cdot S \\ C_2 = C_R + C_L \cdot S \end{cases} \quad \text{or} \quad \begin{cases} C_L = \dfrac{C_1 - C_2 S}{1 - S^2} \\ C_R = \dfrac{C_2 - C_1 S}{1 - S^2} \end{cases} \tag{3.2.19}$$

This leads to the set of matrix equations

$$\left[\frac{1}{2}\begin{pmatrix} \Delta & -\Delta_0 \\ -\Delta_0 & -\Delta \end{pmatrix} + u_{ii}\frac{\gamma}{2}\begin{pmatrix} 1 & 0 \\ 0 & -1 \end{pmatrix}\right]\begin{pmatrix} C_1 \\ C_2 \end{pmatrix} = \left(E - \frac{1}{2}\hbar\omega_0\right)\begin{pmatrix} C_1 \\ C_2 \end{pmatrix} \tag{3.2.20}$$

where

$$\Delta_0 \approx \hbar\omega_0 \exp\left(-\frac{1}{2}\frac{d}{\hbar}\sqrt{2MV}\right) \tag{3.2.21}$$

Equation (3.2.19), as well as the inequalities $\hbar\omega_0 \gg \Delta$ and $S \ll 1$, are taken into account in the latter relationship.

Let us introduce a unitary matrix

$$\hat{U} = \hat{U}^{-1} = \frac{\Delta_0}{\sqrt{2\varepsilon \cdot (\varepsilon - \Delta)}} \begin{pmatrix} 1 & -\frac{\varepsilon-\Delta}{\Delta_0} \\ -\frac{\varepsilon-\Delta}{\Delta_0} & -1 \end{pmatrix} \tag{3.2.22}$$

where

$$\varepsilon = \sqrt{\Delta^2 + \Delta_0^2} \tag{3.2.23}$$

which is typically $\ll \hbar\omega_0$.

On the left-hand side of Eq. (3.2.20), the term that is independent of the deformation u_{ik} can be diagonalized with this matrix:

$$\hat{U}\begin{pmatrix} \Delta & -\Delta_0 \\ -\Delta_0 & -\Delta \end{pmatrix}\hat{U}^{-1} = \begin{pmatrix} \varepsilon & 0 \\ 0 & -\varepsilon \end{pmatrix} \tag{3.2.24}$$

Now we consider the case of undeformed vicinity of the 'tunneling' states, i.e. $u_{ii} = 0$. Then we can diagonalize the left-hand side of Eq. (3.2.20) by multiplying the equation by $\hat{U}$:

$$\frac{1}{2}\begin{pmatrix} \hbar\omega_0 + \varepsilon & 0 \\ 0 & \hbar\omega_0 - \varepsilon \end{pmatrix}\begin{pmatrix} C_1' \\ C_2' \end{pmatrix} = E\begin{pmatrix} C_1' \\ C_2' \end{pmatrix} \tag{3.2.25}$$

where

$$\begin{pmatrix} C_1' \\ C_2' \end{pmatrix} = \hat{U}\begin{pmatrix} C_1 \\ C_2 \end{pmatrix} = \frac{\Delta_0}{\sqrt{2\varepsilon(\varepsilon - \Delta)}} \begin{pmatrix} C_1 & -\frac{\varepsilon-\Delta}{\Delta_0}C_2 \\ -\frac{\varepsilon-\Delta}{\Delta_0}C_1 & -C_2 \end{pmatrix} \tag{3.2.26}$$

The matrix equation (3.2.25) is equivalent to the system of equations

$$\begin{cases} \frac{1}{2}(\hbar\omega_0 + \varepsilon)C_1' = EC_1' \\ \frac{1}{2}(\hbar\omega_0 + \varepsilon)C_2' = EC_2' \end{cases}$$

This system has non-trivial solutions under the following conditions:
(a) if $C_2' \neq 0$ it is necessary that

$$E = \frac{1}{2}(\hbar\omega_0 - \varepsilon) \quad \text{and} \quad C_1' = 0;$$

(b) if $C_1' \neq 0$ then

$$E = \frac{1}{2}(\hbar\omega_0 + \varepsilon) \quad \text{and} \quad C_2' = 0.$$

Thus, due to tunneling, the following pair of nearby energy levels are set up:

$$\begin{cases} E_\mathrm{g} = \dfrac{1}{2}(\hbar\omega_0 - \varepsilon) \\ E_\mathrm{e} = \dfrac{1}{2}(\hbar\omega_0 + \varepsilon) \end{cases} \tag{3.2.27}$$

The condition $C_1' = 0$ corresponds to the ground state E_g or, according to Eq. (3.2.26),

$$C_1 = -\frac{\varepsilon - \Delta}{\Delta_0} C_2$$

Substituting this expression into Eq. (3.2.19) and employing the normalization condition (3.2.14) we find the coefficients C_L and C_R, and subsequently the wave function of the ground state

$$\varphi_\mathrm{g}(x) = \frac{\left(\dfrac{\varepsilon - \Delta}{\Delta_0} - S\right)\varphi_\mathrm{L}(x) + \left(1 - \dfrac{\varepsilon - \Delta}{\Delta_0} S\right)\varphi_\mathrm{R}(x)}{\sqrt{\left[1 - 2\dfrac{\varepsilon - \Delta}{\Delta_0} S + \left(\dfrac{\varepsilon - \Delta}{\Delta_0}\right)^2\right](1 - S^2)}} \tag{3.2.28}$$

The condition $C_2' = 0$ corresponds to the excited state E_e or

$$C_1 = -\frac{\varepsilon + \Delta}{\Delta_0} C_2$$

It can easily be shown with this relation that the wave function of the excited state is:

$$\varphi_\mathrm{e}(x) = \frac{-\left(\dfrac{\varepsilon + \Delta}{\Delta_0} - +S\right)\varphi_\mathrm{L}(x) + \left(1 + \dfrac{\varepsilon + \Delta}{\Delta_0} S\right)\varphi_\mathrm{R}(x)}{\sqrt{\left[1 + 2\dfrac{\varepsilon + \Delta}{\Delta_0} S + \left(\dfrac{\varepsilon + \Delta}{\Delta_0}\right)^2\right](1 - S^2)}} \tag{3.2.29}$$

Two important consequences follow from the above. First and foremost, the energy splitting ε cannot be less than Δ_0 (see Eq. (3.2.23)). Therefore, because we want ε and thus Δ_0 to become small, there has to be a barrier of finite height V (see Eq. (3.2.21)) to produce sufficiently closely spaced energy levels. Further, the wave functions

$\varphi_g(x)$ and $\varphi_e(x)$ are linear combinations of $\varphi_L(x)$ and $\varphi_R(x)$ and, therefore, the states corresponding to these functions are localized in neither of the wells.

Now we turn to the general case when $u_{ii} \neq 0$ and is time independent. Let us multiply the left- and right-hand sides of Eq. (3.2.20) by the matrix $\hat{U}$ taking into consideration that

$$\hat{U}\begin{pmatrix} 1 & 0 \\ 0 & -1 \end{pmatrix}\hat{U}^{-1} = \frac{\Delta}{\varepsilon}\begin{pmatrix} 1 & 0 \\ 0 & -1 \end{pmatrix} + \frac{\Delta}{\varepsilon}\begin{pmatrix} 0 & -1 \\ -1 & -0 \end{pmatrix}$$

So this equation transforms to

$$\hat{\mathbf{H}}\Psi = E\Psi \tag{3.2.30}$$

where

$$\hat{\mathbf{H}} = \frac{1}{2}\begin{pmatrix} \hbar\omega_0 + \varepsilon & 0 \\ 0 & \hbar\omega_0 - \varepsilon \end{pmatrix} + \frac{u_{ii}}{2}D\begin{pmatrix} 1 & 0 \\ 0 & -1 \end{pmatrix} + u_{ii}M\begin{pmatrix} 0 & -1 \\ -1 & 0 \end{pmatrix} \tag{3.2.31}$$

$$D = \frac{\Delta}{\varepsilon}\gamma; \quad M = \frac{1}{2}\frac{\Delta_0}{\varepsilon}\gamma; \quad \Psi = \begin{pmatrix} C_1' \\ C_2' \end{pmatrix} \tag{3.2.32}$$

If u_{ii} is time dependent, instead of Eq. (3.2.30), the time-dependent Schrödinger equation has to be solved:

$$\hat{\mathbf{H}}\Psi(x) = -i\hbar\frac{\partial\Psi}{\partial t} \tag{3.2.33}$$

The second term in the Hamiltonian (3.2.31) is a diagonal matrix describing the modulation of energy levels E_g and E_e under deformation; the third summand represents a non-diagonal interaction and, therefore, results in direct transitions between energy levels. For a given ε the variable M ranges from a maximal value $1/2\gamma$ a in the symmetrical case ($\Delta = 0;\ \varepsilon = |\Delta_0|$) to a minimum, which is zero in the case of a high barrier ($\Delta_0 = 0;\ \varepsilon = |\Delta|$). Thus it is clear that experimentally significant states are characterized by *a distribution of barriers* where the minimal barrier height is determined by the requirement for the smallness of energy differences, and the maximal height by observability (i.e. by the possibility of interlevel transitions).

To complete the model, explicit assumptions have to be made on the distribution of the parameters Δ, Δ_0 and γ over the entire potential energy surface. The common conventions (Jäckle, 1972; Halperin, 1976) are that the distribution function:

(a) is independent of Δ;

(b) depends on Δ_0 only through the overlapping parameter λ, uniformly distributed and having a sharp peak in the symmetrical case ($\Delta = 0;\ \varepsilon = |\Delta_0|$).

These assumptions are in agreement, at least qualitatively, with the majority of experiments at temperatures below 2 K. As a result we have a useful expression for the partial density of 'tunneling' states with splitting ε and non-diagonal interaction $r = \frac{\Delta_0^2}{\varepsilon^2}$

$$p(\varepsilon, r) = \frac{1}{2r}\bar{p}\frac{1}{\sqrt{1-r}} \tag{3.2.34}$$

where $0 \le r \le 1$, and $\bar{p}$ is the experimental density of states of the order of 10^{19}–$10^{20}\,\mathrm{eV^{-1}cm^{-3}}$.

Let us consider the inferences from the above concepts as applied to low-temperature experiments on glasses. In our case a rapidly relaxing subsystem is the two-level system that comes into thermodynamic equilibrium within times that are much shorter than the characteristic times of the experiment. For the given atomic configuration and with no deformation (that is Δ and Δ_0 are fixed, and $u_{ii}=0$) the Hamiltonian of the two-level system is (see Eq. (3.2.31))

$$\hat{\mathbf{H}}_0 = \frac{1}{2}\begin{pmatrix} \hbar\omega_0 + \varepsilon & 0 \\ 0 & \hbar\omega_0 - \varepsilon \end{pmatrix} = \begin{pmatrix} E_g & 0 \\ 0 & E_e \end{pmatrix} \tag{3.2.35}$$

and the average energy is determined by the expression

$$\langle \hat{\mathbf{H}}_0 \rangle = \frac{E_g e^{-E_g/k_B T} + E_e e^{-E_e/k_B T}}{e^{-E_g/k_B T} + e^{-E_e/k_B T}}$$

The heat capacity of the system is readily found with the relation

$$\langle C_V \rangle = \frac{\partial \langle \hat{\mathbf{H}}_0 \rangle}{\partial T} = \frac{\varepsilon^2}{k_B T^2}\frac{1}{4}\,\mathrm{sech}^2\left(\frac{\varepsilon}{2k_B T}\right) \tag{3.2.36}$$

To obtain the observable contribution given by the 'tunneling' states to the total specific heat of the glass, Eq. (3.2.36) has to be further averaged over all possible realizations of the variables Δ and Δ_0 $\left(\text{or } \varepsilon = \sqrt{\Delta^2 + \Delta_0^2} \text{ and } r = \frac{\Delta_0^2}{\varepsilon^2}\right)$:

$$C_V = \overline{\langle C_V \rangle} = k_B \int_0^\infty d\varepsilon \int_{r_0}^1 p(\varepsilon, r)\frac{\varepsilon^2}{k_B^2 T^2}\frac{1}{4}\,\mathrm{sech}^2\left(\frac{\varepsilon}{2k_B T}\right) dr$$

$$= \frac{\pi^2}{6}k_B^2 T \cdot \bar{p} \cdot \frac{1}{2}\int_{r_0}^1 \frac{dr}{r(1-r)^{1/2}} \tag{3.2.37}$$

Here r_0 is the maximal cut-off parameter of the barrier height, which can be a function of the characteristic time of the experiment (Black, 1978).

An almost linear term in the heat capacity, $C_p \propto T$ at temperatures below 2 K, has been observed in many non-metallic glasses. It is well known that in metals a similar contribution would be conventional and simply come from the metallic electrons, but in non-metallic glasses such a contribution is unconventional, and definitely due to atomic tunneling states, which are strongly anharmonic low-energy structural excitations of the system, as described above.

Also in the so-called '*spin glasses*' a pronounced linear specific heat contribution $C_H \propto T$ arises. However, there only the 'magnetic part' of the total heat capacity is concerned, i.e. the part that can be suppressed by strong magnetic fields. Typical 'spin glasses' are *crystalline*-diluted magnetic alloys like $Au_{1-x}Fe_x$ or $Cu_{1-x}Mn_x$, with typical values of $x \approx 0.01$, and a 'glassy *magnetic* behavior'. The reason for the glassy magnetic behavior in these particular systems is the *simultaneous* presence of (i) *strong positional disorder* of the magnetic Fe rsp. Mn impurities, and (ii) *competing interactions*, i.e. whether the interaction between a pair of spins favors parallel or antiparallel mutual orientation of the spins, depends sensitively on the mutual random distance between the atoms considered, with equal probability for both cases. As a consequence, there is a strong '*spin frustration*' in these systems, leading to the existence of many metastable magnetic configurations of the system. These model systems, which are comprehensively reviewed by Binder and Young (1986), are in fact in many aspects analogous, concerning their *magnetic* behavior, to the glassy *structural* behavior of the conventional glasses, which is also partially due to 'structural frustration' in the system; however, this analogy is only mentioned here without further explanation.

In any case, in spin glasses the linear behavior $C_H \propto T$ of the relevant heat capacity is partially explained by (anharmonic) tunneling states of the *spin system* (Halperin, 1976), although there is an alternative explanation involving harmonic spin excitations, a type of spin wave, in the disordered spin system (Krey, 1980, 1981, 1985). So, after all, this 'magnetic contribution' $C_H \propto T$ to the specific heat in spin glasses is not unexpected, following the corresponding contribution $C_p \propto T$ from 'structural tunneling excitations' in conventional silicate glasses that was found by Zeller and Pohl (1971).

In fact, the first experimental discovery of the contribution $C_p \propto T$ in silicate glasses by Zeller and Pohl (1971) came as a surprise. What is more, the value of the density of tunneling states obtained in this way (the product of $\bar{p}$ times the integral over r in Eq. (3.2.37)) varies at most by a factor of 5 in passing from one glass to another, so it is rather 'universal'. In fact, the observed density of states implies that the tunneling' states are concentrated in an energy region with $\varepsilon < 1$ K, amounting to roughly between 1 and 10 tunneling states per million atoms.

In addition to the heat capacity, most of the principal results from the tunneling model in silicate glasses, and in metallic glasses (Black and Gyorffy, 1978), involve interactions between the 'tunneling' states and phonons. Suppose a longitudinal sound wave propagating in the glass with frequency $\omega = \varepsilon/\hbar$ and wavevector **k**. Let the propagation direction be the x-axis. Then the components of the displacement vector are:

$$u_x = u_{0x} \cdot \cos(kx - \omega t)\,; \; u_y = u_z = 0$$

and the time-averaged energy per unit volume of the sound wave is determined by the relationship

$$\bar{E}_{\text{wave}} = \rho \cdot \frac{1}{T} \int_0^T \dot{u}^2(x,t)\mathrm{d}t = \frac{1}{2}\rho\omega^2 u_{0x}^2 \qquad (3.2.38)$$

where $T = \frac{2\pi}{\omega}$ is the period of the wave and ρ is the mass density of the substance.

Only one of the components of the deformation tensor $u_{ik} = \frac{1}{2}\left(\frac{\partial u_i}{\partial x_k} + \frac{\partial u_k}{\partial x_i}\right)$ will be non-zero, namely:

$$u_{xx} = -\frac{1}{2} u_{0x} \frac{\omega}{V} \sin\left(kx - \omega t\right) \qquad (3.2.39)$$

Here $k = \frac{\omega}{V}$ is taken into account, where v is the velocity of a longitudinal acoustical wave.

In quantum mechanical terms this wave is a parallel flux of phonons with $\omega = \varepsilon/\hbar$. Passing through a two-level system, phonons with this frequency can be absorbed – exciting the system – or else can stimulate the release of an additional phonon of the same frequency ω with wavevector $\mathbf{k}$, if the system is excited.

Let us find the mean free path of a phonon specified by these processes. According to the principles of quantum mechanics (e.g. Landau and Lifshitz, 1963; Davydov, 1963) the interlevel transition rate induced by a periodical disturbance $M \cdot u_{ii}(t) = \frac{1}{2} M u_{0x} \frac{\omega}{V} \sin\left(kx - \omega t\right)$ is given by *Fermi's Golden Rule*, namely

$$d\omega_j = \frac{2\pi}{\hbar}\left|\frac{1}{2} M\, u_{0x} \frac{\omega}{v}\right|^2 \delta(E_g - E_e + \hbar\omega)\mathrm{d}E_j$$

With $\rho(E_j)$ denoting the density of final states, the total transition rate, i.e. the probability per time unit of a transition of the system to the state E_j (j = g, e), given an initial state with an energy E_i, which is higher or lower in energy by $\hbar\omega$, is

$$\omega_j = \frac{2\pi}{\hbar}\left|\frac{M}{2} u_{0x} \frac{\omega}{v}\right|^2 \rho(E_j) \qquad (3.2.40)$$

given that

$$E_e - E_g = \varepsilon = \hbar\omega$$

In our case the density of final states is a product of the partial density of the tunneling states $p\left(\varepsilon, r\right)$ times the thermodynamic population of the levels:

$$\rho(E_j) = p(\hbar\omega, r) \cdot \frac{e^{-E_j/k_B T}}{e^{-E_g/k_B T} + e^{-E_e/k_B T}}$$

Because in passing from the level E_g to level E_e the system *absorbs* energy $\hbar\omega$, and going from E_e to E_g *emits* $\hbar\omega$, the average energy absorbed by the two-level system in a unit of time is:

$$\hbar\omega\,(w_{\mathrm{g}} - w_{\mathrm{e}}) = 2\pi\omega\left|\frac{M}{2}u_{0x}\frac{\omega}{v}\right|^2[p(E_{\mathrm{g}}) - p(E_{\mathrm{e}})]$$
$$= \frac{\pi}{2}\frac{\omega^3}{v^2}u_{0x}^2\frac{\gamma^2}{4}\tanh\left(\frac{\hbar\omega}{2k_{\mathrm{B}}T}\right)\cdot\gamma\cdot p\,(\hbar\omega, r)$$

The following integration of this expression, reversed in sign, integrated over the parameter *r*, gives the energy that the acoustic wave of frequency ω loses in a unit time due to interactions with the two-level system:

$$\bar{\dot{E}}_{\mathrm{res}} = -\frac{\pi}{2}\frac{\omega^3}{v^2}u_{0x}^2\frac{\gamma^2}{4}\cdot\bar{p}\cdot\tanh\left(\frac{\hbar\omega}{2k_{\mathrm{B}}T}\right)$$

From now on thermodynamic quantities are taken per unit volume.

The phonon's mean free path time is found with a known formula (Anderson *et al.*, 1972; Phillips, 1972):

$$\frac{1}{\tau(\omega)} = \frac{\bar{\dot{E}}_{\mathrm{res}}}{\bar{E}_{\mathrm{wave}}} = \frac{\pi\omega}{\rho v^2}\frac{\gamma^2}{4}\cdot\bar{p}\cdot\tanh\left(\frac{\hbar\omega}{2k_{\mathrm{B}}T}\right) \tag{3.2.41}$$

Hence the reciprocal of the phonon-free path length (or twice the coefficient of the acoustic resonance absorption) is, according to Anderson *et al.* (1972) and Phillips (1972):

$$\frac{1}{l(\omega)} = \frac{1}{\tau(\omega)\cdot v} = \frac{\pi\omega}{\rho v^3}\frac{\gamma^2}{4}\cdot\bar{p}\cdot\tanh\left(\frac{\hbar\omega}{2k_{\mathrm{B}}T}\right) \tag{3.2.42}$$

The hyperbolic tangent in this equation results in a negative temperature coefficient of sound attenuation, i.e. by enhancing *T* the sound attenuation is reduced. Experiments by Golding *et al.* (1967) with the dielectric glass SiO_2 have shown close agreement with Eq. (3.2.42).

Moreover, this free path length expression permits calculation of the phonon contribution to the thermal conduction (the localized 'tunneling' states in themselves are non-thermoconductive): the longitudinal phonons of frequency ω account for the heat capacity

$$C_{\mathrm{ph}}(\omega) = \frac{\hbar^2\omega^2}{k_{\mathrm{B}}T^2}\left[\mathrm{e}^{\hbar\omega/k_{\mathrm{B}}T} - 1\right]^{-1}\cdot[1 - \mathrm{e}^{-\hbar\omega/k_{\mathrm{B}}T}]^{-1}$$

and the total contribution of longitudinal phonons to the heat capacity is therefore:

$$C_{\mathrm{ph}} = \int C_{\mathrm{ph}}(\omega)\frac{\mathrm{d}^3k}{(2\pi)^3}$$

(Here no consideration is given to the transversal phonons because they do not interact with the two-level systems.)

As a consequence, the coefficient of thermal conductivity due to the considered phonons can be found from the simple kinetic formula:

$$\kappa(\omega) = \frac{1}{3} C_{\mathrm{ph}}(\omega) \cdot v \cdot l(\omega)$$

Because thermal conduction involves longitudinal phonons of all wavelengths, the total contribution to the phonon thermoconductivity in glasses is therefore determined by the expression (Anderson *et al.*, 1972; Phillips, 1972):

$$\kappa = \int \kappa(\omega) \frac{d^3k}{(2\pi)^3} = \frac{1}{6\pi} \frac{\rho v k_B^3}{\bar{p}\frac{\gamma^2}{4}\hbar^2} T^2 \qquad (3.2.43)$$

Here the T^2 dependence of the thermal conductivity κ is in perfect agreement with experiments for a great variety of glasses (Stephens, 1973).

The *sound velocity* is the other characteristic quantity associated with phonon propagation in glasses. The resonance interaction leading to the mean free path length, Eq. (3.2.42), is responsible for a measurable and temperature-dependent shift of the sound velocity in glasses. Performing a Kramers-Kronig analysis of the scattering law in Eq. (3.2.42) we can derive the following logarithmic expression for the temperature dependence of the relative shift of the longitudinal sound velocity (Golding *et al.*, 1976):

$$\frac{\Delta v}{v} = \frac{\bar{p}}{\rho v^2} \frac{\gamma^2}{4} \cdot \ln\left(\frac{T}{T_0}\right) \qquad (3.2.44)$$

where T_0 is some unspecified initial temperature. A velocity shift of this kind has been observed in many glasses (Golding *et al.*, 1976). Note that the magnitude of $\bar{p} \cdot \gamma^2$ can be obtained either by measurement of sound attenuation or by sound velocity measurements. The values given by both methods correlate with an accuracy of 50%.

It should be noted here that these results also apply essentially to metallic glasses; in these systems, however, tunneling levels and longitudinal phonons also interact with the metallic electrons, which gives rise to some additional effects and modifications not discussed here (Black, 1981, Section 8.3).

Among the most impressive successes of the tunneling model is its explanation of the *saturation of the ultrasound resonance attenuation*. The 'saturation', as shown in Fig. 3.1.3 above, manifests itself in a decrease in the absorption coefficient with increasing phonon flux. Responsible for this is the effect of the phonon beam to balance the population of states in the two-level system. If the population is equalized, the processes of resonance phonon absorption and emission will completely compensate each other, and the phonon absorption as a whole will be absent. But practically there are always various processes seeking to restore the thermodynamically equilibrium population values in the two-level system and thus counteracting the tendency to equalize. Calculations demonstrate that the competition between these two tendencies determines the saturation threshold. Usually the threshold is determined in such a way that the attenuation length l of the phonon energy flux is enhanced by a factor of $\sqrt{2}$ compared to its value without saturation.

Data from ultrasonic experiments are in good agreement with the tunneling model. Hunklinger *et al.* (1972) and Golding *et al.* (1973) observed resonance absorption of the kind shown in Fig. 3.1.3, depending on the intensity of the ultrasound beam; in this way the saturation threshold in SiO_2 and borosilicate glasses could be determined in reasonable agreement with theoretical calculations.

The resonance interaction between two-level systems and phonons brings about another important effect, referred to as the *phonon echo*. The phonon echo is a spontaneous generation of a coherent phonon pulse by the tunneling levels, after they have been previously excited with two (or more) ultrasonic pulses. This emitted pulse results from a simultaneous 'phase-coherent tunneling' of a multitude of excited two-level systems.

Golding and Graebner (1976) were the first to observe the phonon echo in SiO_2 glass at temperatures below 100 mK. These experiments provided extensive and comprehensive information on the magnitude of γ – the constant of interaction between the two-level system and phonons.

And, finally, consider one more interaction process of the 'tunneling' states with phonons. The case in point is the relaxational phonon attenuation caused by the 'tunneling' systems (Jäckle *et al.*, 1976). This effect is brought about because the elastic stresses related to the phonons produce a periodic modulation of the energy splitting in the two-level systems (periodic with phonon frequency ω), which is predicted by the term proportional to D in the Hamiltonian (3.2.31). As before, assume that a monochromatic longitudinal acoustic wave of frequency ω propagates through the glass in the x-direction. Then the u_{xx}-component of the deformation tensor will be non-zero, see Eq. (3.2.39). Let us dwell on the case

$$k_B T > \varepsilon \gg \hbar\omega \gg \frac{\hbar}{\tau_1} \tag{3.2.45}$$

where τ_1 is the relaxation time specifying the direct return to equilibrium of the population in the two-level system due to the interaction with thermal phonons.

Here the first strong inequality, $\varepsilon \gg \hbar\omega$, implies that the acoustic wave does not induce transitions between the system levels, and therefore in the Hamiltonian (3.2.31) the term describing the non-diagonal interaction may be neglected, i.e.

$$\hat{\mathbf{H}}(t) = \hat{\mathbf{H}}_0 + \Delta\hat{\mathbf{H}}(t) \tag{3.2.46}$$

where

$$\Delta\hat{\mathbf{H}}(t) = \frac{D}{2} u_{ii}(t) \begin{pmatrix} 1 & 0 \\ 0 & -1 \end{pmatrix} \tag{3.2.47}$$

and $\hat{\mathbf{H}}_0$ has the form of Eq. (3.2.35).

From the second strong inequality ($\omega \gg \frac{1}{\tau_1}$ or $\tau \ll \tau_1$) it follows that within times $t \sim \tau$ equilibrium has become settled in the system only locally and thermodynamic variables may be calculated locally in terms of equilibrium statistics. To obtain thermodynamic quantities describing the system in complete thermodynamic equilibrium, the calculated expressions have to be further time-averaged. Because $k_B T > \varepsilon > \langle \Delta \hat{\mathbf{H}}(t) \rangle$ we employ the thermodynamic perturbation theory to represent the free energy in the following form:

$$\langle F(t) \rangle = \langle F \rangle_0 + \langle \Delta \hat{\mathbf{H}}(t) \rangle_0 - \frac{1}{2k_B T}\left[\langle \Delta \hat{\mathbf{H}}^2(t) \rangle_0 - \langle \Delta \hat{\mathbf{H}}(t) \rangle_0^2\right] \tag{3.2.48}$$

where $\langle F \rangle_0$ denotes the 'undisturbed' free energy calculated for $\Delta \bar{\mathbf{H}} = 0$; and $\langle \ldots \rangle_0$ is the averaging with an undisturbed Gibbs distribution. Using Eqs (3.2.35) and (3.2.48) we can easily find that

$$\langle \Delta \hat{\mathbf{H}}(t) \rangle = \frac{D}{2} u_{ii}(t) \cdot \coth\left(\frac{\varepsilon}{2k_B T}\right)$$

$$\langle \Delta \hat{\mathbf{H}}^2(t) \rangle = \frac{D^2}{4} u_{ii}^2(t)$$

Substituting these expressions into Eq. (3.2.49) and time averaging by

$$\langle F \rangle = \frac{1}{\tau} \int_0^\tau \langle F(t) \rangle \mathrm{d}t$$

gives the free energy of the two-level system in thermodynamic equilibrium with the environment. The average energy of the system is found with the well-known relationship

$$\langle \hat{\mathbf{H}} \rangle = \langle F \rangle - T \frac{\partial}{\partial T} \langle F \rangle$$

or

$$\langle \hat{\mathbf{H}} \rangle = \langle \hat{\mathbf{H}}_0 \rangle_0 + \langle \Delta E \rangle_0 \tag{3.2.49}$$

where

$$\langle \Delta E \rangle_0 = \frac{1}{k_B T} \frac{D^2}{8} u_{0x}^2 \frac{\omega^2}{v^2} \frac{1}{4} \operatorname{sech}^2\left(\frac{\varepsilon}{2k_B T}\right) \tag{3.2.50}$$

It can be seen from these equations that the average energy of the two-level system exposed to the sonic wave of frequency ω is increased by $\langle \Delta E \rangle_0$. Consequently, the sonic wave loses the same energy in a time τ_1. Hence, the average dissipation of energy of the wave (i.e. its energy loss in a unit of time) is

$$\bar{\dot{E}}_{\text{rel}} = \frac{\langle \Delta E \rangle_0}{\tau_1}$$

The sound absorption coefficient $\langle a \rangle$ is defined as the ratio of the average energy dissipation to twice the average (over time) of the energy flux of the sound wave (see Eq. (3.2.39)):

$$\langle a \rangle = \frac{\bar{\dot{E}}_{\text{rel}}}{2v\bar{E}_{\text{wave}}} = \frac{D^2}{8k_B T\rho v^3}\frac{1}{4}\,\text{sech}^2\left(\frac{\varepsilon}{2k_B T}\right)\cdot\frac{1}{\tau_1} \tag{3.2.51}$$

This expression was derived for a given atomic arrangement $\left(\text{or with fixed parameters } \varepsilon \text{ and } r = \frac{\Delta_0^2}{\varepsilon^2}\right)$. To obtain the observable a, Eq. (3.2.52) must still be averaged over all values of ε and r (Jäckle *et al.*, 1976):

$$a = \overline{\langle a \rangle} = \int_0^\infty d\varepsilon \int_{r_0}^1 dr p(\varepsilon, r)\frac{D^2}{8k_B T\rho v^3}\frac{1}{4}\,\text{sech}^2\left(\frac{\varepsilon}{2k_B T}\right)\cdot\frac{1}{\tau_1(\varepsilon, r)} \tag{3.2.52}$$

The relaxation time $\tau_1(\varepsilon, r)$, in which thermodynamic equilibrium between the two-level system and its environment is reached, remains to be found. Assume this equilibrium is reached through the absorption or emission of thermal phonons of energy ε. To find the number $\dot{N}$ of such events in a unit of time and within a unit volume we can use Eq. (3.2.41) with $\rho(\varepsilon)$ in the form:

$$\rho(\varepsilon) = \int \frac{d^3k'}{(2\pi)^3}[N_{k'} + (N_{k'} + 1)]\delta(\varepsilon' - \varepsilon)$$

where $N_{k'} = \dfrac{1}{e^{\varepsilon'/k_B T} - 1}$ is the equilibrium distribution of thermal phonons, and $\varepsilon' = \hbar\omega' = \hbar v k'$

Then

$$\dot{N} = \frac{1}{\pi}\frac{M^2}{4}\frac{u_{0x}^2\,\varepsilon^4}{\hbar^6 v^5}\coth\left(\frac{\varepsilon}{2k_B T}\right)$$

Dividing $\dot{N}$ now by the average concentration of phonons of energy ε, i.e. by

$$N = \frac{\bar{E}_{\text{wave}}}{\varepsilon} = \frac{1}{2}\rho\, u_{0x}^2\frac{\varepsilon}{\hbar^2}$$

we find the desired relaxation time (Kovalenko and Krasny, 1990):

$$\frac{1}{\tau_1} = \frac{\dot{N}}{N} = \frac{1}{2\pi} \frac{M^2 \varepsilon^3}{\rho \hbar^4 v^5} \coth\left(\frac{\varepsilon}{2k_B T}\right) \tag{3.2.53}$$

Substitution of τ_1^{-1} into Eq. (3.2.53) gives the sound absorption coefficient for $\omega\tau_1 \gg 1$:

$$\begin{aligned} a &= \int_0^\infty d\varepsilon \int_{r_0}^1 p(\varepsilon, r) dr \frac{D^2}{8k_B T \rho v^3} \frac{1}{4} \operatorname{sech}^2\left(\frac{\varepsilon}{2k_B T}\right) \cdot \frac{1}{2\pi} \frac{M^2 \varepsilon^3}{\rho \hbar^4 v^5} \coth\left(\frac{\varepsilon}{2k_B T}\right) \\ &= \frac{1}{4\pi} \frac{\gamma^4 \bar{p} (k_B T)^3}{\rho^2 v^7 \hbar^4} \int_0^\infty dx\, x^3 \frac{1}{4} \operatorname{sech}^2(x) \cdot \coth(x) \frac{1}{2} \int_{r_0}^1 dr \sqrt{1-r} \end{aligned} \tag{3.2.54}$$

It is clear from this expression that relaxational contributions to the sound absorption coefficient are frequency independent for $\omega\tau_1 \gg 1$ and at low temperatures approach zero as T^3. These relationships agree well with experiment (see Fig. 3.1.2).

If the condition $\omega\tau_1 \gg 1$ does not hold, the expression (3.2.53) can be generalized, and the frequency dependence is described by the familiar Debye formula (Jäckle *et al.*, 1976):

$$a = \int_0^\infty d\varepsilon \int_{r_0}^1 dr\, p(\varepsilon, r) \frac{D^2}{8\rho v^3 k_B T} \frac{1}{4} \operatorname{sech}^2\left(\frac{\varepsilon}{2k_B T}\right) \frac{\omega^2 \tau_1}{1 + \omega^2 \tau_1^2} \tag{3.2.55}$$

At this point our discussion of the tunneling model and its experimental consequences for glasses is complete. In conclusion it should be noted that the tunneling model adequately explains the low-temperature anomalous properties of glasses; this is borne out by a multitude of experiments. As already mentioned, the theory has a certain amount of 'universality' applying both to non-metallic and metallic glasses, although in metallic glasses excitations of the two-level systems not only relax by interaction with phonons but also by electronic and magnetic degrees of freedom. The interested reader can find more details concerning these final points in the review article by J. L. Black (1981) in the book of Güntherodt and Beck (1981).

3.3 A Quasi-phonon Model for Amorphous Metals; Heat Capacity at Moderately Low Temperatures

We now come to the investigation of harmonic vibrational excitations in amorphous solids. Let us consider a topologically disordered system of N identical atoms distributed within a volume V. The Hamiltonian of such a system can be represented as follows:

$$\hat{H} = \sum_{i=1}^{N} \frac{M\hat{\mathbf{v}}_i^2}{2} + U(\mathbf{R}_1 \ldots \mathbf{R}_N) \tag{3.3.1}$$

where $\hat{\mathbf{v}}_i$ is the velocity operator and M denotes the mass of the ith atom. The first and second terms on the r.h.s. of Eq. (3.3.1) are the kinetic and potential energy operators, respectively. The pair-interaction approximation will now be employed, whereby

$$U(\mathbf{R}_1 \ldots \mathbf{R}_N) = \frac{1}{2} \sum_{i \neq j} \Phi(|\mathbf{R}_i - \mathbf{R}_j|) \tag{3.3.2}$$

In this case the potential $\Phi(\mathbf{R})$ is assumed to have a well-defined Fourier transform. Such an approximation adequately describes the ion interaction in simple metals.

The coordinates of the atoms may be written as

$$\mathbf{R}_i = \mathbf{R}_{0i} + \vec{\xi}_i \tag{3.3.3}$$

where $\mathbf{R}_{0i}$ are their fixed quasi-equilibrium positions and $\vec{\xi}_i \equiv \vec{\xi}(\mathbf{R}_{0i})$ refer to the displacements of the atoms due to thermal vibrations.

In amorphous solids the positions $\mathbf{R}_{0i}$ do not form a regular crystalline lattice. Let us expand the potential energy U into a series in the small displacements $\vec{\xi}_i$. The composite force exerted on the ith atom in a quasi-equilibrium position $\mathbf{R}_{0i}$ we put equal to zero. In the potential energy expansion we restrict ourselves to a term quadratic in $\vec{\xi}_i$, and in the kinetic energy $\mathbf{v}_j = \dot{\vec{\xi}}_j$ is taken into consideration. In a quantum approach the variables ξ_j^a and $\dot{\xi}_j^\beta (a, \beta = x, y, z)$ are matched by the operators $\hat{\xi}_j^a, \hat{\dot{\xi}}_j^\beta$ satisfying the standard commutation relations for canonically conjugate coordinates and momenta:

$$\begin{cases} \left[\hat{\xi}_j^a; \hat{\xi}_{j'}^\beta\right] = \left[\hat{\dot{\xi}}_j^a; \hat{\dot{\xi}}_{j'}^\beta\right] = 0 \\ \left[\hat{\xi}_j^a; \hat{\dot{\xi}}_{j'}^\beta\right] = \frac{i\hbar}{M} \delta_{jj'} \cdot \delta_{a\beta} \end{cases} \tag{3.3.4}$$

where $\delta_{a\beta}$ is the Kronecker delta and i denotes the imaginary unit.

Thus in a harmonic approximation the Hamiltonian Eq. (3.3.1) becomes:

$$\hat{H} = U_0 + \sum_{i=1}^{N} \frac{M}{2} \hat{\dot{\xi}}_i^2 + \frac{1}{2} \sum_{i,j} \sum_{a,\beta} \hat{\xi}_i^a \hat{\xi}_j^\beta \frac{\partial^2 U_0}{\partial R_{0i}^a \partial R_{0j}^\beta} \tag{3.35}$$

where $U_0 = U(\mathbf{R}_{01} \ldots \mathbf{R}_{0N})$.

If the system is taken as a cube, with edge length L and with the usual Born-von Karman periodic boundary conditions imposed at the faces, the displacement vectors, as well as their derivatives, and the potential $\Phi(\mathbf{R})$, can be expanded into the Fourier series:

$$\begin{cases} \hat{\vec{\xi}} = \dfrac{1}{\sqrt{N}} \sum\limits_{\mathbf{k}} \hat{\mathbf{q}}(\mathbf{k}) \cdot \exp(i\mathbf{k}\mathbf{R}_{0j}) \\ \hat{\dot{\vec{\xi}}} = \dfrac{1}{\sqrt{N}} \sum\limits_{\mathbf{k}} \hat{\dot{\mathbf{q}}}(\mathbf{k}) \cdot \exp(i\mathbf{k}\mathbf{R}_{0j}) \\ \Phi(\mathbf{R}) = \dfrac{1}{\sqrt{N}} \sum\limits_{\mathbf{k}} \tilde{\Phi}(\mathbf{k}) \cdot \exp(i\mathbf{k}\mathbf{R}) \end{cases} \tag{3.3.6}$$

where

$$\begin{cases} k_a = \dfrac{2\pi}{L} n_a = \dfrac{2\pi}{a} \dfrac{n_a}{N_a}; \; N_a = \sqrt[3]{N} \\ a^3 = v_0 = \dfrac{V}{N}; \; n_a = 0, \pm 1, \pm 2 \ldots; a = x, y, z \end{cases} \tag{3.3.7}$$

and $\mathbf{kR}$ rsp. $\mathbf{kR}_0$ mean as usual the dot products $\mathbf{k} \cdot \mathbf{R}$ rsp. $\mathbf{k} \cdot \mathbf{R}_0$.

Because the operators $\hat{\xi}_j^a$ and $\hat{\dot{\xi}}_j^\beta$ are self-adjoint, and from the commutation relations in Eq. (3.3.4), it follows that the Fourier transforms $\hat{q}^a(\mathbf{k})$ and $\hat{\dot{q}}^a(\mathbf{k})$ must satisfy the equations:

$$\begin{cases} \hat{q}^a(\mathbf{k}) = [\hat{q}^a(-\mathbf{k})]^+; \; \hat{\dot{q}}^a(\mathbf{k}) = [\hat{\dot{q}}^a(-\mathbf{k})]^+ \\ [\hat{q}^a(\mathbf{k}); \hat{q}^\beta(\mathbf{k}')] = [\hat{\dot{q}}^a(\mathbf{k}); \hat{\dot{q}}^\beta(\mathbf{k}')] = 0 \\ [\hat{q}^a(\mathbf{k}); \hat{\dot{q}}^\beta(\mathbf{k}')] = \dfrac{i\hbar}{M} \delta_{\mathbf{k}_1 - \mathbf{k}'} \cdot \delta_{a\beta} \end{cases} \tag{3.3.8}$$

Employing Eqs (3.3.6) and (3.3.8) we can easily rewrite the Hamiltonian (3.3.5) as follows:

$$\hat{\mathrm{H}} = U_0 + \hat{\mathrm{H}}_0 + \hat{\mathrm{H}}_{\mathrm{int}} \tag{3.3.9}$$

where

$$\hat{\mathrm{H}}_0 = \frac{M}{2} \sum_{\mathbf{k}} \hat{\dot{\mathbf{q}}}^+(\mathbf{k}) \cdot \hat{\dot{\mathbf{q}}}(\mathbf{k}) \tag{3.3.10}$$

$$\begin{aligned} \hat{\mathrm{H}}_{\mathrm{int}} = & \frac{M}{2} \sum_{\mathbf{k},\mathbf{k}'} \hat{\dot{\mathbf{q}}}(\mathbf{k}) \cdot \hat{\dot{\mathbf{q}}}(\mathbf{k}') \left\{ \frac{1}{N} \sum_{j=1}^{N} \mathrm{e}^{i(\mathbf{k}+\mathbf{k}') \cdot \mathbf{R}_{0j}} - \delta_{\mathbf{k},-\mathbf{k}'} \right\} \\ & + \frac{1}{2} \sum_{\mathbf{k},\mathbf{k}'} \sum_{a,\beta} \hat{q}^a(\mathbf{k}) \Gamma^{a\beta}(\mathbf{k},\mathbf{k}') \hat{q}^\beta(\mathbf{k}') \end{aligned} \tag{3.3.11}$$

$$\begin{aligned} \Gamma^{a\beta}(\mathbf{k},\mathbf{k}') = & \frac{1}{N^2} \sum_{\mathbf{k}_1} K_1^a K_1^\beta \, \tilde{\Phi}(\mathbf{k}_1) \\ & \times \sum_{j \neq j'} \left\{ \mathrm{e}^{i(\mathbf{k}_1+\mathbf{k}) \cdot \mathbf{R}_{0j}} \cdot \mathrm{e}^{i(-\mathbf{k}_1+\mathbf{k}') \cdot \mathbf{R}_{0j'}} - \mathrm{e}^{i(\mathbf{k}_1+\mathbf{k}+\mathbf{k}') \cdot \mathbf{R}_{0j}} \cdot \mathrm{e}^{-i\mathbf{k}_1 \cdot \mathbf{R}_{0j'}} \right\} \end{aligned} \tag{3.3.12}$$

Because the thermodynamic equilibrium of vibrational states of atoms is reached in a time far less than the characteristic time of the experiment, the system's free energy, according to Eq. (3.3.1), can be expressed as:

$$F = \bar{U}_0 - k_B T \cdot \overline{\ln\left\{ \mathrm{Tr}\left[\exp\left(-\frac{\hat{H}_0 + \hat{H}_{int}}{k_B T} \right) \right] \right\}} \tag{3.3.13}$$

where the trace operation Tr applies to the atoms' displacement operators $\hat{\xi}_j^a$ and to their derivatives $\dot{\hat{\xi}}_j^a$ (or their Fourier transforms). The overscribed bar denotes the averaging over all configurations of atomic quasi-equilibrium arrangements. The calculation of Eq. (3.3.13) involves a "frozen liquid" model. Considering the fact that the operator $\hat{H}_0$ does not depend on the atomic coordinates $\mathbf{R}_{0j}$, Eq. (3.3.13) may be written in the form (see Appendix B):

$$F = \bar{U}_0 - k_B T \cdot \ln\left\{ \mathrm{Tr}\left[\exp\left(-\frac{\hat{H}_0 + \tilde{\hat{H}}_{int}}{k_B T} \right) \right] \right\} \tag{3.3.14}$$

where

$$\begin{aligned}
\tilde{\hat{H}}_{int} = \hat{\hat{H}}_{int} - e^{-\hat{H}_0 \beta} \Bigg\{ & \frac{(-1)^2}{2!} \int_0^\beta d\tau_1 \overline{\hat{T}_\tau \left[\Delta\hat{H}_{int}(\beta) \cdot \Delta\hat{H}_{int}(\tau_1) \right]} \\
& + \frac{(-1)^3}{3!} \int_0^\beta d\tau_1 \int_0^\beta d\tau_2 \overline{\hat{T}_\tau \left[\Delta\hat{H}_{int}(\beta) \cdot \Delta\hat{H}_{int}(\tau_1) \cdot \Delta\hat{H}_{int}(\tau_2) \right]} \\
& + \frac{(-1)^4}{4!} \int_0^\beta d\tau_1 \int_0^\beta d\tau_2 \int_0^\beta d\tau_3 \overline{\hat{T}_\tau \left[\Delta\hat{H}_{int}(\beta) \cdot \Delta\hat{H}_{int}(\tau_1) \cdot \Delta\hat{H}_{int}(\tau_2) \cdot \Delta\hat{H}_{int}(\tau_3) \right]} \\
& - 3 \cdot \overline{\Delta\hat{H}_{int}(\beta) \cdot \Delta\hat{H}_{int}(\tau_1)} \cdot \overline{\Delta\hat{H}_{int}(\tau_2) \cdot \Delta\hat{H}_{int}(\tau_3)} + \ldots \Bigg\} e^{\hat{H}_0 \beta}
\end{aligned}$$

$$\Delta\hat{H}_{int}(\tau) = e^{-\hat{H}_0 \tau} \left(\hat{H}_{int} - \hat{\hat{H}}_{int} \right) \cdot e^{\hat{H}_0 \tau} \tag{3.3.15}$$

with $\beta = \frac{1}{k_B T}$, and $\hat{T}_\tau$ being the time ordering operator.

The second and following 'residual' summands on the r.h.s. of Eq. (3.3.15) describe the phonon scattering from the material density fluctuations: being proportional to (no less than) *fourth*-power products of $\hat{\xi}_j^a$ and $\dot{\hat{\xi}}_j^a$, because $\Delta\hat{H}_{int}(\tau)$ contains second-power products of these operators, the residual terms mentioned above describe therefore free-energy contributions from processes involving at least four phonons. In a quadratic approximation for the free energy these summands can therefore be neglected. So the free energy expression takes the form:

$$\Phi = \bar{U}_0 - k_B T \cdot \ln\{\mathrm{Tr}\exp[-\beta(\hat{H}_0 + \hat{\hat{H}}_{int})]\} \tag{3.3.16}$$

According to the frozen-liquid model:

$$\begin{cases} \overline{\frac{1}{N}\sum_{j=1}^{N}\exp[i(\mathbf{k}+\mathbf{k}')\mathbf{R}_{0j}]} = \delta_{\mathbf{k},-\mathbf{k}'} \\ \overline{\frac{1}{N}\sum_{j\neq j'}^{N}\exp[i\mathbf{k}_1\mathbf{R}_{0j} + i\mathbf{k}_2\mathbf{R}_{0j'}]} = N\cdot\delta_{\mathbf{k}_1,0}\cdot\delta_{\mathbf{k}_2,0} + [a(\mathbf{k}_1)-1]\cdot\delta_{\mathbf{k}_1,-\mathbf{k}_2} \end{cases} \tag{3.3.17}$$

where the so-called 'structure factor' $a(\mathbf{k})$ is related to the pair distribution function $g(\mathbf{r})$ as usual:

$$a(\mathbf{k}) - 1 = \frac{1}{v_0}\int d^3r\{g(\mathbf{r})-1\}\cdot\exp(-i\mathbf{k}\mathbf{r})$$

Consequently

$$\Gamma^{\alpha\beta}(\mathbf{k},\mathbf{k}') = \Gamma^{\alpha\beta}(\mathbf{k})\cdot\delta_{\mathbf{k},-\mathbf{k}'} \tag{3.3.18}$$

where

$$\Gamma^{\alpha\beta}(\mathbf{k}) = k^\alpha k^\beta\Phi(\mathbf{k}) + \frac{1}{N}\sum_{\mathbf{k}_1} k_1^\alpha\cdot k_1^\beta\,\tilde{\Phi}(\mathbf{k}_1)[a(\mathbf{k}_1+\mathbf{k}) - a(\mathbf{k}_1)] \tag{3.3.19}$$

If the z-axis is aligned with the vector $\mathbf{k}$ so that $\mathbf{k}=(0, 0, k)$, only the non-diagonal components of the tensor $\Gamma^{\alpha\beta}(\mathbf{k})$ will be non-zero, and

$$\begin{cases} \Gamma^{xx}(\mathbf{k}) = \Gamma^{yy}(\mathbf{k}) = \frac{1}{N}\sum_{\mathbf{k}_1} k_1^x\cdot k_1^x\,\tilde{\Phi}(\mathbf{k}_1)[a(k_1^x, k_1^y, k_1^z + k) - a(k_1)] \\ \Gamma^{zz}(\mathbf{k}) = k^2\,\tilde{\Phi}(\mathbf{k}) + \frac{1}{N}\sum_{\mathbf{k}_1} k_1^z\cdot k_1^z\,\tilde{\Phi}(\mathbf{k}_1)[a(k_1^x, k_1^y, k_1^z + k) - a(k_1)] \end{cases} \tag{3.3.20}$$

Hence

$$\hat{\text{H}}_{\text{int}} = \frac{1}{2}\sum_{\alpha,\beta}\sum_{\mathbf{k}_1}\hat{q}^{+\alpha}(\mathbf{k})\Gamma^{\alpha\beta}(\mathbf{k})\hat{q}^\beta(\mathbf{k}) \tag{3.3.21}$$

This expression can be diagonalized with the substitution

$$\hat{\mathrm{q}}(\mathbf{k}) = \sum_{\lambda=1}^{3}\left(\frac{\hbar}{2M\omega_\lambda(\mathbf{k})}\right)^{1/2}(\hat{c}_{\mathbf{k},\lambda} + \hat{c}^{+}_{-\mathbf{k},\lambda})\cdot\vec{\varepsilon}(\mathbf{k},\lambda) \tag{3.3.22}$$

if the components of the vector $\vec{\varepsilon}(\mathbf{k},\lambda)$ satisfy the equations:

$$\sum_\beta \Gamma^{\alpha\beta}(\mathbf{k})\varepsilon^\beta(\mathbf{k},\lambda) = M\omega_\lambda^2(\mathbf{k})\cdot\varepsilon^\alpha(\mathbf{k},\lambda) \tag{3.3.23}$$

Because solutions of homogeneous equations are determined up to a constant factor, the solutions of the Eq. (3.3.23) are sought so that they satisfy the normalization condition

$$\sum_{\alpha} \varepsilon^{\alpha}(\mathbf{k},\lambda)\varepsilon^{\alpha}(\mathbf{k},\lambda) = \vec{\varepsilon}(\mathbf{k},\lambda)\cdot\vec{\varepsilon}(\mathbf{k},\lambda) = 1$$

From Eq. (3.3.23) follows the orthogonality of vectors with different λ. To prove this point Eq. (3.3.23) should be multiplied by $\varepsilon^{\alpha}(\mathbf{k},\lambda')$ and summed over α:

$$\sum_{\alpha,\beta} \varepsilon^{\alpha}(\mathbf{k},\lambda')\cdot\Gamma^{\alpha\beta}(\mathbf{k})\cdot\varepsilon^{\beta}(\mathbf{k},\lambda) = M\omega_{\lambda}^{2}(\mathbf{k})\sum_{\alpha}\varepsilon^{\alpha}(\mathbf{k},\lambda')\cdot\varepsilon^{\alpha}(\mathbf{k},\lambda)$$

In a similar way we can obtain the relation

$$\sum_{\alpha,\beta} \varepsilon^{\alpha}(\mathbf{k},\lambda)\cdot\Gamma^{\alpha\beta}(\mathbf{k})\cdot\varepsilon^{\beta}(\mathbf{k},\lambda') = M\omega_{\lambda'}^{2}(\mathbf{k})\sum_{\alpha}\varepsilon^{\alpha}(\mathbf{k},\lambda)\cdot\varepsilon^{\alpha}(\mathbf{k},\lambda')$$

Subtracting these equations from one another term by term and taking into account that $\Gamma^{\alpha\beta}(\mathbf{k}) = \Gamma^{\beta\alpha}(\mathbf{k})$ and $\omega_{\lambda}^{2}(\mathbf{k}) \neq \omega_{\lambda'}^{2}(\mathbf{k})$ when $\lambda \neq \lambda'$, we find

$$\sum_{\alpha} \varepsilon^{\alpha}(\mathbf{k},\lambda')\varepsilon^{\alpha}(\mathbf{k},\lambda) = \vec{\varepsilon}(\mathbf{k},\lambda')\cdot\vec{\varepsilon}(\mathbf{k},\lambda) = 0$$

Thus the vectors $\vec{\varepsilon}(\mathbf{k},\lambda)$ satisfy the condition:

$$\vec{\varepsilon}(\mathbf{k},\lambda)\cdot\vec{\varepsilon}(\mathbf{k},\lambda') = \delta_{\lambda,\lambda'} \tag{3.3.24}$$

Likewise it can be proved that

$$\sum_{\lambda} \varepsilon^{\alpha}(\mathbf{k},\lambda)\cdot\varepsilon^{\beta}(\mathbf{k},\lambda) = \delta_{\alpha\beta}$$

The functions $\omega_{\lambda}(\mathbf{k})$ are specified by the condition that the system in Eq. (3.3.23) possesses nontrivial solutions, obtained by the vanishing determinant

$$|\Gamma^{\alpha\beta}(\mathbf{k}) - M\omega_{\lambda}^{2}(\mathbf{k})\cdot\delta_{\alpha\beta}| = 0$$

or

$$[\Gamma^{xx}(\mathbf{k}) - M\omega_{\lambda}^{2}(\mathbf{k})]^{2}\cdot[\Gamma^{zz}(\mathbf{k}) - M\cdot\omega_{\lambda}^{2}(\mathbf{k})] = 0$$

It follows from this equation that when $\lambda = 1$ (Kovalenko and Krasny, 1990):

$$\omega_1^2(\mathbf{k}) = \omega_l^2(\mathbf{k}) = \frac{1}{M}\Gamma^{zz}(\mathbf{k}) = \frac{1}{M}k^2\,\tilde{\Phi}(\mathbf{k}) + \frac{v_0}{M}\int\frac{d^3k_1}{(2\pi)^3}(k_1^z)^2\,\tilde{\Phi}(\mathbf{k}_1)[a(k_1^x, k_1^y, k_1^z + k) - a(k_1)] \tag{3.3.25}$$

and when $\lambda = 2, 3$:

$$\omega_2^2(\mathbf{k}) = \omega_3^2(\mathbf{k}) = \omega_t^2(\mathbf{k}) = \frac{1}{M}\Gamma^{xx}(\mathbf{k}) = \frac{v_0}{M}\int\frac{d^3k_1}{(2\pi)^3}(k_1^x)^2\,\tilde{\Phi}(\mathbf{k}_1)[a(k_1^x, k_1^y, k_1^z + k) - a(k_1)] \tag{3.3.26}$$

In the above expressions we passed to the limit $V, N \to \infty\left(v_0 = \frac{V}{N} = \text{const}\right)$

The operators $\hat{c}^+_{\mathbf{k},\lambda}, \hat{c}_{\mathbf{k},\lambda}$ in Eq. (3.3.22) are chosen to fit the Bose commutation relations:

$$\begin{cases} [\hat{c}_{\mathbf{k},\lambda};\ \hat{c}_{\mathbf{k}',\lambda'}] = [\hat{c}^+_{\mathbf{k},\lambda};\ \hat{c}^+_{\mathbf{k}',\lambda'}] = 0 \\ [\hat{c}_{\mathbf{k}\lambda};\ \hat{c}^+_{\mathbf{k}'\lambda'}] = \delta_{\mathbf{k},\mathbf{k}'}\cdot\delta_{\lambda,\lambda'} \end{cases} \tag{3.3.27}$$

Then from the commutation relations Eq. (3.3.8) it immediately follows that the operator $\hat{\mathbf{q}}(\mathbf{k})$ should be defined as:

$$\hat{\mathbf{q}}(\mathbf{k}) = \sum_{\lambda=1}^{3}\left(\frac{\hbar\omega_\lambda(\mathbf{k})}{2M}\right)^{1/2}\cdot i(\hat{c}_{\mathbf{k},\lambda} - \hat{c}^+_{-\mathbf{k},\lambda})\cdot\vec{\varepsilon}(\mathbf{k},\lambda) \tag{3.3.28}$$

Taking into account the expansions Eqs (3.3.8) and (3.3.28) with commutation relations Eq. (3.3.27) one can show that

$$\hat{\mathrm{H}}_0 + \hat{\mathrm{H}}_{\text{int}} = \sum_{\lambda=1}^{3}\sum_{\mathbf{k}}\hbar\omega_\lambda(\mathbf{k})\cdot\left[\hat{c}^+_{\mathbf{k},\lambda}\hat{c}_{\mathbf{k},\lambda} + \frac{1}{2}\right] \tag{3.3.29}$$

It is evident from this equation that $\hat{c}^+_{\mathbf{k},\lambda}$ and $\hat{c}_{\mathbf{k},\lambda}$ act as operators of creation and destruction, respectively, of 'homogeneous phonons' from phonon branch λ with wavenumber $\mathbf{k}$, i.e. the energy $\hbar\omega_\lambda(\mathbf{k})$. In this case longitudinal phonons correspond to the acoustic branch $\lambda = 1$, and transversal ones to the branches $\lambda = 2, 3$.

In the long-wave limit, asymptotically for $k \to 0$, we get (Kovalenko and Krasny, 1990):

$$\begin{cases} \omega_1^2(\mathbf{k}) = v_1^2\cdot k^2 \\ v_1^2 = \dfrac{1}{M}\,\tilde{\Phi}(0) + \dfrac{v_0}{2M}\displaystyle\int\frac{d^3k_1}{(2\pi)^3}(k_1^z)^2\,\tilde{\Phi}(k_1)\frac{\partial^2 a(k_1)}{\partial k_1^{z^2}} \end{cases} \tag{3.3.30}$$

$$\begin{cases} \omega_t^2(\mathbf{k}) = v_t^2 \cdot k^2 \\ v_t^2 = \dfrac{v_0}{2M} \displaystyle\int \frac{d^3 k_1}{(2\pi)^3} (k_1^x)^2 \, \tilde{\Phi}(k_1) \frac{\partial^2 a(k_1)}{\partial k_1^{z^2}} \end{cases} \tag{3.3.31}$$

Here v_l and v_t are the *longitudinal and transverse sound velocities.*

Expressions similar to Eqs (3.3.32) and (3.3.33) for the sound velocities *in crystals* have been obtained by Brovman and Kagan (1974). The difference is just that, instead of a summation with respect to the reciprocal lattice vectors, in Eqs (3.3.32) and (3.3.33) an integration over the wavevector **k** with the weight function $[a(\mathbf{k})-1]$ is carried out.

We stress that the true harmonic eigenvibrations of an amorphous sample are typically not at all plane-wave-like except at long wavelengths, i.e. the decomposition (3.3.29) applies only to the averaged effective *quasi-crystalline* Hamiltonian entering the calculation of the free energy. Therefore the approximate result (3.3.16) for the free energy will be called the 'quasi-phonon approximation'.

Equations (3.3.25), (3.3.26), (3.3.32) and (3.3.33) make it possible to calculate analytically the 'quasi-phonon' dispersion curves and sound velocities, given the Fourier transform of the interatomic interaction and the structure factor of a material. For instance, in simple amorphous metals $\tilde{\Phi}(\mathbf{k})$ is relatively simple, see Eq. (2.5.43) and Harrison (1966), namely:

$$\tilde{\Phi}(\mathbf{k}) = \frac{1}{v_0} \left[\frac{z^2 \cdot 4\pi e^2}{k^2} - \frac{\pi(k)|w(k)|^2}{1 + \dfrac{4\pi e^2}{k^2}\pi(k)} \right]$$

where $\pi(k)$ is the polarization function of the electron gas of the metal; z is the valency, and $w(k)$ denotes the pseudopotential of the electron-ion interaction.

Figure 3.3.1 and Tab. 3.3.1 show the dispersion curves and sound velocities calculated for amorphous Mg and Zn (Kovalenko and Krasny, 1990) [2)]. The calculations involved the hard-sphere structure factor in the Percus-Yevick approximation with a dimensionless density $\eta = 0.45$ (Kovalenko and Fisher, 1972). The Ashcroft pseudopotential (Harrison, 1966, and Chap. 2) was selected as $w(k)$.

From Fig. 3.3.1 we can see that in the dispersion curves $\omega_\lambda(k)$ there are specific minima indicative of "soft phonons" (called "*rotons*" on historical grounds) in the short-wave region. This point is directly borne out by experiments with inelastic neutron scattering (see e.g. Suck and Rudin, 1983), where of course besides the dispersion curves of the phonons, their scattering lifetimes are also measured (see Section 3.5, and Von Heimendahl 1979).

In addition to our *calculated* sound velocities in amorphous Mg and Zn, in Tab. 3.3.1 (taken from Kovalenko and Krasny, 1990) the corresponding *experimen-*

2) At room temperature only amorphous *alloys* of the type $Mg_{1-x}Zn_x$ with $x \approx 0.3$ are stable; but from the calculations for the pure amorphous metals one can already draw important conclusions also for these alloys (see below).

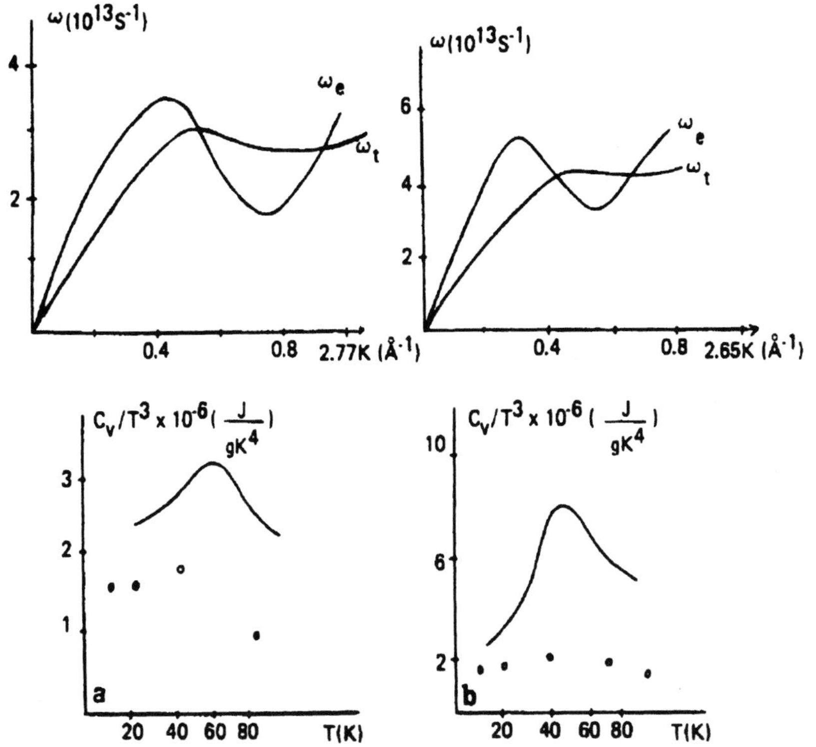

Fig. 3.3.1 Dispersion curves for the longitudinal and transverse 'quasi-phonon excitations' in amorphous Mg (upper left plot) and amorphous Zn (upper right). For the longitudinal excitations, the pronounced *roton minimum* at short wavelengths is steeper. The lower left and lower right plots represent the reduced heat capacity C_V/T^3 as a function of the temperature *T*. Here the corresponding results for crystals are presented by the open circles (from Kovalenko and Krasny, 1990)

Tab. 3.3.1 Sound velocities for Mg and Zn

Metal	*a* (Å)	*b* (Å)	$\bar{a}$ (Å)	c_l (10^5 cm/s)	c_t (10^5 cm/s)
Mg (cryst) *	3.2	5.21	–	6.157	4.90
Mg (am) **	–	–	2.8	5.09	3.39
Zn (cryst) ***	4.9	3.66	–	5.01	4.135
Zn (am) **	–	–	2.75	4.05	2.33

* Frantsevich *et al.* (1982)

** The calculated sound-velocities for amorphous Mg and Zn in the table can also be compared to the values calculated by computer simulations for amorphous $Mg_{70}Zn_{30}$, which is an amorphous alloy stable at room temperature; for this alloy, the sound velocities can be read off from Fig. 8 in Von Heimendahl (1979). See also Kovalenko *et al.* (1991), and below

*** Kikoin (1976)

tal values for the analogous crystals are given. Additionally the lattice constants for the crystals are given, and the average interatomic distances $\bar{a}$ for the amorphous case.

The calculated sound velocities agree satisfactorily with experimental values. Besides, the calculations suggest – and this is supported by experiments – that the transition from the crystalline to the amorphous state leads to a *significant reduction of the sound velocity.*

The obtained results can be generalized for a multicomponent compound. Among other things the free energy of such a system, as for the single-component substance, is determined by Eq. (3.3.16) where (Kovalenko *et al.*, 1990)

$$\hat{\mathrm{H}}_0 + \hat{\mathrm{H}}_{\mathrm{int}} = \sum_{\lambda=1}^{3s} \sum_{\mathbf{k}} \hbar\omega_\lambda(\mathbf{k}) \cdot \left[\hat{c}^{+}_{\mathbf{k},\lambda} \hat{c}_{\mathbf{k},\lambda} + \frac{1}{2} \right] \tag{3.3.32}$$

and s is the number of components in the compound.

In the case of $s \geq 2$, apart from the acoustic branch there are $3s$–3 'optical' vibrational branches. The dispersion law for these branches can be determined through the diagonalization of quadratic forms like Eq. (3.3.21). To illustrate, for a two-component compound the dispersion curves are determined by the following expressions (Kovalenko *et al.*, 1990):

$$\begin{cases} \omega_l^2(\mathbf{k}) = \dfrac{1}{2}\left(\dfrac{\Gamma_{11}^{zz}}{c_1 M_1} + \dfrac{\Gamma_{22}^{zz}}{c_2 M_2} \right) \pm \left\{ \dfrac{1}{4}\left(\dfrac{\Gamma_{11}^{zz}}{c_1 M_1} - \dfrac{\Gamma_{22}^{zz}}{c_2 M_2} \right)^2 + \dfrac{(\Gamma_{12}^{zz})^2}{c_1 M_1 c_2 M_2} \right\}^{1/2} \\ \omega_t^2(\mathbf{k}) = \dfrac{1}{2}\left(\dfrac{\Gamma_{11}^{xx}}{c_1 M_1} + \dfrac{\Gamma_{22}^{xx}}{c_2 M_2} \right) \pm \left\{ \dfrac{1}{4}\left(\dfrac{\Gamma_{11}^{xx}}{c_1 M_1} - \dfrac{\Gamma_{22}^{xx}}{c_2 M_2} \right)^2 + \dfrac{(\Gamma_{12}^{xx})^2}{c_1 M_1 c_2 M_2} \right\}^{1/2} \end{cases} \tag{3.3.33}$$

where

$$\begin{aligned} \Gamma_{ll'}^{aa} \equiv \Gamma_{ll'}^{aa}(\mathbf{k}) &= k^a \cdot k^a \cdot c_l c_{l'} \tilde{\Phi}_{ll'}(\mathbf{k}) \\ &+ \frac{v_0}{(2\pi)^3} \int \mathrm{d}^3 k_1 (k_1^a)^2 \left\{ c_l c_{l'} \tilde{\Phi}_{ll'}(\mathbf{k}_1) a_{ll'}(\mathbf{k}_1 + \mathbf{k}) - \delta_{ll'} \sum_{l_1} c_l c_{l_1} \tilde{\Phi}_{ll_1}(\mathbf{k}_1) a_{ll_1}(\mathbf{k}_1) \right\} \end{aligned} \tag{3.3.34}$$

Here $\mathbf{k} = (0, 0, k)$; $c_l = \frac{N_l}{N}$; N_l; M_l stand respectively for concentration, number and mass of the atoms in the lth component; $a_{ll'}(\mathbf{k})$ are the partial structure factors.

When in Eq. (3.3.33) the minus is chosen we have acoustic wave frequencies, if plus is taken, the frequencies of optic waves. It can easily be shown that in the long-wave limit ($k \to 0$) for the acoustic sound waves

$$\omega_\lambda(\mathbf{k}) \to c_\lambda k$$

and for the optic sound waves

$\omega_\lambda(\mathbf{k}) \to \omega_\lambda(0) \neq 0$

Expressions similar to Eq. (3.3.33) have also been obtained in a paper by Vakarchuk *et al.* (1988), where a Green's function method was employed. This method not only permits determination of *dispersion curves* but also investigation of the *attenuation* of phonon excitations due to the phonon scattering from the material density fluctuations. Analytically, the attenuation could be evaluated by this method in the long-wave limit. It turns out that for acoustic phonons the ratio of attenuation to frequency when $k \to 0$ tends to zero as k^3. With this in mind one can state that at long wavelengths, i.e. for $|(k \cdot r_0)| \ll 1$, where k is the wavenumber and r_0 the average nearest-neighbor distance in the amorphous sample, acoustical phonons in amorphous solids are well-defined excitations.

The dispersion curves for the amorphous *alloy* $Mg_{70}Zn_{30}$ have been numerically calculated by Kovalenko *et al.* (1990) within the hard-sphere approximation (Fig. 3.3.2). This calculation demonstrates that the dispersion law for acoustic waves has hardly changed in character; in fact, in the short-wave region the dispersion curves again show the pronounced 'roton minimum'.

There are also minima in the optical-wave dispersion curves: the occurrence of a minimum in a short-wave region in the dispersion curves of phonon excitations makes it possible to explain the anomalies observed in the temperature depen-

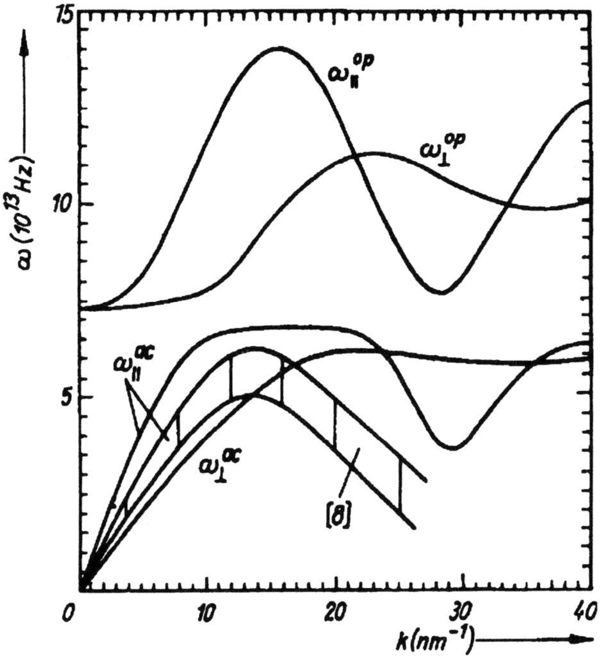

Fig. 3.3.2 Quasi-phonon dispersion law of amorphous $Mg_{70}Zr_{30}$ alloys, according to Kovalenko *et al.* (1991)

dence of glass heat capacity at moderately low temperatures. Within the approximation Eq. (3.3.32) the free energy is readily calculated by Eq. (3.3.16) and takes the form:

$$F = \bar{U}_0 + k_B T \cdot V \cdot \sum_{\lambda=1}^{3s} \int \ln\left[1 - \exp\left(-\frac{\hbar\omega_\lambda(\mathbf{k})}{k_B T}\right)\right] \frac{d^3k}{(2\pi)^3} \tag{3.3.35}$$

The heat capacity of the system can be found using the thermodynamic relation

$$C_V = \left(\frac{\partial E}{\partial T}\right)_{V,C_l} = -\frac{\partial}{\partial T}\left(T^2 \frac{\partial}{\partial T}\frac{F}{T}\right)_{V,C_l}$$

Assuming that the structure factors, $a_{ll'}(\mathbf{k})$, of the material are temperature independent it is easy to show that

$$C_V = V \cdot \frac{\hbar^2}{k_B T^2} \sum_{\lambda=1}^{3s} \int \frac{d^3k}{(2\pi)^3} \frac{\omega_\lambda^2(\mathbf{k})}{\left[\exp\left(\frac{\hbar\omega_\lambda(\mathbf{k})}{k_B T}\right) - 1\right]\left[1 - \exp\left(-\frac{\hbar\omega_\lambda(\mathbf{k})}{k_B T}\right)\right]} \tag{3.3.36}$$

We can see from this equation that owing to the minima i-n the dispersion curves $\omega_\lambda(\mathbf{k})$ at short wavelengths, the graph of the $C_V(T)$ curve at low temperatures deviates from the Debye law, $C_V(T) \propto T^3$. This deviation is easy to verify. Let us consider, for simplicity, a single-component compound ($s=1$). Then for small k the function $\omega_\lambda(\mathbf{k})$ can be approximated as

$$\omega_\lambda(k) = v_\lambda \cdot k \tag{3.3.37}$$

and in the vicinity of a roton minimum by the expression

$$\omega_\lambda(k) = \Delta_\lambda + \frac{\hbar(k - k_{0\lambda})^2}{2m_\lambda^*} \tag{3.3.38}$$

Thus the integral over k in Eq. (3.3.36) can be decomposed into a pair of integrals:

$$C_V = V \cdot \frac{\hbar^2}{k_B T^2} \sum_{\lambda=1}^{3s} \left\{ \int_0^{k_{1\lambda}} \frac{k^2 dk}{2\pi^2} \frac{\omega_\lambda^2(k) \exp\left(-\frac{\hbar\omega_\lambda(k)}{k_B T}\right)}{\left[1 - \exp\left(-\frac{\hbar\omega_\lambda(k)}{k_B T}\right)\right]^2} + \int_{k_{1\lambda}}^{k_{max}} \frac{k^2 dk}{2\pi^2} \frac{\omega_\lambda^2(k) \exp\left(-\frac{\hbar\omega_\lambda(k)}{k_B T}\right)}{\left[1 - \exp\left(-\frac{\hbar\omega_\lambda(k)}{k_B T}\right)\right]^2} \right\}$$

where $k_{1\lambda}$ is the wavenumber at the peak of $\omega_\lambda(k)$. In the first integral we carry out the substitution Eq. (3.3.37) and let the upper limit tend to infinity. In the second integral we make the substitution Eq. (3.3.38) and the limits of integration turn $\pm\infty$. Then the integrals can be calculated analytically and the low-temperature heat capacity for $k_B T \ll \Delta_\lambda$ takes the form:

$$\frac{C_V}{(k_B T)^3} = V \cdot A + V \cdot \sum_\lambda B_\lambda (k_B T)^{-5/2} \cdot \exp\left(-\frac{\hbar\Delta_\lambda}{k_B T}\right) \times \left[\frac{3}{4} + \frac{\hbar\Delta_\lambda}{k_B T} + \left(\frac{\hbar\Delta_\lambda}{k_B T}\right)^2\right] \tag{3.3.39}$$

where

$$A = \frac{2\pi^2}{15}\frac{k_B}{\hbar^3}\sum_\lambda v_\lambda^{-3}$$

$$B_\lambda = \frac{2}{(2\pi)^{3/2}} \cdot \frac{k_B}{\hbar} k_{0\lambda}^2 (m_\lambda^*)^{1/2}$$

The calculations show (see Fig. 3.3.2) that $\Delta_1 < \Delta_2 = \Delta_3$. So the term with $\lambda=1$ gives the major contribution to the sum over λ. This term peaks at $\frac{\hbar\Delta_1}{k_B T} \approx 4.15$. In amorphous Mg and Zn, where $\Delta_1 \sim (2-3) \cdot 10^{13}\ \mathrm{s}^{-1}$, the function $\frac{C_V}{(k_B T)^3}$ has a maximum at $T \sim 30 - 50$ K. This is also corroborated by direct numerical integration of Eq. (3.3.36) for amorphous Mg and Zn, Fig. 3.3.1, (Browman and Kagan, 1974). It is in this temperature range that the peak of the $\frac{C_V}{T^3}$ function was also observed in experiments with amorphous $Mg_{70}Zn_{30}$ (Panova *et al.*, 1985).

3.4 The Quasi-phonon Contribution to the Heat Conductivity and Sound Absorption in Amorphous Solids at Moderately Low Temperatures

The physics behind the *kinetic phenomena* (heat conductivity, absorption of sound, etc.) in amorphous dielectrics (or the phonon contribution to these effects in amorphous materials) is contained in the transport processes of thermal phonons. At very low temperatures (T<2 K) the interaction between phonons and 'tunneling states' is crucial, whereas at higher temperatures the interaction between phonons themselves provides the decisive contributions to these effects. In amorphous metals, particularly in superconducting materials, the electron-phonon interaction would also play a role, but this will not be considered here.

The concept of free phonons originates from quantization of the vibrational motion of atoms in solids in a *harmonic* approximation, i.e. when only the bilinear terms (bilinear in the displacements of the atoms w.r.t. their quasi-equilibrium positions) in the Hamiltonian are considered (see Section 3.3). But the various interactions between phonons will be described if the Hamiltonian $\hat{H}$ includes the terms of the following orders of smallness, namely the anharmonic terms of the third- and higher-order in the displacements $\hat{\vec{\xi}}$. If we restrict our consideration to a system of N identical atoms of mass M within a volume V, the Hamiltonian of such a system can be written as follows (Lifshitz and Pitaevski, 1978; Krasny and Kovalenko, 1990; Kovalenko *et al.*, 1993):

$$\hat{\mathrm{H}} = \hat{\mathrm{H}}_2 + \hat{\mathrm{H}}_3 + \ldots \tag{3.4.1}$$

where

$$\hat{\mathrm{H}}_2 = \mathrm{U}_0 + \sum_{j=1}^{N} \frac{M}{2} \hat{\dot{\xi}}_j^2 + \frac{1}{2} \sum_{j_1, j_2} \sum_{\alpha,\beta=1}^{3} \hat{\xi}_{j_1}^{\alpha} \hat{\xi}_{j_2}^{\beta} \frac{\partial^2 \mathrm{U}_0}{\partial R_{0j_1}^{\alpha} \partial R_{0j_2}^{\beta}} \tag{3.4.2}$$

is the Hamiltonian of vibrating atoms in harmonic approximation;

$$\mathrm{U}_0 = \mathrm{U}(\vec{\mathrm{R}}_{01}, \vec{\mathrm{R}}_{02}, \ldots \vec{\mathrm{R}}_{0N}) \tag{3.4.3}$$

is the potential energy of the atoms in a quasi-equilibrium position;

$$\hat{\mathrm{H}}_3 = \frac{1}{3!} \sum_{j_1, j_2, j_1} \sum_{\alpha,\beta,\gamma} \hat{\xi}_{j_1}^{\alpha} \hat{\xi}_{j_2}^{\beta} \hat{\xi}_{j_3}^{\gamma} \frac{\partial^3 \mathrm{U}_0}{\partial R_{0j_1}^{\alpha} \partial R_{0j_2}^{\beta} \partial R_{0j_3}^{\gamma}} \tag{3.4.4}$$

is the third-order anharmonic term, etc.

It is assumed that the quasi-equilibrium atomic positions $\vec{R}_{0j}$ do not make up a crystal lattice.

From now on in the expression (3.4.1) we shall restrict ourselves only to the anharmonic terms of third order, considering them as small perturbations responsible for the transitions between the states of the basic Hamiltonian $\hat{\mathrm{H}}_2$. According to the fundamentals of quantum mechanics (e.g. Landau and Lifshitz, 1963; Davydov, 1976) the probability rate of transitions of the system from the state $|i\rangle$ with energy ε_i to the state $|f\rangle$ of energy ε_f under the perturbation $\hat{\mathrm{H}}_3$ is given by *Fermi's Golden Rule*:

$$P_{i\to f} = \frac{2\pi}{\hbar} \left|\langle i|\hat{\mathrm{H}}_3|f\rangle\right|^2 \delta(\varepsilon_f - \varepsilon_i) \tag{3.4.5}$$

Here for the unperturbed harmonic part of the Hamiltonian it is $\hat{\mathrm{H}}_2|f\rangle = \varepsilon_f|f\rangle$. Because initially the system can be in any quasi-equilibrium state, i.e. near any one of the minima of the potential energy, Eq. (3.4.5) should be averaged over all configurations of quasi-equilibrium positions of atoms, i.e.

$$\overline{P_{i\to f}} = \frac{2\pi}{\hbar} \overline{|\langle i\hat{\mathrm{H}}_3|f\rangle|^2 \delta(\varepsilon_f - \varepsilon_i)} \tag{3.4.6}$$

Let the Hamiltonian $\hat{\mathrm{H}}_2$ be represented in the form:

$$\hat{\mathrm{H}}_2 = \hat{\bar{\mathrm{H}}}_2 + \Delta\hat{\mathrm{H}}_2$$

The second summand describes the energy fluctuations when the system passes from one configuration of quasi-equilibrium positions to another. Similarly

$$\begin{cases} \varepsilon_f &= \bar{\varepsilon}_f + \Delta\varepsilon_f \\ |f\rangle &= |\bar{f}\rangle + |\Delta f\rangle \end{cases}$$

where $\bar{\varepsilon}_f$ and $|\bar{f}\rangle$ are the eigenvalues and eigenfunctions of the operator $\hat{\bar{H}}_2$.

In terms of quasi-phonons the above-mentioned fluctuations arise from quasi-phonon scattering by the material density fluctuations. These effects – as was argued in the previous section – correspond to members of the fourth and higher orders of smallness in the atoms' displacements from quasi-equilibrium, concerning the evaluation of contributions to averages of free energies. Therefore, in Eq. (3.4.6) we put:

$$|f\rangle \approx |\bar{f}\rangle \quad \text{and} \quad \varepsilon_f \approx \bar{\varepsilon}_f \tag{3.4.7}$$

Moving to the second-quantized representation with respect to the corresponding quasi-phonon *creation* and *annihilation* operators $\hat{C}^{+}_{\vec{k}\lambda}$ and $\hat{C}_{\vec{k}\lambda}$, we can write the operators $\hat{\bar{H}}_2$ and $\hat{\vec{\xi}}_i$ as follows (see Eqs (3.4.27), (3.4.6) and (3.4.20)):

$$\hat{\bar{H}}_2 = \bar{U}_0 + \sum_{\vec{k},\lambda} \hbar\omega_\lambda(\vec{k}) \left[\hat{C}^{+}_{\vec{k}\lambda}\hat{C}_{\vec{k}\lambda} + \frac{1}{2}\right] \tag{3.4.8}$$

$$\hat{\vec{\xi}}_j = \sum_{\vec{k},\lambda} \left(\frac{\hbar}{2N \cdot M \cdot \omega_\lambda(\vec{k})}\right)^{1/2} \left(\hat{C}_{\vec{k}\lambda} + \hat{C}^{+}_{-\vec{k}\lambda}\right) \vec{\varepsilon}\,(\vec{k},\lambda)\, e^{i\vec{k}\cdot\vec{R}_{0j}} \tag{3.4.9}$$

Here $\vec{\varepsilon}(\vec{k},\lambda)$ is the polarization vector of quasi-phonons of branch λ, with momentum $\hbar\vec{k}$ and energy $\hbar\omega_\lambda(\vec{k})$ determined by Eqs (3.4.22) and (3.4.23). The anharmonic term $\hat{H}_3$ in the same representation, according to Eq. (3.4.3), takes the form:

$$\begin{aligned} \hat{H}_3 = &\sum_{\vec{k}_1\lambda_1;\,\vec{k}_2\lambda_2;\,\vec{k}_3\lambda_3} \left(\frac{\hbar}{2NM}\right)^{3/2} \left[\omega_{\lambda_1}(\vec{k}_1)\cdot\omega_{\lambda_2}(\vec{k}_2)\cdot\omega_{\lambda_3}(\vec{k}_3)\right]^{-1/2} \\ &\times \left(\hat{C}_{\vec{k}_1\lambda_1} + \hat{C}^{+}_{-\vec{k}_1\lambda_1}\right)\left(\hat{C}_{\vec{k}_2\lambda_2} + \hat{C}^{+}_{-\vec{k}_2\lambda_2}\right)\left(\hat{C}_{\vec{k}_3\lambda_3} + \hat{C}^{+}_{-\vec{k}_3\lambda_3}\right)\cdot \Lambda_{\vec{k}_1\lambda_1,\,\vec{k}_2\lambda_2,\,\vec{k}_3\lambda_3} \end{aligned} \tag{3.4.10}$$

where

$$\begin{aligned} \Lambda_{\vec{k}_1\lambda_1,\,\vec{k}_2\lambda_2,\,\vec{k}_3\lambda_3} = &\sum_{j_1,j_2,j_1}\sum_{a,\beta,\gamma} \varepsilon^{a}(\vec{k}_1\lambda_1)\cdot\varepsilon^{\beta}(\vec{k}_2\lambda_2)\cdot\varepsilon^{\gamma}(\vec{k}_3\lambda_3) \\ &\times \frac{\partial^3 U_0}{\partial R^{a}_{0j_1}\,\partial R^{\beta}_{0j_2}\,\partial R^{\gamma}_{0j_3}} \cdot \exp\left[i\left(\vec{k}_1\cdot\vec{R}_{0j_1} + \vec{k}_2\cdot\vec{R}_{0j_2} + \vec{k}_3\cdot\vec{R}_{0j_3}\right)\right] \end{aligned} \tag{3.4.11}$$

Substituting this expression into Eq. (3.4.6) and taking into account the above approximations we find for the averaged transition rate $P_{i\to f}$ $(= \bar{P}_{i\to f})$:

$$P_{i\to f} = \frac{2\pi}{\hbar}\delta(\bar{\varepsilon}_f - \bar{\varepsilon}_f) \sum_{\substack{\vec{k}_1\lambda_1,\vec{k}_2\lambda_2,\\ \vec{k}_3\lambda_3}} \sum_{\substack{\vec{k}'_1\lambda'_1,\vec{k}'_2\lambda'_2,\\ \vec{k}'_3\lambda'_3}} \left(\frac{\hbar}{2NM}\right)^3$$
$$\times [\omega_{\lambda_1}(\vec{k}_1)\cdot\omega_{\lambda_2}(\vec{k}_2)\cdot\omega_{\lambda_3}(\vec{k}_3)\cdot\omega_{\lambda'_1}(\vec{k}'_1)\cdot\omega_{\lambda'_2}(\vec{k}'_2)\cdot\omega_{\lambda'_3}(\vec{k}'_3)]^{-1/2}$$
$$\times \langle\bar{i}|(\hat{C}_{\vec{k}_1\lambda_1} + \hat{C}^+_{-\vec{k}_1\lambda_1})\cdot(\hat{C}_{\vec{k}_2\lambda_2} + \hat{C}^+_{-\vec{k}_2\lambda_2})\cdot(\hat{C}_{\vec{k}_3\lambda_3} + \hat{C}^+_{-\vec{k}_3\lambda_3})|\bar{f}\rangle$$
$$\times \langle\bar{f}|(\hat{C}_{\vec{k}'_1\lambda'_1} + \hat{C}^+_{-\vec{k}'_1\lambda'_1})\cdot(\hat{C}_{\vec{k}'_2\lambda'_2} + \hat{C}^+_{-\vec{k}'_2\lambda'_2})\cdot(\hat{C}_{\vec{k}'_3\lambda'_3} + \hat{C}^+_{-\vec{k}'_3\lambda'_3})|\bar{i}\rangle$$
$$\times \overline{\Lambda_{\vec{k}_1\lambda_1,\vec{k}_2\lambda_2,\vec{k}_3\lambda_3}\cdot\Lambda^*_{\vec{k}'_1\lambda'_1,\vec{k}'_2\lambda'_2,\vec{k}'_3\lambda'_3}} \qquad (3.4.12)$$

In the second quantized representation the eigenfunctions $|\bar{f}\rangle$ of the operator $\hat{\bar{H}}_2$ depend on the occupation numbers $n^r_{\vec{k}\lambda}$ of the quasi-phonons. Here the creation and annihilation operators transform the wave function by increasing or decreasing the occupation numbers by one:

$$\hat{C}|n\rangle = \sqrt{n}\cdot|n-1\rangle\,; \qquad \hat{C}^+|n-1\rangle = \sqrt{n}\cdot|n\rangle$$

Consequently, only the following matrix elements of production or destruction operators will be non-zero:

$$\langle n|\hat{C}^+|n-1\rangle = \langle n-1|\hat{C}|n\rangle = \sqrt{n} \qquad (3.4.13)$$

Equation (3.4.12) involves matrix elements of three-fold products of $\hat{C}$ and $\hat{W}$ operators in various combinations. The matrix element of the product of three-phonon annihilation operators $\hat{C}_{\vec{q}_1\lambda_1}\hat{C}_{\vec{q}_2\lambda_2}\hat{C}_{\vec{q}_3\lambda_3}$ is non-zero when

$$\begin{cases} |\bar{i}\rangle = |n_{\vec{q}_1\lambda_1}\,;\ n_{\vec{q}_2\lambda_2}\,;\ n_{\vec{q}_3\lambda_3}\rangle \\ |\bar{f}\rangle = |n_{\vec{q}_1\lambda_1}-1\,;\ n_{\vec{q}_2\lambda_2}-1\,;\ n_{\vec{q}_3\lambda_3}-1\rangle \end{cases}$$

Because the matrix element mentioned above describes the process of annihilation of three quasi-phonons, the total energy of the system decreases by

$$\bar{\varepsilon}_f - \bar{\varepsilon}_i = -\hbar\omega_{\lambda_1}(\vec{q}_1) - \hbar\omega_{\lambda_2}(\vec{q}_2) - \hbar\omega_{\lambda_3}(\vec{q}_3) < 0$$

This happens therefore with *zero probability*, because according to Eq. (3.4.12), $\overline{P_{i\to f}}$ is non-zero only when the law of conservation of energy holds, i.e. when $\bar{\varepsilon}_f - \bar{\varepsilon}_i = 0$. In a similar manner, the matrix element corresponding to a product of three creation operators $\hat{C}^+_{\vec{q}_1\lambda_1}\hat{C}^+_{\vec{q}_2\lambda_2}\hat{C}^+_{\vec{q}_3\lambda_3}$ describes the simultaneous creation of three quasi-phonons, where the law of conservation of energy would also be violated:

$$\bar{\varepsilon}_f - \bar{\varepsilon}_i = -\hbar\omega_{\lambda_1}(\vec{q}_1) + \hbar\omega_{\lambda_2}(\vec{q}_2) - \hbar\omega_{\lambda_3}(\vec{q}_3) > 0$$

The probability of such a process is also zero.

The only processes possible are those in which two interacting quasi-phonons are destroyed producing a third one, or vice versa. The terms with products $\hat{C}_{\vec{q}_1\lambda_1}\hat{C}_{\vec{q}_2\lambda_2}\hat{C}^{+}_{\vec{q}_3\lambda_3}$ or $\hat{C}^{+}_{\vec{q}_1\lambda_1}\hat{C}^{+}_{\vec{q}_2\lambda_2}\hat{C}_{\vec{q}_3\lambda_3}$, correspond to these processes, which are called '*confluence*' or '*splitting*' processes of quasi-phonons, respectively (see the cases → and ← below). Symbolically such three-phonon transitions will be represented below as

$$(\vec{q}_1\lambda_1) + (\vec{q}_2\lambda_2) \rightleftarrows (\vec{q}_3\lambda_3)$$

allowing us to show explicitly the values of wave and polarization vectors for each of the three quasi-phonons participating in the process.

The law of conservation of energy in this case can be nontrivially satisfied:

$$\bar{\varepsilon}_f - \bar{\varepsilon}_i = \pm[\hbar\omega_{\lambda_1}(\vec{q}_1) + \hbar\omega_{\lambda_2}(\vec{q}_2) - \hbar\omega_{\lambda_3}(\vec{q}_3)] = 0$$

The matrix elements of such transitions are

$$\langle n_{\vec{q}_1\lambda_1} - 1\,;\, n_{\vec{q}_2\lambda_2} - 1\,;\, n_{\vec{q}_3\lambda_3} + 1|\hat{C}_{\vec{q}_1\lambda_1}\hat{C}_{\vec{q}_2\lambda_2}\hat{C}^{+}_{\vec{q}_3\lambda_3}|n_{\vec{q}_1\lambda_1}\,;\, n_{\vec{q}_2\lambda_2}\,;\, n_{\vec{q}_3\lambda_3}\rangle$$
$$= \sqrt{n_{\vec{q}_1\lambda_1}\cdot n_{\vec{q}_2\lambda_2}\cdot (n_{\vec{q}_3\lambda_3}+1)}$$

$$\langle n_{\vec{q}_1\lambda_1} + 1\,;\, n_{\vec{q}_2\lambda_2} + 1\,;\, n_{\vec{q}_3\lambda_3} - 1|\hat{C}^{+}_{\vec{q}_1\lambda_1}\hat{C}^{+}_{\vec{q}_2\lambda_2}\hat{C}_{\vec{q}_3\lambda_3}|n_{\vec{q}_1\lambda_1}\,;\, n_{\vec{q}_2\lambda_2}\,;\, n_{\vec{q}_3\lambda_3}\rangle$$
$$= \sqrt{(n_{\vec{q}_1\lambda_1}+1)\,(n_{\vec{q}_2\lambda_2}+1) n_{\vec{q}_3\lambda_3}}$$

Hence the probabilities of transitions in which two quasi-phonons $(\vec{q}_1\lambda_1)$ and $(\vec{q}_2\lambda_2)$ are annihilated and a quasi-phonon $(\vec{q}_3\lambda_3)$ is created, or conversely, are as follows (Krasny *et al.*, 1990; Kovalenko *et al.*, 1993):

$$\overline{P_{\vec{q}_3\lambda_3\to\vec{q}_1\lambda_1;\,\vec{q}_2\lambda_2}} = \frac{1}{N^2}\frac{(n_{\vec{q}_1\lambda_1}+1)(n_{\vec{q}_2\lambda_2}+1)n_{\vec{q}_3\lambda_3}}{\omega_{\lambda_1}(\vec{q}_1)\cdot\omega_{\lambda_2}(\vec{q}_2)\cdot\omega_{\lambda_3}(\vec{q}_3)}$$
$$\times\, \mathrm{I}_{\vec{q}_1\lambda_1;\,\vec{q}_2\lambda_2;\,\vec{q}_3\lambda_3}\cdot\delta\left(\hbar\omega_{\lambda_1}(\vec{q}_1) + \hbar\omega_{\lambda_2}(\vec{q}_2) - \hbar\omega_{\lambda_3}(\vec{q}_3)\right)$$
$$\times\, \overline{P_{\vec{q}_1\lambda_1;\,\vec{q}_2\lambda_2\to\vec{q}_3\lambda_3}} = \frac{1}{N^2}\frac{n_{\vec{q}_1\lambda_1}\cdot n_{\vec{q}_2\lambda_2}\cdot(n_{\vec{q}_3\lambda_3}+1)}{\omega_{\lambda_1}(\vec{q}_1)\cdot\omega_{\lambda_2}(\vec{q}_2)\cdot\omega_{\lambda_3}(\vec{q}_3)}$$
$$\times\, \mathrm{I}_{\vec{q}_1\lambda_1;\,\vec{q}_2\lambda_2;\,\vec{q}_3\lambda_3}\cdot\delta\left(\hbar\omega_{\lambda_1}(\vec{q}_1) + \hbar\omega_{\lambda_2}(\vec{q}_2) - \hbar\omega_{\lambda_3}(\vec{q}_3)\right)$$
$$\times\, \mathrm{I}_{\vec{q}_1\lambda_1;\,\vec{q}_2\lambda_2;\,\vec{q}_3\lambda_3} = \mathrm{I}_{-\vec{q}_1\lambda_1;\,-\vec{q}_2\lambda_2;\,-\vec{q}_3\lambda_3}$$
$$= \left(\frac{\hbar}{2M}\right)^3 \frac{1}{N}\frac{2\pi}{\hbar}\,\overline{\Lambda_{\vec{q}_1\lambda_1;\,\vec{q}_2\lambda_2;\,\vec{q}_3\lambda_3}\cdot\Lambda^{*}_{\vec{q}_1\lambda_1;\,\vec{q}_2\lambda_2;\,\vec{q}_3\lambda_3}} \qquad (3.4.14)$$

Assuming the quasi-phonon gas is rarified and taking into account all scattering events we construct kinetic equations for every kind of quasi-phonon (Lifshitz and

Pitaevski, 1978; Krasny *et al.*, 1990; Kovalenko *et al.*, 1993). Let $n_{\vec{q}_1\lambda_1} = n_{\vec{q}_1\lambda_1}(\vec{r}, t)$ be the distribution function of quasi-phonons of sort λ_1, then the kinetic equations take the form:

$$\begin{aligned}\frac{\partial n_{\vec{q}_1\lambda_1}}{\partial t} + \vec{v}_{\vec{q}_1\lambda_1}\frac{\partial n_{\vec{q}_1\lambda_1}}{\partial \vec{r}} = &+\sum_{\lambda_2\lambda_3}\frac{v_0^2}{(2\pi)^6}\int d^3q_2\int d^3q_3 \\ &\times\Big\{[(1+n_{\vec{q}_1\lambda_1})(1+n_{\vec{q}_2\lambda_2})n_{\vec{q}_3\lambda_3} - n_{\vec{q}_1\lambda_1}n_{\vec{q}_2\lambda_2}(1+n_{\vec{q}_3\lambda_3})] \\ &\times\frac{\mathrm{I}_{\vec{q}_1\lambda_1;\vec{q}_2\lambda_2;\vec{q}_3\lambda_3}\cdot\delta(\hbar\omega_{\lambda_1}(\vec{q}_1)+\hbar\omega_{\lambda_2}(\vec{q}_2)-\hbar\omega_{\lambda_3}(\vec{q}_3))}{\omega_{\lambda_1}(\vec{q}_1)\cdot\omega_{\lambda_2}(\vec{q}_2)\cdot\omega_{\lambda_3}(\vec{q}_3)} \\ &+\frac{1}{2}[(1+n_{\vec{q}_1\lambda_1})n_{\vec{q}_2\lambda_2}n_{\vec{q}_3\lambda_3} - n_{\vec{q}_1\lambda_1}(1+n_{\vec{q}_2\lambda_2})(1+n_{\vec{q}_3\lambda_3})] \\ &\times\frac{\mathrm{I}_{\vec{q}_3\lambda_3;\vec{q}_2\lambda_2;\vec{q}_1\lambda_1}\cdot\delta(\hbar\omega_{\lambda_1}(\vec{q}_1)-\hbar\omega_{\lambda_2}(\vec{q}_2)-\hbar\omega_{\lambda_3}(\vec{q}_3))}{\omega_{\lambda_1}(\vec{q}_1)\cdot\omega_{\lambda_2}(\vec{q}_2)\cdot\omega_{\lambda_3}(\vec{q}_3)}\Big\}\end{aligned} \tag{3.4.15}$$

Here

$$\vec{v}_{\vec{q}\lambda} = \frac{\partial\omega_\lambda(\vec{q})}{\partial\vec{q}}\,;\; v_0 = \frac{V}{N} \tag{3.4.16}$$

The first term in curly brackets on the right-hand side of Eq. (3.4.15) corresponds to the following confluence and splitting processes, respectively:

$$(\vec{q}_1\lambda_1) + (\vec{q}_2\lambda_2) \rightleftarrows (\vec{q}_3\lambda_3)$$

The second term, with the prefactor 1/2, describes the processes

$$(\vec{q}_1\lambda_1) \rightleftarrows (\vec{q}_2\lambda_2) + (\vec{q}_3\lambda_3)$$

The factor 1/2 in this term takes into consideration that, in view of identity, to avoid double counting, the summation must be carried out only over half of the final states. It should be pointed out that in the integrand the terms involving $n^{\mathrm{r}}_{q_1\lambda_1}\cdot n^{\mathrm{r}}_{q_2\lambda_2}\cdot n^{\mathrm{r}}_{q_3\lambda_3}$ cancel out.

The collision integral (the right-hand side of Eq. (3.4.15)) vanishes identically for the equilibrium distribution of phonons

$$n^{(0)}_{\vec{q}\lambda} = \left[\exp\left(\frac{\hbar\omega_\lambda(\vec{q})}{k_B T}\right) - 1\right]^{-1} \tag{3.4.17}$$

This can easily be checked: the multiplication immediately gives

$$n^{(0)}_{\vec{q}_1\lambda_1}(n^{(0)}_{\vec{q}_2\lambda_2}+1)(n^{(0)}_{\vec{q}_3\lambda_3}+1)$$
$$=(n^{(0)}_{\vec{q}_1\lambda_1}+1)\cdot n^{(0)}_{\vec{q}_2\lambda_2}\cdot n^{(0)}_{\vec{q}_3\lambda_3}\cdot\exp\left[\frac{\hbar\omega_{\lambda_1}(\vec{q}_1)+\hbar\omega_{\lambda_2}(\vec{q}_2)-\hbar\omega_{\lambda_3}(\vec{q}_3)}{k_B T}\right] \tag{3.4.18}$$

and on the basis of the energy conservation law the exponential factor on the r.h.s. reduces to unity. If the sum of quasi-momenta of phonons were conserved, too, not only the function (3.4.17) would be in equilibrium, but also the function

$$n^{(0)}_{\vec{q}\lambda}=\left[\exp\left(\frac{\hbar\omega_\lambda(\vec{q})-\vec{q}\cdot\vec{V}}{k_B T}\right)-1\right]^{-1} \tag{3.4.19}$$

which corresponds to the translational motion (a drift) of the quasi-phonon gas as a whole relative to the lattice with velocity $\vec{V}$. This can be verified directly: the function (3.4.19) substituted as $n^{(0)}_{\text{r}_{q\lambda}}$ into the r.h.s. of Eq. (3.4.18) gives yet another factor

$$\exp\left[\frac{\vec{N}\cdot(\vec{q}_1+\vec{q}_2-\vec{q}_3)}{k_B T}\right]$$

reducing to unity when $\vec{q}_1+\vec{q}_2-\vec{q}_3=0$. The distribution (3.4.19), however, brings about a non-zero energy flux. So, if the law of conservation of phonon quasi-momenta were fulfilled, thermal transmission at constant temperature along the whole specimen would be possible, i.e. the solid would show an infinite heat conduction. The *finite* thermal conduction is thus a consequence of the *absence* of quasi-momentum conservation, i.e. in crystals it is a consequence of the existence of the so-called 'Umklapp processes', where $\vec{q}_1+\vec{q}_2-\vec{q}_3=\vec{G}$, where in crystals $\vec{G}$ is a vector of the reciprocal lattice, whereas in amorphous systems $\vec{G}$ is arbitrary in any case.

To calculate the heat conductivity we have to write down a kinetic equation with slow temperature variation throughout the bulk of the body. We look for a quasi-phonon distribution function in the form of the following *ansatz*:

$$n_{\vec{q}\lambda}=n^{(0)}_{\vec{q}\lambda}+\delta n_{\vec{q}\lambda}=n^{(0)}_{\vec{q}\lambda}-\Phi_{\vec{q}\lambda}\frac{\partial n^{(0)}_{\vec{q}\lambda}}{\partial\hbar\omega_\lambda(\vec{q})} \tag{3.4.20}$$

restricting ourselves to terms linear in $\vec{\nabla}T$. On the l.h.s. of Eq. (3.4.15) we retain the term

$$(\vec{v}_{\vec{q}_1\lambda_1}\cdot\vec{\nabla}T)\frac{\partial n^{(0)}_{\vec{q}_1\lambda_1}}{\partial T}$$

The r.h.s. of Eq. (3.4.15) can be simplified by considering

$$\delta n_{\vec{q}\lambda} = -\Phi_{\vec{q}\lambda} = \frac{\partial n^{(0)}_{\vec{q}\lambda}}{\partial \hbar\omega_\lambda(\vec{q})} = \frac{n^{(0)}_{\vec{q}\lambda}\left(1+n^{(0)}_{\vec{q}\lambda}\right)}{k_B T}\,\Phi_{\vec{q}\lambda}$$

$$= \delta\frac{n_{\vec{q}\lambda}}{1+n_{\vec{q}\lambda}} = \frac{n^{(0)}_{\vec{q}\lambda}}{1+n^{(0)}_{\vec{q}\lambda}}\cdot\frac{\Phi_{\vec{q}\lambda}}{k_B T}$$

The expressions in square brackets on the r.h.s. of Eq. (3.4.15) can be rewritten as follows:

$$[(1+n_{\vec{q}_1\lambda_1})(1+n_{\vec{q}_2\lambda_2})n_{\vec{q}_3\lambda_3} - n_{\vec{q}_1\lambda_1}n_{\vec{q}_2\lambda_2}(1+n_{\vec{q}_3\lambda_3})]$$

$$= (1+n_{\vec{q}_1\lambda_1})(1+n_{\vec{q}_2\lambda_2})(1+n_{\vec{q}_3\lambda_3})\cdot\left[\frac{-n_{\vec{q}_1\lambda_1}}{1+n_{\vec{q}_1\lambda_1}}\cdot\frac{n_{\vec{q}_2\lambda_2}}{1+n_{\vec{q}_2\lambda_2}}+\frac{n_{\vec{q}_3\lambda_3}}{1+n_{\vec{q}_3\lambda_3}}\right]$$

$$\times[(1+n_{\vec{q}_1\lambda_1})\cdot n_{\vec{q}_2\lambda_2}\cdot n_{\vec{q}_3\lambda_3} - n_{\vec{q}_1\lambda_1}(1+n_{\vec{q}_2\lambda_2})(1+n_{\vec{q}_3\lambda_3})]$$

$$= (1+n_{\vec{q}_1\lambda_1})(1+n_{\vec{q}_2\lambda_2})(1+n_{\vec{q}_3\lambda_3})\cdot\left[\frac{n_{\vec{q}_2\lambda_2}}{1+n_{\vec{q}_2\lambda_2}}\cdot\frac{n_{\vec{q}_3\lambda_3}}{1+n_{\vec{q}_3\lambda_3}}-\frac{n_{\vec{q}_1\lambda_1}}{1+n_{\vec{q}_1\lambda_1}}\right]$$

In the factors outside the square brackets we can put $n_{\vec{q}\lambda} = n^0_{\vec{q}\lambda}$. As for the difference in the square brackets, it gives

$$\begin{cases} \dfrac{1}{k_B T}\dfrac{n^0_{\vec{q}_3\lambda_3}}{1+n^0_{\vec{q}_3\lambda_3}}\,[\Phi_{\vec{q}_3\lambda_3}-\Phi_{\vec{q}_1\lambda_1}-\Phi_{\vec{q}_2\lambda_2}] \\ \dfrac{1}{k_B T}\dfrac{n^0_{\vec{q}_1\lambda_1}}{1+n^0_{\vec{q}_1\lambda_1}}\,[\Phi_{\vec{q}_2\lambda_2}+\Phi_{\vec{q}_3\lambda_3}-\Phi_{\vec{q}_1\lambda_1}] \end{cases}$$

Here we take into account that in the first relationship

$$\frac{n^0_{\vec{q}_1\lambda_1}}{1+n^0_{\vec{q}_1\lambda_1}}\cdot\frac{n^0_{\vec{q}_2\lambda_2}}{1+n^0_{\vec{q}_2\lambda_2}} = \frac{n^0_{\vec{q}_3\lambda_3}}{1+n^0_{\vec{q}_3\lambda_3}}$$

and in the second relation

$$\frac{n^0_{\vec{q}_2\lambda_2}}{1+n^0_{\vec{q}_2\lambda_2}}\cdot\frac{n^0_{\vec{q}_3\lambda_3}}{1+n^0_{\vec{q}_3\lambda_3}} = \frac{n^0_{\vec{q}_1\lambda_1}}{1+n^0_{\vec{q}_1\lambda_1}}$$

Thus the kinetic equation (3.4.15) takes the following form:

$$-(\vec{v}_{\vec{q}_1\lambda_1}\cdot\vec{\nabla}T)\,\frac{\partial n^0_{\vec{q}_1\lambda_1}}{\partial T}=\frac{1}{k_B T}+\frac{v_0^2}{(2\pi)^6}\sum_{\lambda_2\lambda_3}\int d^3q_2\int d^3q_3$$
$$\times\left\{n^0_{\vec{q}_1\lambda_1}\cdot n^0_{\vec{q}_2\lambda_2}\cdot(1+n^0_{\vec{q}_3\lambda_3})\cdot(\Phi_{\vec{q}_1\lambda_1}+\Phi_{\vec{q}_2\lambda_2}-\Phi_{\vec{q}_3\lambda_3})\right.$$
$$\times\,\mathrm{I}_{\vec{q}_1\lambda_1;\vec{q}_2\lambda_2;\vec{q}_3\lambda_3}\frac{\delta\left(\hbar\omega_{\lambda_1}(\vec{q}_1)+\hbar\omega_{\lambda_2}(\vec{q}_2)-\hbar\omega_{\lambda_3}(\vec{q}_3)\right)}{\omega_{\lambda_1}(\vec{q}_1)\cdot\omega_{\lambda_2}(\vec{q}_2)\cdot\omega_{\lambda_3}(\vec{q}_3)}$$
$$+\frac{1}{2}n^0_{\vec{q}_1\lambda_1}(1+n^0_{\vec{q}_2\lambda_2})(1+n^0_{\vec{q}_3\lambda_3})\cdot(\Phi_{\vec{q}_1\lambda_1}-\Phi_{\vec{q}_2\lambda_2}-\Phi_{\vec{q}_3\lambda_3})$$
$$\left.\times\,\mathrm{I}_{\vec{q}_3\lambda_3;\vec{q}_2\lambda_2;\vec{q}_3\lambda_3}\frac{\delta\left(\hbar\omega_{\lambda_1}(\vec{q}_1)-\hbar\omega_{\lambda_2}(\vec{q}_2)-\hbar\omega_{\lambda_3}(\vec{q}_3)\right)}{\omega_{\lambda_1}(\vec{q}_1)\cdot\omega_{\lambda_2}(\vec{q}_2)\cdot\omega_{\lambda_3}(\vec{q}_3)}\right\}\tag{3.4.21}$$

We will solve this equation using the variational principle, namely, the functional

$$\frac{1}{\kappa}=\left\{\frac{v_0^2}{(2\pi)^6}\sum_{\lambda_1\lambda_2\lambda_3}\int d^3q_1\int d^3q_2\int d^3q_3[\Phi_{\vec{q}_1\lambda_1}+\Phi_{\vec{q}_2\lambda_2}-\Phi_{\vec{q}_3\lambda_3}]^2\right.$$
$$\left.\times\,n^{(0)}_{\vec{q}_1\lambda_1}\cdot n^{(0)}_{\vec{q}_2\lambda_2}\cdot(1+n^{(0)}_{\vec{q}_3\lambda_3})\frac{\mathrm{I}_{\vec{q}_1\lambda_1;\vec{q}_2\lambda_2;\vec{q}_3\lambda_3}\,\delta\left[\hbar\omega_{\lambda_1}(\vec{q}_1)+\hbar\omega_{\lambda_2}(\vec{q}_2)-\hbar\omega_{\lambda_3}(\vec{q}_3)\right]}{\omega_{\lambda_1}(\vec{q}_1)\cdot\omega_{\lambda_2}(\vec{q}_2)\cdot\omega_{\lambda_3}(\vec{q}_3)}\right\}$$
$$:\left\{2k_B T^2\left|\sum_{\lambda}\int\frac{d^3q}{(2\pi)^3}\,\vec{v}_{\vec{q}\lambda}\cdot\frac{\partial n^{(0)}_{\vec{q}\lambda}}{\partial T}\right|^2\right\}\tag{3.4.22}$$

reaches its minimum when the function $\Phi_{\vec{q}_\lambda}$ is a solution of the kinetic equation (3.4.22) (Lifshitz and Pitaevski, 1978; Krasny *et al.*, 1990; Kovalenko *et al.*, 1993). Here κ stands for the heat conductivity of the medium.

An analogous expression for heat conductivity also results from a direct solution of Eq. (3.4.21) when the function

$$\Phi_{\vec{q}\lambda}=A\cdot(\vec{q}\cdot\vec{u})\tag{3.4.23}$$

is chosen as a solution, where $\vec{u}=\frac{\vec{\nabla}_T}{|\vec{\nabla}_T|}$ is a unit vector aligned with the temperature gradient. Then

$$\Phi_{\vec{q}_1\lambda_1}+\Phi_{\vec{q}_2\lambda_2}-\Phi_{\vec{q}_3\lambda_3}=A(\vec{q}_1+\vec{q}_2-\vec{q}_3)\cdot\vec{u}\tag{3.4.24}$$

The potential energy $U_0=U(\vec{R}_{01},\vec{R}_{02},..\vec{R}_{0N})$ is generally characterized by a parameter r_0 determining the radius of interatomic interaction, and can be expressed in terms of dimensionless variables $\vec{r}_j=\vec{R}_{0j}/r_0$. For a spatially uniform and isotropic amorphous medium the function $\mathrm{I}_{\vec{q}_1\lambda_1;\vec{q}_2\lambda_2;\vec{q}_3\lambda_3}$, upon averaging over all possible configurations of atomic equilibrium positions, must be real and depend only on the absolute values of the vectors $\vec{q}_i$, i.e.

$$\mathrm{I}_{\vec{q}_1\lambda_1;\vec{q}_2\lambda_2;\vec{q}_3\lambda_3}=\mathrm{I}_{-\vec{q}_1\lambda_1;\vec{q}_2\lambda_2;\vec{q}_3\lambda_3}=\mathrm{I}_{\vec{q}_1\lambda_1;-\vec{q}_2\lambda_2;\vec{q}_3\lambda_3}=\mathrm{I}_{\vec{q}_1\lambda_1;\vec{q}_2\lambda_2;-\vec{q}_3\lambda_3}$$

This is possible on the condition that

$$\overline{\sum_{\substack{j_1 \cdot j_2 \cdot j_1 \\ j_1' \cdot j_2' \cdot j_1'}} \frac{\partial^3 U_0}{\partial R_{0j_1}^{\alpha}\, \partial R_{0j_2}^{\beta}\, \partial R_{0j_3}^{\gamma}} \frac{\partial^3 U_0}{\partial R_{0j_1'}^{\alpha'}\, \partial R_{0j_2'}^{\beta'}\, \partial R_{0j_3'}^{\gamma'}}}$$

$$\overline{\times \exp\left[i\vec{q}_1 \cdot (\vec{R}_{0j_1} - \vec{R}_{0j_1'}) + i\vec{q}_2 \cdot (\vec{R}_{0j_2} - \vec{R}_{0j_2'}) - i\vec{q}_3 \cdot (\vec{R}_{0j_3} - \vec{R}_{0j_3'})\right]}$$

$$= \frac{\delta_{\alpha\alpha'}\, \delta_{\beta\beta'}\, \delta_{\gamma\gamma'} \cdot r_0^{-6}}{27} \overline{\sum_{\substack{j_1 \cdot j_2 \cdot j_3 \\ j_1' \cdot j_2' \cdot j_3'}} \sum_{\alpha_1 \beta_1 \gamma_1} \frac{\partial^3 U_0}{\partial r_{j_1}^{\alpha_1}\, \partial r_{j_2}^{\beta_1}\, \partial r_{j_3}^{\gamma_1}} \cdot \frac{\partial^3 U_0}{\partial r_{j_1'}^{\alpha_1}\, \partial r_{j_2'}^{\beta_1}\, \partial r_{j_3'}^{\gamma_1}}}$$

$$\overline{\times\, e^{i\vec{q}_1 r \cdot (\vec{r}_{j_1} - \vec{r}_{j_1'}) + i\vec{q}_2 r_0 \cdot (\vec{r}_{j_2} - \vec{r}_{j_2'}) - i\vec{q}_3 r_0 \cdot (\vec{r}_{j_3} - \vec{r}_{j_3'})}}$$

Because $\vec{\varepsilon}(\vec{q}, \lambda) \cdot \vec{\varepsilon}(\vec{q}, \lambda) = 1$, it follows that $I_{\vec{q}_1\lambda_1;\vec{q}_2\lambda_2;\vec{q}_3\lambda_3}$ is independent of λ_i and takes the form:

$$I_{\vec{q}_1\lambda_1;\vec{q}_2\lambda_2;\vec{q}_3\lambda_3} \equiv I(q_1 r_0; q_2 r_0; q_3 r_0)$$

$$= \left(\frac{\hbar}{2M}\right)^3 \frac{2\pi}{\hbar} \frac{r_0^{-6}}{27N} \overline{\sum_{\substack{j_1 \cdot j_2 \cdot j_3 \\ j_1' \cdot j_2' \cdot j_3'}} \sum_{\alpha,\beta,\gamma} \frac{\partial^3 U_0}{\partial r_{j_1}^{\alpha}\, \partial r_{j_2}^{\beta}\, \partial r_{j_3}^{\gamma}} \cdot \frac{\partial^3 U_0}{\partial r_{j_1'}^{\alpha}\, \partial r_{j_2'}^{\beta}\, \partial r_{j_3'}^{\gamma}}}$$

$$\overline{\times\, e^{i\vec{q}_1 r_0 \cdot (\vec{r}_{j_1} - \vec{r}_{j_1'}) + i\vec{q}_2 r_0 \cdot (\vec{r}_{j_2} - \vec{r}_{j_2'}) - i\vec{q}_3 r_0 \cdot (\vec{r}_{j_3} - \vec{r}_{j_3'})}} \tag{3.4.25}$$

Because $r_0 \approx 1/q_D$, and because the integration in Eq. (3.4.22) is performed over the region $0 \le q \le q_D = \left(\frac{6\pi^2}{v_0}\right)^{1/3}$, the function $I(q_1 r_0; q_2 r_0; q_3 r_0)$ under the integrals can be expanded into a Taylor series, and only the first term of the expansion must be retained:

$$I(q_1 r_0; q_2 r_0; q_3 r_0) \approx I(0, 0, 0) = I_0$$

$$= \left(\frac{\hbar}{2M}\right)^3 \frac{2\pi}{\hbar} \frac{r_0^{-6}}{27N} \overline{\sum_{\substack{j_1 \cdot j_2 \cdot j_3 \\ j_1' \cdot j_2' \cdot j_3'}} \sum_{\alpha,\beta,\gamma} \frac{\partial^3 U_0}{\partial r_{j_1}^{\alpha}\, \partial r_{j_2}^{\beta}\, \partial r_{j_3}^{\gamma}} \cdot \frac{\partial^3 U_0}{\partial r_{j_1'}^{\alpha}\, \partial r_{j_2'}^{\beta}\, \partial r_{j_3'}^{\gamma}}} \tag{3.4.26}$$

Substituting Eqs (3.4.23), (3.4.24) and (3.4.26) into (3.4.22) we have the numerator of the resulting fraction expressed in the form:

$$\frac{v_0^2}{(2\pi)^3}\sum_{\lambda_1\lambda_2\lambda_3}\int_0^{q_D} q_1^2 dq_1 \int_0^{q_D} q_2^2 dq_2 \int_0^{q_D} q_3^2 dq_3 \, \frac{n^{(0)}_{\vec{q}_1\lambda_1}\cdot n^{(0)}_{\vec{q}_2\lambda_2}\cdot(1+n^{(0)}_{\vec{q}_3\lambda_3})}{\omega_{\lambda_1}(q_1)\cdot\omega_{\lambda_2}(q_2)\cdot\omega_{\lambda_3}(q_3)}$$
$$\times\,\delta\left(\hbar\omega_{\lambda_1}(q_1)+\hbar\omega_{\lambda_2}(q_2)-\hbar\omega_{\lambda_3}(q_3)\right)\cdot A^2\int d\Omega_1\int d\Omega_2\int d\Omega_3$$
$$\times\,\mathrm{I}_{\vec{q}_1\lambda_1;\vec{q}_2\lambda_2;\vec{q}_3\lambda_3}\cdot[(\vec{q}_1+\vec{q}_2-\vec{q}_3)\vec{u}]^2$$
$$\approx\frac{A^2}{3}\frac{v_0^2}{\pi^3}\,\mathrm{I}_0\sum_{\lambda_1\lambda_2\lambda_3}\int_0^{q_D} q_1^2 dq_1\int_0^{q_D} q_2^2 dq_2\int_0^{q_D} q_3^2 dq_3\cdot(q_1^2+q_2^2+q_3^2)$$
$$\times\,\frac{n^{(0)}_{\vec{q}_1\lambda_1}\cdot n^{(0)}_{\vec{q}_2\lambda_2}\cdot(1+n^{(0)}_{\vec{q}_3\lambda_3})}{\omega_{\lambda_1}(q_1)\cdot\omega_{\lambda_2}(q_2)\cdot\omega_{\lambda_3}(q_3)}\,\delta[\hbar\omega_{\lambda_1}(q_1)+\hbar\omega_{\lambda_2}(q_2)-\hbar\omega_{\lambda_3}(q_3)] \qquad (3.4.27)$$

The expression in the denominator is calculated in a similar way:

$$\frac{1}{(2\pi)^3}\sum_\lambda\int d^3q\vec{v}_{\vec{q}\lambda}\cdot\Phi_{\vec{q}\lambda}\cdot\frac{\partial n^{(0)}_{\vec{q}\lambda}}{\partial T}$$
$$=\frac{A\vec{u}}{6\pi^2}\frac{\hbar}{k_B T^2}\sum_\lambda\int_0^{q_D} q^3 dq\,\frac{\partial\omega_\lambda(q)}{\partial q}\,\omega_\lambda(q)\cdot\frac{e^{\hbar\omega_\lambda(q)/k_B T}}{[e^{\hbar\omega_\lambda(q)/k_B T}-1]^2} \qquad (3.4.28)$$

Thus the phonon contribution to the heat conductivity becomes (Krasny *et al.*, 1990; Kovalenko *et al.*, 1993):

$$\kappa=\left\{\frac{2\hbar^2}{k_B T^2(6\pi^2)^2}\left[\sum_\lambda\int_0^{q_D} q^3 dq\,\frac{\partial\omega_\lambda(q)}{\partial q}\,\omega_\lambda(q)\cdot\frac{\exp\left(\hbar\omega_\lambda(q)/k_B T\right)}{\left[\exp\frac{\hbar\omega_\lambda(q)}{k_B T}-1\right]^2}\right]^2\right\}$$
$$:\left\{\frac{v_0^2}{3\pi^3}\,I_0\sum_{\lambda_1\lambda_2\lambda_3}\int_0^{q_D} q_1^2 dq_1\int_0^{q_D} q_2^2 dq_2\int_0^{q_D} q_3^2 dq_3\,\frac{n^{(0)}_{\vec{q}_1\lambda_1}\cdot n^{(0)}_{\vec{q}_2\lambda_2}(1+n^{(0)}_{\vec{q}_3\lambda_3})}{\omega_{\lambda_1}(q_1)\cdot\omega_{\lambda_2}(q_2)\cdot\omega_{\lambda_3}(q_3)}\right.$$
$$\left.\times\,(q_1^2+q_2^2+q_3^2)\,\delta\left[\hbar\omega_{\lambda_1}(q_1)+\hbar\omega_{\lambda_2}(q_2)-\hbar\omega_{\lambda_3}(q_3)\right]\right\} \qquad (3.4.29)$$

This relationship demonstrates that temperature behavior of the thermal conductivity coefficient κ is determined by the dispersion law $\omega_\lambda(q)$ of the quasi-phonons.

Experiments as well as direct calculations (see Section 3.4) show maxima and minima in the dispersion curves $\omega_\lambda(q)$. The presence of pronounced minima implies the existence of so-called 'soft quasi-phonons', which – as already mentioned – are also called 'rotons' on historical grounds, in the short-wave region, i.e. at $q \approx 2\pi/r_0$, where r_0 is the average nearest-neighbor distance between the ions. In the long-wave limit ($q \to 0$), $\omega_\lambda(q)$ is found to be proportional to:

$$\omega_\lambda(q) = v_\lambda \cdot q$$

where v_λ is the corresponding sound velocity.

To evaluate the temperature dependence of the *heat conductivity* numerically, the dispersion curves are approximated by polygonal lines connecting special points on the dispersion curves (the initial point, maximum and minimum positions, and the end-point). Figure 3.4.1 demonstrates these approximating lines (the dotted ones) along with the calculated full dispersion curves (solid lines) for amorphous Mg and Zn (Krasny *et al.*, 1990). As already mentioned above, at room temperature, only amorphous *alloys*, e.g. $Mg_{30}Zn_{70}$, are stable, however from the computed results for amorphous Mg and Zn we can already draw conclusions for the alloy system.

The above-mentioned polygon approximation allows the analytical calculation of the integrals in Eq. (3.4.29). In a moderately low temperature region ($k_B T \lesssim \hbar\Delta_l$ where $\hbar\Delta_l = \hbar\omega_l(q_{min})$, typically $\approx 30\,k_B$ K) the result for the heat conductivity κ depends on T as (Krasny *et al.*, 1990; Kovalenko *et al.*, 1993)

$$\kappa = \kappa_0 \frac{1}{\varepsilon} \frac{1 + (B_0 + B_1\varepsilon + B_2\varepsilon^2 + B_3\varepsilon^3 + B_4\varepsilon^4)\,e^{-\varepsilon}}{1 + (D_0 + D_1\varepsilon + D_2\varepsilon^2 + D_3\varepsilon^3 + D_4\varepsilon^4)\,e^{-\varepsilon}} \tag{3.4.30}$$

where

$$\varepsilon = \frac{\hbar\Delta_l}{k_B T}; \; \kappa_0 = \frac{C_t^5 \hbar\Delta_l \cdot k_B \cdot C}{6\pi v_0^2 I_0 \cdot A_0}$$

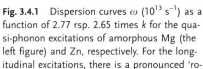

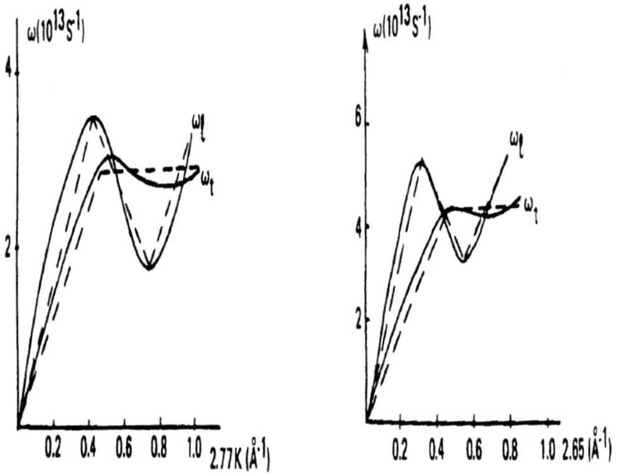

Fig. 3.4.1 Dispersion curves ω (10^{13} s^{-1}) as a function of 2.77 rsp. 2.65 times k for the quasi-phonon excitations of amorphous Mg (the left figure) and Zn, respectively. For the longitudinal excitations, there is a pronounced 'roton minimum' already mentioned several times. The dashed line represents the 'polygon approximation' used in the calculation of the thermal conductivity κ (taken from Krasny *et al.*, 1990)

and the dimensionless constants A_0, C, B_i and $D_i D$ can be expressed in terms of specific parameters of the dispersion curves, namely $(v_t, v_l, \hbar\Delta_l, \hbar\omega_l(q_{\max}), \hbar\omega_\lambda(q_D))$.

Computations for amorphous Mg and Zn gave the results shown in Fig. 3.4.2, taken from Krasny *et al.* (1990).

It turns out that within the temperature range where the function C_V/T^3 has a maximum (i.e. when $2\text{K} < T \lesssim 100\,\text{K}$) there is a *plateau* in the curve $\kappa(T)$, which results from the 'roton minimum' in Fig. 3.4.1. Outside this region (when $T > 100$ K) the heat conductivity is linear in T.

The interaction between phonons determines the temperature dependence not only of the *heat conductivity*, but of *sound absorption* in a material as well. Suppose that a sound wave with the wavevector $\vec{q}$ and frequency ω travels through the material. Let the x-axis be in alignment with the propagating wave. In quantum terms, the wave considered is a parallel flux of phonons. Due to the interaction with *thermal* phonons the concentration of ultra-sound phonons, N, becomes dependent on x and satisfies the kinetic equation (3.4.15). Assuming a steady-state flux of ultra-sound phonons, we have

$$n_{\vec{q}_1\lambda_1}(\vec{r}, t) = N(x) \tag{3.4.31}$$

Then

$$\frac{\partial n_{\vec{q}_1\lambda_1}}{\partial t} = \frac{\partial N(x)}{\partial t} = 0$$

$$\vec{v}_{\vec{q}_1\lambda_1} \frac{\partial n_{\vec{q}_1\lambda_1}}{\partial r} = v_\lambda \frac{\partial N}{\partial x}$$

where v_λ is the sound velocity of the quasi-phonons considered.

From here on we put

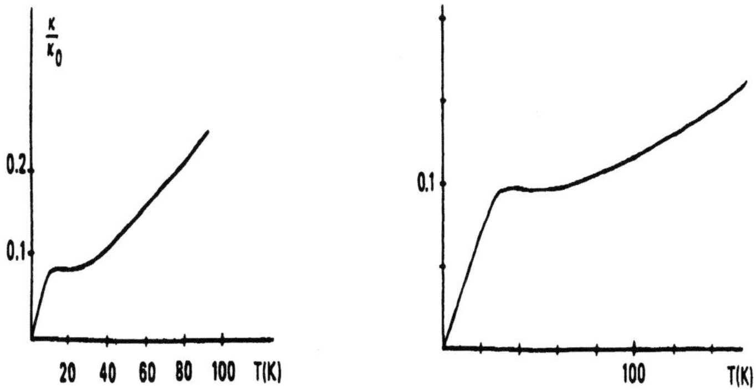

Fig. 3.4.2 The behavior of the thermal conductivity κ as a function of the Kelvin temperature T (K) for amorphous Mg (left plot) and amorphous Zn (right plot) (from Krasny *et al.*, 1990)

$$\hbar\omega \ll k_B T$$

Actually, in experiments (see Section 3.1) waves of frequency $\omega \sim 10^9$ s^{-1} dominate. So the energy of ultra-sound phonons is

$$\hbar\omega \sim 10^{-18}\,\text{erg} \ll k_B T \sim 10^{-15} - 10^{-14}\,\text{erg}$$

Besides, these frequencies are matched by wavelengths

$$\lambda = \frac{2\pi}{\omega} v_\lambda \sim 10^{-3}\,\text{cm} \gg a$$

where a is the mean interatomic distance. The dispersion law for the sound waves considered is therefore well approximated by the expression

$$\omega = v_\lambda \cdot q$$

where v_λ is constant.

The energy of thermal phonons is of the order of $k_B T$:

$$\hbar\omega_{\lambda_1}(q_1);\ \hbar\omega_{\lambda_2}(q_2) \sim k_B T$$

Thus the inequality

$$\hbar\omega \ll \hbar\omega_{\lambda_1}(q_1) + \hbar\omega_{\lambda_2}(q_2) \sim k_B T$$

always holds. (On the r.h.s. of Eq. (3.4.15) we change the subscripts and integration variables: $\lambda_2 \to \lambda_1, \lambda_3 \to \lambda_2; q_2 \to q_1, q_3 \to q_2$). Hence the second summand on the r.h.s. of the kinetic equation (3.4.15) vanishes identically, i.e. processes are impossible in which an ultra-sound phonon splits into two thermal ones. Instead, processes in which sonic quanta are *absorbed* by thermal phonons play a leading part in sound absorption.

A macroscopic sound wave corresponds to a very large occupation number $N(x) \gg 1$. The occupation numbers of thermal phonons are equal to the equilibrium distribution function:

$$n_{\vec{q}\lambda} \approx n^0_{\vec{q}\lambda} = \left[\exp\left(\frac{\hbar\omega_\lambda(q)}{k_B T}\right) - 1\right]^{-1}$$

Finally, assuming $\mathrm{I}_{\mathrm{r}_{q_1\lambda_1};\mathrm{r}_{q_2\lambda_2};\mathrm{r}_{q_3\lambda_3}} \approx \mathrm{I}_0$, we have, upon integration over all possible angles, the kinetic equation (3.4.15) for sonic phonons as follows:

$$\frac{dN}{dx} = -\frac{1}{l(\omega)} N(x) \tag{3.4.32}$$

Here $l(\omega)$ is referred to as the free path length of the sonic phonon; its reciprocal equals twice the coefficient of sound-wave absorption, a (Lifshitz and Pitaevski, 1978; Krasny *et al.*, 1990), and is determined by the relationship

$$\frac{1}{l(\omega)} = \frac{v_0^2}{(2\pi^2)^2 v_\lambda} \cdot \int_0^{q_D} q_1^2 dq_1 \int_0^{q_D} q_2^2 dq_2 \Psi[\omega_{\lambda_1}(q_1); \omega_{\lambda_2}(q_2)] \times \delta[\omega + \omega_{\lambda_1}(q_1) - \omega_{\lambda_2}(q_2)] \tag{3.4.33}$$

with

$$\Psi(\omega_{\lambda_1}(q_1); \omega_{\lambda_2}(q_2)) = -\Psi(\omega_{\lambda_2}(q_2); \omega_{\lambda_1}(q_1)) = \frac{I_0}{\hbar \cdot \omega \cdot \omega_{\lambda_1}(q_1) \cdot \omega_{\lambda_2}(q_2)} \times \left\{ \left[\exp\left(\frac{\hbar\omega_{\lambda_1}(q_1)}{k_B T}\right) - 1\right]^{-1} - \left[\exp\left(\frac{\hbar\omega_{\lambda_2}(q_2)}{k_B T}\right) - 1\right]^{-1} \right\} \tag{3.4.34}$$

For thermal phonons, asymptotically for $q \to 0$, the function $\omega_\lambda(q)$ takes the form

$$\omega_\lambda(q) = v_\lambda q \tag{3.4.35}$$

In the vicinity of a 'roton minimum' the function $\omega_\lambda(q)$ can be approximated by a 'parabolic fit' with three fit-parameters $\Delta_\lambda, q_{0\lambda}$ and m_λ^*:

$$\omega_\lambda(q) \approx \omega_\lambda^{(0)}(q) \cong \Delta_\lambda + \frac{\hbar(q - q_{0\lambda})^2}{2m_\lambda^*} \tag{3.4.36}$$

Here $\frac{\hbar^2(q - q_{0\lambda})^2}{2m_\lambda^*}$ is a quantity of second-order smallness compared to $\hbar\Delta_\lambda$, which is of the order of $k_B T$. From here on $\hbar\omega$ is assumed to be a quantity of first-order smallness, compared with $k_B T$, i.e. the inequalities hold:

$$\frac{\hbar^2(q - q_{0\lambda})^2}{2m_\lambda^*} \ll \hbar\omega \ll \hbar\Delta_\lambda \sim k_B T \tag{3.4.37}$$

To evaluate numerically the integrals in Eq. (3.4.33) the following approximation is made. Let the function $\omega_\lambda(q)$ peak at $q = q_{1\lambda}$. Then under the integral over the interval $(0, q_{1\lambda})$ we can put $\omega_\lambda(q) = v_\lambda q$, and under that over $(q_{1\lambda}, q_D)$ the function $\omega_\lambda(q)$ can be replaced with $\omega_\lambda^{(0)}(q)$. Hence the quantity $1/l(\omega)$ can be written in the form:

$$\frac{1}{l(\omega)}=\frac{2v_0^2}{(2\pi^2)^2\cdot v_\lambda}\cdot\sum_{\lambda_1\lambda_2}\left\{\int_0^{q_{1\lambda_1}}q_1^2\mathrm{d}q_1\int_0^{q_{1\lambda_2}}q_2^2\mathrm{d}q_2\cdot\Psi(v_{\lambda_1}q_1;v_{\lambda_2}q_2)\cdot\delta(\omega+v_{\lambda_1}q_1-v_{\lambda_2}q_2)\right.$$
$$+\int_{q_{1\lambda_1}}^{q_D}q_1^2\mathrm{d}q_1\int_0^{q_{1\lambda_2}}q_2^2\mathrm{d}q_2\Psi(\omega_{\lambda_1}^{(0)}(q_1);v_{\lambda_2}q_2)\cdot\delta(\omega+\omega_{\lambda_1}^{(0)}(q_1)-v_{\lambda_2}q_2)$$
$$+\int_0^{q_{1\lambda_1}}q_1^2\mathrm{d}q_1\int_{q_{1\lambda_2}}^{q_D}q_2^2\mathrm{d}q_2\cdot\Psi[v_{\lambda_1}q_1;\omega_{\lambda_2}^{(0)}(q_2)]\cdot\delta[\omega+v_{\lambda_1}q_1-\omega_{\lambda_2}^{(0)}(q_2)]$$
$$\left.+\int_{q_{1\lambda_1}}^{q_D}q_1^2\mathrm{d}q_1\int_{q_{1\lambda_2}}^{q_D}q_2^2\mathrm{d}q_2\cdot\Psi[\omega_{\lambda_1}^{(0)}(q_1);\omega_{\lambda_2}^{(0)}(q_2)]\cdot\delta[\omega+\omega_{\lambda_1}^{(0)}(q_1)-\omega_{\lambda_2}^{(0)}(q_2)]\right\} \tag{3.4.38}$$

The last summand in this expression vanishes because

$$\omega+\omega_{\lambda_1}^{(0)}(q_1)-\omega_{\lambda_2}^{(0)}(q_2)\neq 0 \tag{3.4.39}$$

Indeed, when $\lambda_1=\lambda_2$, the l.h.s. of the inequality takes the form

$$\omega+\frac{\hbar(q_1-q_{0\lambda_1})^2}{2m_{\lambda_1}^*}-\frac{\hbar(q_2-q_{0\lambda_2})^2}{2m_{\lambda_1}^*}$$

and this is non-zero as a consequence of inequality (3.4.37). If $\lambda_1\neq\lambda_2$ the l.h.s. of Eq. (3.4.39) becomes

$$\omega+\frac{\hbar(q_1-q_{0\lambda_1})^2}{2m_{\lambda_1}^*}-\frac{\hbar(q_2-q_{0\lambda_2})^2}{2m_{\lambda_2}^*}+\Delta_{\lambda_1}-\Delta_{\lambda_2}$$

which likewise does not vanish in view of the inequality Eq. (3.4.37).

In the first and second terms of Eq. (3.4.38) we integrate with respect to q_2, in the third one with respect to q_1, and take into account that

$$\hbar\omega\ll\hbar\omega_\lambda(q)\sim k_BT$$

i.e.

$$\pm\Psi(\omega_\lambda;\omega_\lambda\pm\omega)\cdot(\omega_\lambda\pm\omega)^2=\frac{I_0}{\hbar}\frac{\partial n_{q\lambda}^{(0)}}{\partial\omega_\lambda}\left\{1\pm\left[\frac{1}{2}\left(\frac{\partial n_{q\lambda}^{(0)}}{\partial\omega_\lambda}\right)^{-1}\frac{\partial^2 n_{q\lambda}^{(0)}}{\partial\omega_\lambda^2}+\frac{1}{\omega_\lambda}\right]\omega\right.$$
$$\left.+\left[\frac{1}{6}\left(\frac{\partial n_{q\lambda}^{(0)}}{\partial\omega_\lambda}\right)^{-1}\left(\frac{\partial^3 n_{q\lambda}^{(0)}}{\partial\omega_\lambda^3}\right)+\frac{1}{2}\left(\frac{\partial n_{q\lambda}^{(0)}}{\partial\omega_\lambda}\right)^{-1}\left(\frac{\partial^2 n_{q\lambda}^{(0)}}{\partial\omega_\lambda^2}\right)\frac{1}{\omega_\lambda}\right]\omega^2+\ldots\right\}$$

Finally, using the expression for the equilibrium distribution function and Eqs (3.4.35) and (3.4.36), we arrive at the following result for $1/l(\omega)$:

$$\frac{1}{l(\omega)} = \frac{v_0^2}{(2\pi^2)^2 v_\lambda} \sum_{\lambda_1\lambda_2} \int_0^\infty \frac{q^2 dq}{v_{\lambda_2}^3} \frac{I_0}{k_B T} \frac{\exp(-\hbar v_{\lambda_1} q/k_B T)}{\left[1 - \exp\left(\frac{-\hbar v_{\lambda_1} q}{k_B T}\right)\right]^2}$$

$$\times \left\{1 - \left[\frac{1}{2}\frac{1+\exp(-\hbar v_{\lambda_1} q/k_B T)}{1-\exp(-\hbar v_{\lambda_1} q/k_B T)} - \frac{k_B T}{\hbar v_{\lambda_1} q_1}\right]\frac{\hbar\omega}{k_B T} + O\left(\frac{\hbar^2\omega^2}{k_B^2 T^2}\right)\right\}$$

$$+ \frac{2v_0^2}{(2\pi^2)^2 v_\lambda} \sum_{\lambda_1\lambda_2} \int_{-\infty}^{\infty} \frac{q^2 dq}{v_{\lambda_2}^3} \frac{I_0}{k_B T} \left\{1 + O\left(\frac{\hbar^2\omega^2}{k_B^2 T^2}\right)\right\}$$

$$\times \frac{\exp\left[-\left(\hbar\Delta_{\lambda_1} + \frac{\hbar^2 (q-q_{0\lambda_1})^2}{2m_{\lambda_1}^*}\right)^2 \Big/ k_B T\right]}{\left\{1 - \exp\left[-\left(\hbar\Delta_{\lambda_1} + \frac{\hbar^2 (q-q_{0\lambda_1})^2}{2m_{\lambda_1}^*}\right)^2 \Big/ k_B T\right]\right\}^2}$$

In the first integral we change the variable

$$x = \frac{\hbar v_{\lambda_1} q}{k_B T}$$

and in the second one we make the substitution

$$x^2 = \frac{\hbar^2 (q-q_{0\lambda_1})^2}{2m_{\lambda_1}^* \cdot k_B T}$$

then the result can be integrated to give the expression

$$\frac{1}{l(\omega)} = \sum_{\lambda_1} A_{\lambda_1} \left(\frac{k_B T}{\hbar\Delta_{\lambda_1}}\right)^2 \times \left\{1 + B_{\lambda_1} \cdot \exp\left(-\frac{\hbar\Delta_{\lambda_1}}{k_B T}\right)\right.$$

$$\left. \times \sum_{n=0}^{\infty} \left[\sqrt{n+1}\left(\frac{\hbar\Delta_{\lambda_1}}{k_B T}\right)^{5/2} + \frac{D_{\lambda_1}}{\sqrt{n+1}}\left(\frac{\hbar\Delta_{\lambda_1}}{k_B T}\right)^{3/2}\right] \cdot e^{\frac{-n\hbar\Delta_{\lambda_1}}{k_B T}}\right.$$

where

$$\begin{cases} A_{\lambda_1} = \sum_{\lambda_2} \dfrac{v_0^2}{(2\pi^2)^2 v_\lambda} I_0 \dfrac{(\hbar\Delta_{\lambda_1})^2}{\hbar^3 v_{\lambda_1}^3 v_{\lambda_2}^3} \\ B_{\lambda_1} = \dfrac{2}{\pi^{3/2}} \left(\dfrac{\hbar v_{\lambda_1} q_{0\lambda_1}}{\hbar\Delta_{\lambda_1}}\right)^3 \sqrt{\dfrac{2m_{\lambda_1}^* \cdot \hbar\Delta_{\lambda_1}}{\hbar^2 q_{0\lambda_1}^2}} \\ D_{\lambda_1} = \dfrac{m_{\lambda_1}^* \cdot \hbar\Delta_{\lambda_1}}{\hbar^2 q_{0\lambda_1}^2} \end{cases}$$

Because $\dfrac{\hbar\Delta_t}{k_B T} > \dfrac{\hbar\Delta_l}{k_B T} \gtrsim 1$, the major contribution to the sound absorption comes from the interaction of acoustic phonons with longitudinal thermal ones, i.e.

$$\frac{1}{l(\omega)} \approx A_l \left(\frac{k_B T}{\hbar\Delta_l}\right)^2 \left\{1 + f_l\left(\frac{\hbar\Delta_l}{k_B T}\right) - \frac{3}{2\pi^2}\frac{\hbar\omega}{k_B T} + O\left(\frac{\hbar^2\omega^2}{k_B^2 T^2}\right)\right\} \tag{3.4.41}$$

where

$$f_l(\varepsilon) = B_l \cdot e^{-\varepsilon} \sum_{n=0}^{\infty} \left[\sqrt{n+1}\,\varepsilon^{5/2} + \frac{D_l}{\sqrt{n+1}}\varepsilon^{3/2}\right] \cdot e^{-n\varepsilon} \approx B_l \cdot e^{-\varepsilon}\left[\varepsilon^{5/2} + D_l \varepsilon^{3/2}\right]$$

It is evident from this expression that the phonon contribution to the sound absorption goes down as the frequency increases. Moreover, the frequency dependence tends to diminish with increasing temperature.

The temperature dependence of the absorption coefficient a is determined by the behavior of $f(\varepsilon)$. This function approaches zero for $\varepsilon \to 0$ and $\varepsilon \to \infty$, and it peaks at $\varepsilon = \dfrac{1}{2}\left(\dfrac{5}{2} - D_l\right) + \sqrt{\dfrac{D_l^2}{4} + \dfrac{1}{4}D_l + \dfrac{25}{16}}$. This means that neglecting the frequency dependence at very low ($T < 2$ K) and at sufficiently high ($T \gtrsim T_D$) temperatures we get $a \sim T^2$.

At moderately low temperatures ($k_B T \sim \hbar\Delta_l$) the sound absorption a, growing initially faster than T^2 with elevation of temperature, subsequently demonstrates a significant *slowing down* of growth. This is illustrated by Fig. 3.4.3 representing $\dfrac{a}{A_l} = \dfrac{1}{2A_l\, l(\omega)}$ as a function of $\dfrac{k_B T}{\hbar\Delta_l}$ when $B_l \approx 40$ and $D_l \approx \frac{1}{2}$, which corresponds to amorphous Mg and Zn.

It should be noted that upon certain relationships between B_l and D_l the function $a(T)$ could have a maximum in the moderately low temperature range. The obtained frequency and temperature dependencies of the sound absorption coefficient are consistent with experiment (see Section 3.2).

The foregoing proves that at moderately low temperatures the function $1/l$ (or a) resembles in its behavior the function $C_V(T)$ (see Eqs (3.3.41) and (3.4.41)). Because of this, the temperature dependence of the thermal conductivity becomes comprehensible: Using the simplest formula of the kinetic gas theory,

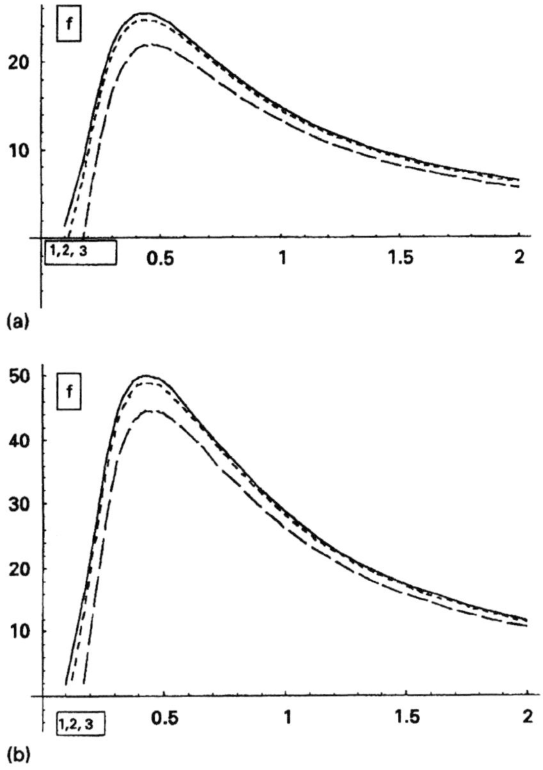

Fig. 3.4.3 (a) Temperature dependence of $f(\varepsilon) = (\hbar\Delta_l/k_B T)^2 l/(\Delta_l l(\omega))$ for amorphous Zn: (1) ν=0, (2) $\nu=10^{1+3}$ Hz, (3) $\nu=5\times10^{13}$Hz; $\varepsilon = k_B T/\hbar\Delta_l$; $\nu = \omega/2\pi$. (b) The same for amorphous Mg (from Krasny *et al.*, 1998)

$$\kappa = \frac{1}{3} C_V \cdot v_\lambda \cdot l$$

we can demonstrate that *outside* the region of moderately low temperatures $C_V \sim T^3, l \sim T^{-2}$, and thus $\kappa \sim T$. In contrast, *within* this temperature region we get $\kappa \sim$ const., because here the functions C_V and $1/l$ are similar in behavior.

Hence, if at very low temperatures the 'anomalous' thermal physics of glasses is related to the 'tunneling states' and to interaction between quasi-phonons and these elementary excitations, at moderately low temperatures the thermal properties arise from the unusual dispersion law of quasi-phonon excitations in the 'roton range', due to interaction between quasi-phonons with different dispersion laws.

3.5
Beyond the Quasi-phonon Approximation

The preceding sections of this chapter were strongly based on the quasi-phonon approximation, in particular

(i) *at long wavelengths* (long compared with the typical nearest-neighbor distance in the metallic glass) on the *identity* of quasi-phonons and long-lived conventional ultra-sound phonons, which could be classified by their polarization λ into *longitudinal* rsp. *transverse* sound waves with well-defined sound velocities v_λ, and

(ii) *at short wavelengths* (i.e. comparable with typical nearest-neighbor distances) on the existence of well-defined quasi-phonons with a pronounced minimum, i.e. the so-called 'roton minimum', in the dispersion curve, for which the 'parabolic fit' in Eq. (3.4.36), i.e. $\omega_\lambda(q) = \Delta_\lambda + \hbar(q - q_{0\lambda})^2/(2m^*)$, is sensible, in particular for the longitudinal polarisation branch. Also in this range of q-vectors it is assumed that the excitations are sufficiently well defined, such that effects of finite lifetime can be neglected.

In the following, these assumptions are critically assessed and essentially confirmed. Such assessment can only be done by *numerical studies*, whereas, as we have seen, it is one of the main virtues of the quasi-phonon approximation that it leads to equations for the thermodynamics of amorphous systems, which are analytically solvable.

Let us describe in the usual adiabatic approximation ('Born-Oppenheimer approximation') the motion of the N *ions* in the pseudopotential $\Phi(\mathbf{R}_1, \mathbf{R}_2, \ldots, \mathbf{R}_N)$ generated by the electrons. We expand this potential around one of the metastable quasi-equilibrium configurations $(\mathbf{R}_1^{(0)}, \mathbf{R}_2^{(0)}, \ldots, \mathbf{R}_N^{(0)})$ of the metallic glass. To describe the positions of all the ions, we have $3N$ cartesian coordinates, and for simplity we enumerate all of them as X_a, with $a = 1, \ldots, 3N$. Then we get in the harmonic order, i.e. by neglecting products of at least three terms in the deviations $u_i := X_i - X_i^{(0)}$, the following Taylor expansion for Φ:

$$\Phi = \Phi^{(0)} + 1/2 \sum_{a,\beta=1}^{3N} u_a u_\beta V_{a\beta} + \ldots \tag{3.5.1}$$

where the dots denote the neglected terms and the non-negative $3N \times 3N$ matrix $V_{a,\beta}$ is given by the derivatives of Φ, $V_{a\beta} = \partial^2\Phi/(\partial X_a \partial X_\beta)$. The classical Hamiltonian H describing the harmonic vibrations of the ions around their quasi-equilibrium positions is thus simply given by

$$\mathrm{H} = \sum_{a=1}^{3N} p_a'^2/2 + 1/2 \sum_{a,\beta=1}^{3N} u_a' u_\beta' V_{a\beta}' \tag{3.5.2}$$

Here we have formally 'gauged away' the masses M_a of the ions by the substitutions $p_a' := p_a M_a^{-1/2}, u_a' := u_a M_a^{1/2}$ and $V_{a\beta}' := V_{a\beta}/(M_a M_\beta)^{1/2}$, which conserve the quantum mechanical commutation rules $[\hat{p}_a, \hat{u}_\beta] = (\hbar/i)\delta_{a\beta}$ for the corresponding operators. Thus the squares of the *eigenfrequencies* $\omega_\nu^2 (\nu = 1, \ldots, 3N)$ are identical to the eigenvalues of the matrix $V_{a\beta}'$, and the corresponding classical eigenvectors are the eigenvectors of this matrix. In an amorphous system, they are *not* plane wave states, i.e. $\mathbf{k}$ is *not* 'a good quantum number'. Numerically we can determine the relevant quantities, i.e. the density $g(\omega_\nu)$ of eigenfrequencies, the corresponding *projected densities of states* $g_{\mathbf{q}}^{(\lambda)}(\omega_\nu)$ for longitudinal rsp. transversal plane waves (see below), and the thermodynamic quantities, for example the *internal vibrational energy*

$$U(T) = \int_0^\infty d\omega\, g(\omega)\hbar\omega/\{\exp[\hbar\omega)/(k_B T)] - 1\} \tag{3.5.3}$$

as a function of the Kelvin temperature T, from a Green's function formalism. This formalism, which is often also called '*resolvent formalism*', proceeds as follows. With a complex number z we define at first the *resolvent matrix* (which is also called formally the *Green's operator*, although here we are actually dealing with classical physics):

$$R(z) = [z - V']^{-1} \tag{3.5.4}$$

The matrix inverse exists, if z does not belong to the spectrum of the matrix V′, which is definitely true if z is written as $z = \omega^2 + i\varepsilon$, with $\varepsilon > 0$ or < 0. The matrix V′ is real-symmetric, actually even positive-semidefinite, because only a homogeneous translation or homogeneous rotation of the whole system leads to vanishing eigenfrequencies; therefore V′ can be diagonalized by a suitable rotation in the $3N$-dimensional space, and in an abstract formalism we can write

$$R(z) = \sum_{\nu=1}^{3N} \frac{|\nu><\nu|}{z - \omega_\nu^2} \tag{3.5.5}$$

where $P_\nu := |\nu><\nu|$ is the projection operator onto the vibrational eigenvector corresponding to the harmonic vibrational mode with index ν. Additionally, we can also form the matrix elements $<\mathbf{q},\lambda|R(z)|\mathbf{q},\lambda>$ with 'plane wave states $|\mathbf{q},\lambda>$ of polarization λ, which are not eigenstates of V′, as already mentioned: if the propagation of the wave is in the Z-direction, we write for *longitudinal* polarization of the plane wave simply $|\mathbf{q},\lambda> := [0, 0, \exp(iqZ_1)/N^{1/2}; 0, 0, \exp(iqZ_2)/N^{1/2}; \ldots; 0, 0, \exp(iqZ_N]]$; whereas for *transverse* polarization we can for example set $|\mathbf{q},\lambda> := [0, \exp(iqZ_1)/N^{1/2}, 0; 0, \exp(iqZ_2)/N^{1/2}, 0; \ldots; 0, \exp(iqZ_N), 0]$ where $Z_1, Z_2, \ldots$ are the z-coordinates of ion number 1, 2,… in the equilibrium position $\mathbf{R}_1^{(0)}, \mathbf{R}_2^{(0)}, \ldots$, respectively.

The scalar product $\langle\mathbf{q},\lambda|R(z)|\mathbf{q},\lambda\rangle$ is a complex function of z. We are particularly interested in the imaginary part of this function, from which the above-mentioned functions $g_{\mathbf{q}}^{(\lambda)}(\omega_\nu)$ can be calculated.

The following equations are obtained with the identity $1/(x - i\varepsilon) = x/(x^2 + \varepsilon^2) + i\varepsilon/(x^2 + \varepsilon^2) \to P(1/x) + \pi i\delta(x)$, where $P(1/x)$ means the so-called 'Cauchy part', or 'Principal part', which is real and defined by the following identity, valid for all smooth 'test functions' $f(x)$ with compact support, and for $\varepsilon \to 0^+$:

$$\int P(1/x)f(x)\mathrm{d}x := \left(\int_{-\infty}^{-\varepsilon} \mathrm{d}x + \int_{+\varepsilon}^{+\infty} \mathrm{d}x \right) f(x)/x\,,$$

whereas the second part $\pi i\delta(x)$ is imaginary and essential for us. It is based on the fact that the function $\pi^{-1}\varepsilon/(\varepsilon^2 + x^2)$ behaves for $\varepsilon \to 0$ as a δ-function, i.e. with a pronounced peak of height $\propto \varepsilon^{-1}$ and width $\propto \varepsilon$ and integral 1. In fact, with this identity we get for $z = \omega^2 - i\varepsilon$ in the limit $\varepsilon \to 0$:

$$\pi^{-1}\mathrm{Im}(\langle \mathbf{q}, \lambda | \mathrm{R}(\mathrm{z}) | \mathbf{q}, \lambda \rangle) = \sum_{\nu=1}^{3N} |\langle \mathbf{q}, \lambda | \nu \rangle|^2 \delta(\omega^2 - \omega_\nu^2) = g_{\mathbf{q}}^{(\lambda)}(\omega^2) \tag{3.5.6}$$

Now the matrix elements $\langle \mathbf{q}, \lambda | \mathrm{R}(\mathrm{z}) | \mathbf{q}, \lambda \rangle$ can be evaluated systematically either (i) by a continued-fraction expansion (Krey *et al.*, 1987), which is quite generally applicable, or (ii) by numerical integration with respect to the time variable of the differential equation for the corresponding Green's function and subsequent Fourier transformation (von Heimendahl, 1979). We do not go into the details here as they can be found in these references, but simply present the results in Fig. 3.5.1.

In Fig. 3.5.1, the functions $S_\lambda(\mathbf{k}, \omega) = g_{\mathbf{k}}^{(\lambda)}(\omega^2)2\omega$, i.e. $S_\lambda(k, \omega)\mathrm{d}\omega = g_k^{(\lambda)}(\omega^2)\,\mathrm{d}\omega^2$, are plotted for longitudinal ($\lambda = 3$) resp. transversal polarization ($\lambda = 1$ or 2 in the above-mentioned example) in part (a) resp. (b).

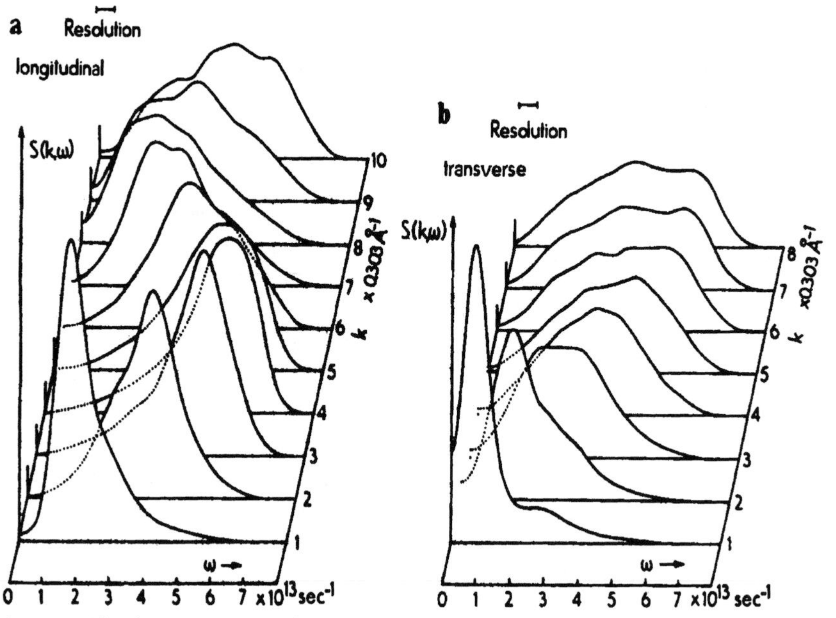

Fig. 3.5.1 The dynamic structure functions $S(\mathbf{k},\omega)$ for the amorphous alloy $Mg_{0.7}Zn_{0.3}$ are presented as a function of the wavenumber $\mathbf{k}$ and the frequency ω for longitudinal (a) and transversal (b) phonon excitations (after von Heimendahl, 1979)

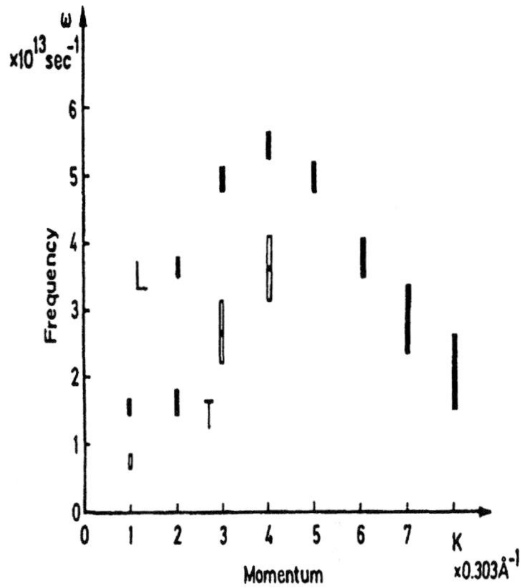

Fig. 3.5.2 The dispersion curves obtained by joining the maximum frequencies ω_{max} in Fig. 3.5.1 as a function of the wavenumber **k** are plotted for the longitudinal (full symbols) rsp. transversal (open symbols) vibrational excitations of the amorphous alloy $Mg_{0.7}Zn_{0.3}$. The width of the bars denotes the linewidth corresponding to half of the maximum.

These so-called *dynamical structure functions* $S_\lambda(\mathbf{k},\omega)$ are directly proportional to the *inelastic double-differential cross-section* $\frac{d^2\sigma}{d\omega\, d\Omega}$ measured in inelastic neutron scattering events for non-magnetic solids, with a momentum transfer of $\hbar\mathbf{k}$ and an energy transfer $\hbar\omega$ from the neutron to the solid. As we can see from the figures, these dynamic structure functions are rather broad, i.e. they are *not* identical to delta functions $\delta(\omega - \omega_\lambda(\mathbf{k}))$. However if we simply join the maxima $\omega_{max}(k)$ of $S_\lambda(\mathbf{k},\omega)$ as a function of **k**, we get effective dispersion curves that are quite similar to those used above for the quasi-phonons.

These effective dispersion curves, which are derived from Fig. 3.5.1, are plotted together with the 'width' of the excitations in Fig. 3.5.2. From this figure we can not only calculate the sound velocities v_λ, which are given by the slopes of the dispersion curves for $\mathbf{k} \to 0$, but we also find the pronounced 'roton minimum' in the dispersion curve for the longitudinal phonons, whereas for the transversal phonons in the 'roton region' the dispersion curve is essentially flat and very broad, as can be seen from Fig. 3.5.1. The quality of the quasi-phonon approach, which was extensively used in the preceding subsections, arises from the facts that (i) in the dispersion curves the 'width' of the curves in Fig. 3.5.2 can be neglected to a good approximation (for the transverse modes, which are non-essential in any case, this statement applies at least for the ascending part of the dispersion relation, i.e. up to $k \approx 1\,\text{Å}^{-1}$). Furthermore, to a good approximation, the total density of eigenfrequencies, namely $g(\omega) = 4\pi \int k^2 dk \Omega_i^* S_\lambda(k,\omega)$, is quite similar to the result that would be obtained from the quasi-phonon approach.

From these results, the quasi-phonon approach appears to be well justified for the calculation of the relevant thermodynamic quantities of amorphous metals.

4
Magnetic Properties of Amorphous Metals

In this chapter, the magnetic properties of metallic glasses are discussed within the framework of *phenomenological models*, e.g. the Heisenberg model and molecular field models, without referring explicitly to the electronic structure of the system. In contrast, in Chapter 5, the *electronic structure* of amorphous metals will be treated, and the magnetism of metallic glasses will be further discussed in that context, i.e. the particular aspects of the so-called *itinerant magnetism* of glassy metals will be considered. Here we start with a short overview of some experimental results.

4.1
Review of Experimental Results

One of the main problems in the theory of amorphous magnets is the question concerning the influence of amorphization on the formation of magnetic order. From the theoretical point of view the possible existence of amorphous ferromagnets was considered first by Gubanov (1960). It was shown that the existence of a crystal structure is not a necessary condition for a ferromagnetic state. A positive exchange interaction, which is mainly short-ranged, is responsible for the ferromagnetic ordering. This is why the *magnitude* and (positive) *sign* of the exchange interaction in the range of typical interatom nearest-neighbor distances is very important.

Soon after the work by Gubanov, some amorphous ferromagnets were obtained by Klein and Brout (1963). The following experimental investigations resulted in the discovery of more amorphous materials with a ferromagnetic phase. It is worth noting that ferromagnetism was discovered both in one-component amorphous metallic systems (Fe, Co or Ni), prepared by Leung and Wright (1974) at He temperatures, and in amorphous alloys, e.g. $Co_{1-x}P_x$, $Fe_{1-x}B_x$, $Pd_{1-x}Pd_x$ systems, which exist in the glassy state at room temperature. (At room temperature, amorphous Fe exists only as thin films of thickness <2.4 nm on amorphous Gd or Y substrates (Handschuh *et al.*, 1993), whereas for greater thicknesses the Fe film transforms to the usual crystalline α-Fe; whether for the thin amorphous Fe films the magnetism is of the ferromagnetic kind or whether we are dealing with a kind of Fe spin glass is unclear at present. However the magnetic moments of these amorphous Fe films are strongly reduced to 1.2 μ_B, whereas, in the conventional bulk α-Fe, the moment per atom is 2.2 μ_B. More details will be found in the second half of Chapter 5.

So it is definite that structural order is not a necessary factor for magnetic ordering to exist. At the same time, however, *structural disorder* in a system has an essential influence on the thermal properties of amorphous magnets. We will now consider such an influence in more detail.

Curie Temperature As numerous experiments show (see, e.g., Luborsky, 1983; Handrich and Kobe, 1980; Hooper and de Graaf, 1973, and the references therein), the Curie temperature of amorphous magnets is usually lower than that of their crystal analogues. Amorphous ferromagnetic metal-metalloid alloys have Curie temperatures that decrease with increasing metalloid concentration. However for some alloys (FeB, CoB, etc.), this concentration dependence is non-monotonic: first T_C increases with metalloid concentration, then it reaches a maximum, and at last begins to diminish sharply (Fig. 4.1.1, Luborsky, 1983).

Temperature Dependence of the Magnetization (see, e.g., Luborsky, 1983; Handrich and Kobe, 1980; Hooper and de Graaf, 1973; Hasegawa and Levy, 1977; Gilman and Leamy, 1978; and references therein). The temperature dependence of the dimensionless magnetization $M(T)/M(0)$ as a function of T/T_C has for amorphous ferromagnets qualitatively the same form as for crystals, i.e. for low temperatures *Bloch's law* applies in both cases, which means that $M(T)/M(0) \approx 1 - \text{const.}\,(T/T_C)^{3/2}$, whereas at high temperatures – near the Curie temperature – a *molecular field approximation* works quite well (see below). However, for amorphous magnets, the corresponding curves are significantly flatter than for crystals (Fig. 4.1.2) (see Gilman and Leamy, 1978).

In the molecular-field approximation (also called 'mean-field approximation') for amorphous solids the magnetization near T_C again has the usual form:

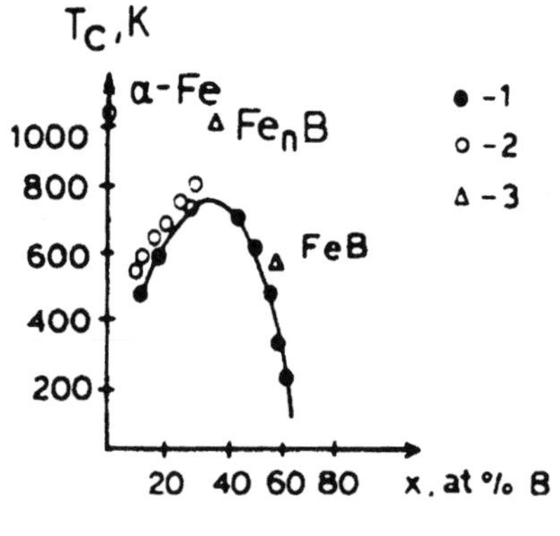

Fig. 4.1.1 Experimental results for the Curie temperature (T_C) of amorphous Fe-B alloys. 1: Sputtered films; 2: ribbons prepared by melt spinning. For comparison, the results for crystalline α-Fe and crystalline Fe-B and Fe_nB with $n=2$ are also shown (after Luborsky, 1983, p. 263)

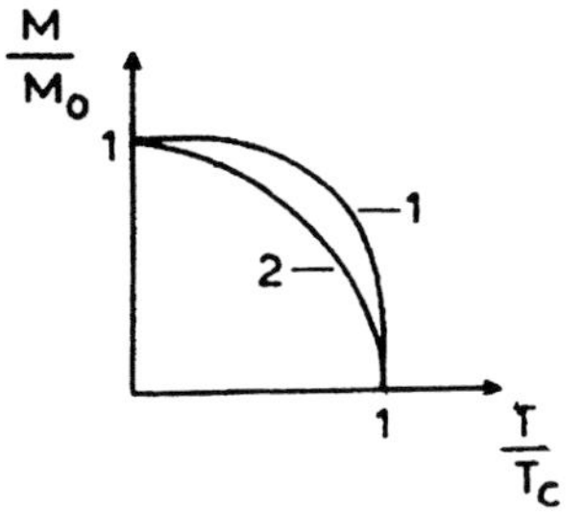

Fig. 4.1.2 Comparison of typical behavior of the magnetization as a function of temperature for crystalline (1) versus amorphous (2) systems

$$\frac{M(T)}{M(0)} = \sqrt{a\left(1 - \frac{T}{T_C}\right)} \tag{4.1.1}$$

however for amorphous solids the coefficient a is significantly smaller than for crystals.

As was noted above, the low-temperature behavior of the magnetization is governed by Bloch's law:

$$\frac{M(T)}{M(0)} = 1 - B \cdot T^{3/2} + \dots \tag{4.1.2}$$

It follows from Fig. 4.1.1 that the coefficient B for amorphous solids is significantly larger than for crystals. Both for crystals and for metallic glasses, the temperature dependence (Eq. (4.1.2)) is a consequence of the existence of ferromagnetic spin-wave excitations with *quadratic dispersion law* in the long-wave domain, i.e. if the product kr_0 is $\ll 1$, where $k = 2\pi/\lambda$ is the wavenumber of the spin wave and r_0 the typical nearest distance between magnetic atoms, then the eigenmodes of the spin waves are plane waves with excitation energy

$$\varepsilon(k) = Dk^2 + \dots \tag{4.1.3}$$

Here D is the so-called *spin-wave stiffness*, and the coefficient B in Eq. (4.1.2) is related to D theoretically, i.e. if the low-energy spin excitations are 'exhausted' by spin waves, then $B = [V/(Ns)]0.0587(k_B/D)^{3/2}$, where (V/N) is the specific volume of the sample and s is the spin quantum number.

Scattering Experiments (Handrich and Kobe, 1980; Hooper and de Graaf, 1973; Hasegawa and Levy, 1977; Gilman and Leamy, 1978; Güntherodt and Beck, 1981; Mook and Tsuei, 1977). Investigations of the long-wave limit of the spin-wave dispersion law in amorphous magnets, with the help of ferromagnetic resonance methods, Brillouin light scattering, and inelastic neutron scattering, confirm the quadratic dependence (Eq. (4.1.3)) of the energy of the magnetic excitation on the wavenumber. However, these 'direct scattering experiments' give a spin-wave stiffness D nearly twice that determined indirectly from the low-temperature magnetization, i.e. in amorphous metals apparently the plane spin waves do not at all 'ex-

Fig. 4.1.3 Dispersion relation of the magnetic excitations for the ferromagnetic amorphous glassy metal Co_4P. Here, apart from the spin-wave branch at small values of q (the dashed line), the pronounced minimum at wavenumbers around $q \approx 3\,\text{Å}^{-1}$ should be noted (after Mook *et al.* 1975)

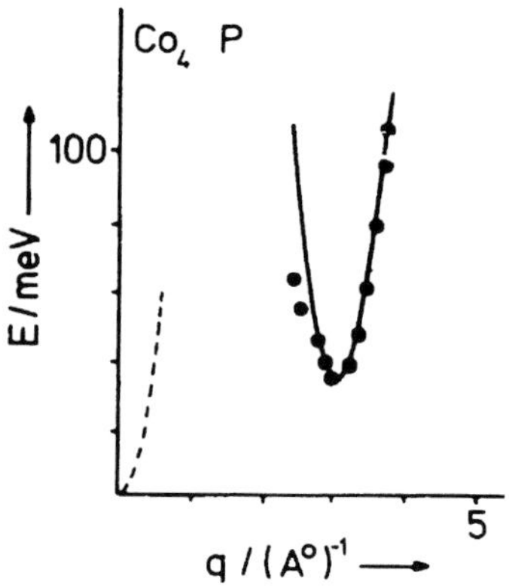

haust' the low-energy magnetic excitations, and, as a rule, D is significantly less than the spin-wave stiffness of crystalline magnets.

Finally, most mysterious at first glance is the 'roton-like' minimum on the dispersion curve for spin waves, which is experimentally found in the short-wave region (Mook *et al.*, 1975; Mook and Tsuei, 1977) (see Fig. 4.1.3).

This is reminiscent of similar 'roton-like minima' in the dispersion curve for the longitudinal quasi-phonon mode treated in Chapter 3. Actually the 'magnetic roton mode' has probably roughly the same explanation as the corresponding longitudinal phonon phenomenon. Namely, in a crystal, if we plotted the dispersion curves $\omega(q)$ *not* in the usual *reduced-zone scheme*, where the dispersion curve ends with the maximum at the boundary of the Brillouin zone, i.e. at $q=\pi/a$) in the edge direction, then we would find that in the *extended-zone scheme* the dispersion curve would come down again with increasing q_x, until at $q_x=2\pi/a$ it would again reach the minimal excitation energy zero. In other propagation directions this would happen at somewhat different values of q; therefore in an *amorphous* system we should observe a 'smeared' pronounced minimum in the dispersion curve at a value of $q_0 \approx 2\pi/r_0$, where r_0 is the average nearest-neighbor distance. The value of q_0 observed in Fig. 4.1.3, namely $q_0 \approx 3.1\ \text{Å}^{-1}$, nicely fits this explanation, and a 'parabolic expansion' around the 'roton minimum', analogously to Eq. (4.4.36), also works here. Mook and Tsuei (1977) obtained similar results also for the amorphous ferromagnet $Fe_{75}P_{15}C_{10}$, and they also found that the 'magneto-roton excitations' have a finite 'damping width' as was also obtained for the 'vibrational rotons' by L. von Heimendahl in 1977 (see e.g. Fig. 3.4.1).

In the following sections of this chapter we give an explanation of the peculiarities of the thermodynamic properties of amorphous magnets. To explain the behavior of magnets near the critical point, the diagrammatic method of Vaks, Larkin and Pikin for crystals is used (Vaks *et al.*, 1967; Barjakhtar *et al.*, 1984), and the low-temperature properties are considered with the help of the so-called Holstein-Primakoff transformation (see below) (Barjakhtar *et al.*, 1984; Akhieser *et al.*, 1967).

4.2 Thermodynamic Properties of Amorphous Ferromagnets Near the Curie Point

Let us consider a system of N magnetic atoms with spin quantum number S. The atoms are fixed at randomly distributed points $\mathbf{R}_1, \ldots, \mathbf{R}_N$ of the volume V. The magnetic interaction in the system is described by Heisenberg's exchange function; the full Hamiltonian is (Barjakhtar *et al.*, 1984; Akhieser *et al.*, 1967; Davydov, 1967):

$$\hat{\mathrm{H}} = \hat{\mathrm{H}}_0 + \hat{\mathrm{H}}_i \tag{4.2.1}$$

$$\hat{\mathrm{H}}_0 = -g\mu_0 \mathrm{H} \sum_{l=1}^{N} \hat{S}_l^z = -\gamma_0 \sum_{l=1}^{N} \hat{S}_l^z \tag{4.2.2}$$

$$\hat{\mathrm{H}}_i = -\frac{1}{2}\sum_{l \neq l'} J_{ll'} \vec{\hat{S}}_l \cdot \vec{\hat{S}}_{l'} = -\frac{1}{2}\sum_{l \neq l'} J_{ll'} [\hat{S}_l^+ \hat{S}_{l'}^- + \hat{S}_l^z \hat{S}_{l'}^z] \tag{4.2.3}$$

where $J_{ll'} \equiv J(\mathbf{R}_l - \mathbf{R}_{l'})$ is the exchange integral between the atoms l and l'; $\vec{\hat{S}}_l = \vec{\hat{S}}(\mathbf{R}_l) = (\hat{S}_l^x; \hat{S}_l^y; \hat{S}_l^z)$ are the spin operators, if the atom l is localized at $\mathbf{R}_l$; $\hat{S}_l^{\pm} = \hat{S}_l^x \pm i\hat{S}_l^y; \gamma_0 = g\mu_0 \mathrm{H}; \mu = \dfrac{e\hbar}{2mc}$ (the Bohr magneton); g is the Landé factor; $\mathbf{H} = (0,0,\mathrm{H})$ is the external magnetic field directed along the z-axis. The spin operators $\hat{S}_l^a$ are governed by the ordinary commutation rules:

$$[\hat{S}_l^+, \hat{S}_{l'}^-] = 2\hat{S}_{l'}^z \delta_{ll'}, \; [\hat{S}_l^z, \hat{S}_{l'}^{\pm}] = \pm \hat{S}_l^{\pm} \delta_{ll'} \tag{4.2.4}$$

Moreover,

$$\vec{\hat{S}}_l^2 = (\hat{S}_l^x)^2 + (\hat{S}_l^y)^2 + (\hat{S}_l^z)^2 = S(S+1) \tag{4.2.5}$$

where S is the spin quantum number of the atom.

Because amorphous magnets are characterized by the absence of structural long-range order, the physical quantities measured in the experiment (such as magnetic moment, free energy, etc.), in addition to Gibbs averaging, $\langle \ldots \rangle$, i.e. with respect to the degrees of freedom of the quickly relaxing spin subsystem, usually also have to be averaged over all possible realizations of random atomic configurations. For instance, the average spin expectation value per atom is determined by the expression

$$S_{\mathrm{av}} = \overline{\frac{1}{N}\sum_{l=1}^{N} \langle \hat{S}_l^z \rangle} \tag{4.2.6}$$

As in the preceding chapters, the line above this formula denotes *configurational averaging* with a normalized function of the distribution probability of random variables, in this case the coordinates of atoms. The choice of this function is based on the idea of the structure of the non-ordered system: a very realistic model describing the topological disorder of a glassy metal is that of a *frozen liquid*. Therefore we will use it further and, in accordance with the liquid theory, we will describe the structure of our amorphous system by means of correlation functions between simultaneous relative positions of several atoms at given points in space. Such functions are defined for N and $V \to \infty/(v_0 = \text{const.})$ by the following expressions, which have already been used in previous chapters:

$$F_s S(\mathbf{r}_1, \ldots, \mathbf{r}_s) = v_0^s \overline{\sum_{l_1} \ldots \sum_{l_s} \delta(\mathbf{r}_1 - \mathbf{R}_{l_s}) \ldots \delta(\mathbf{r}_s - \mathbf{R}_{l_s})}_{(l_1 \neq l_2 \pm \cdots \neq l_s)} \tag{4.2.7}$$

There is no microscopic theory at present that allows us to calculate the functions $F_s(\mathbf{r}_1, \ldots, \mathbf{r}_N)$. Therefore, we will again consider these functions as phenomenological entities, or alternatively we apply models known from the liquid theory, or we employ hard-sphere or soft-sphere models for amorphous systems, which are often used in numerical calculations for $F_2(\mathbf{r}_1, \mathbf{r}_2) = g(\mathbf{r}_1 - \mathbf{r}_2)$.

As usual, Gibbs averaging on the spin variables in Eq. (4.2.6) is denoted by angular brackets:

$$\langle \hat{S}_l^z \rangle = \frac{1}{z} \mathrm{Tr}\{\hat{S}_l^z \exp(-\beta \hat{H})\} \tag{4.2.8}$$

where

$$z = \mathrm{Tr}\{\exp(-\beta \hat{H})\}, \quad \beta = \frac{1}{k_B T}$$

Because the eigenvalues of the operator $\hat{H}_0$ do not depend on the location of the atom, we can use the thermodynamic perturbation theory. To calculate $\langle \hat{S}_l^z \rangle$ we consider $\hat{H}_i$ to be a low value and introduce a temperature scattering matrix, $\hat{\sigma}(\beta)$ (see Akhieser *et al.*, 1967; Abrikosov *et al.*, 1965; and Appendix A):

$$\exp(-\beta \hat{H}) = e^{-\beta \hat{H}_0} \hat{\sigma}(\beta) \tag{4.2.9}$$

where

$$\begin{aligned} \hat{\sigma}(\beta) &= \hat{T}_\tau \exp\left\{ -\int_0^\beta \hat{H}_i(\tau) d\tau \right\} \\ &= \sum_{n=0}^{\infty} \frac{(-1)^n}{n!} \int_0^\beta d\tau_1 \ldots \int_0^\beta d\tau_n \hat{T}_\tau \{\hat{H}_i(\tau_1) \ldots \hat{H}_i(\tau_n)\} \end{aligned} \tag{4.2.10}$$

$$\hat{H}_i(\tau) = e^{\hat{H}_0\tau}\hat{H}_i e^{-\hat{H}_0\tau}$$

and $\hat{T}$ is the time ordering operator. Then Eq. (4.2.8) can be represented as

$$\langle \hat{S}_l^z \rangle = \frac{1}{\langle \hat{\sigma}(\beta) \rangle_0} \langle \hat{S}_l^z \hat{\sigma}(\beta) \rangle_0 = \langle \hat{S}_l^z \hat{\sigma}(\beta) \rangle_{0_c} \, . \tag{4.2.11}$$

where

$$\langle \ldots \rangle_0 = \frac{1}{z_0} \mathrm{Tr}\{\ldots e^{-\beta\hat{H}_0}\} \tag{4.2.12}$$

$$z_0 = \mathrm{Tr}\{\exp(-\beta\hat{H}_0)\}$$

The symbol $\langle \ldots \rangle_{0_c}$ means that only *connected diagrams* should be considered in the series representing $\langle \ldots \rangle_0$. Average values in Eq. (4.2.11) are calculated from Wick's theorem as formulated for spin variables by Vaks *et al.* (1967). For this, let us use the diagram technique proposed by these authors (Vaks *et al.*, 1967; Barjakhtar *et al.*, 1984): Specifically, the operators $\hat{S}^+$ in Eq. (4.2.11) correspond to full points, $\hat{S}^-$ to points with attached rays, i.e. attached solid lines with arrows, and $\hat{S}^z$ to open circles. These points and circles are connected with wavy lines, solid lines with arrows, or dashed lines, respectively (see the example (4.2.15) below).

Each solid line corresponds to a spin Green's function:

$$K_{ll'}^0(\tau - \tau') = \delta_{ll'} e^{-y_0(\tau-\tau')} \times \begin{cases} n_{y_0}, & \tau' > \tau \\ 1 + n_{y_0}, & \tau' < \tau \end{cases} \tag{4.2.13}$$

$$n_{y_0} = [\exp(\beta y_0) - 1]^{-1}$$

Each wavy line corresponds to an exchange interaction J_{ll}. Dashed lines connect only the circles, and a set of circles connected with the same dashed line is called a *block*. Each block which unites ***n*** operators $\hat{S}_l^z$ corresponds to the expression $b^{n-1}(\beta y_0)\delta(\ldots)\ldots\delta(\ldots)$, where the product of ($n$–1) Kronecker symbols takes into account the equality of spatial variables in the block, while

$$b(x) = \left(S + \frac{1}{2}\right) \mathrm{cth}\left(S + \frac{1}{2}\right)x - \frac{1}{2}\mathrm{cth}\left(\frac{x}{2}\right) \tag{4.2.14}$$

is the so-called *Brillouin function* for given S.

Using the above, we can represent the diagram series for $\langle \hat{S}_l^z \rangle$ as

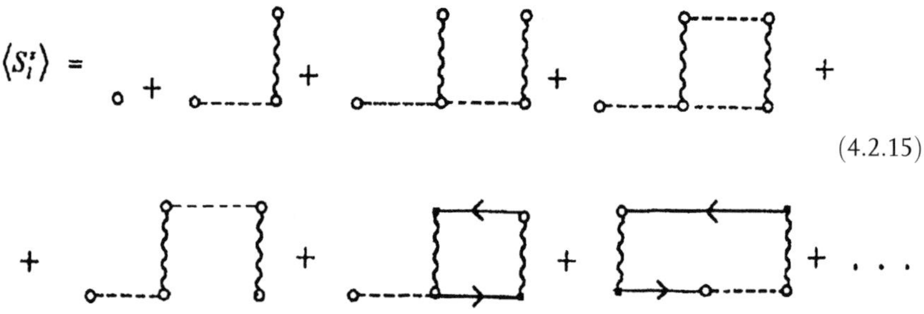

(4.2.15)

This expression includes all the diagrams up to the second-order terms.

Near the Curie point we can use the approximation of a self-consistent field; to do this we introduce an *exchange interaction radius* R_0 in accordance with Vakarchuk *et al.* (1985a,b):

$$R_0 = \left\{ \frac{\int d^3r\, r^2 J(r)}{\int d^3 r J(r)} \right\}^{1/2} \tag{4.2.16}$$

If R_0 is large in comparison with the mean interparticle distance a, i.e. for $(a/R_0) \ll 1$, then the small parameter $(a/R_0)^3$ appears in the theory. In this case all analytical expressions such as

$$\sum_l f(\mathbf{R}_l) J_{ll'} \sim \frac{1}{v_0} \int f(r) J(r) d^3 r = \left(\frac{a}{R_0}\right)^3 \frac{4\pi}{3} \frac{1}{(2\pi)^3} \int \tilde{f}(\mathbf{k}') \tilde{J}(\mathbf{k}') d^3 k' \tag{4.2.17}$$

are proportional to $(a/R_0)^3$. Here we have introduced the abbreviations $\mathbf{k}' := \mathbf{k} R_0$, $f(r)$ is any fast decreasing function; $\tilde{J}(k)$ and $\tilde{f}(k)$ are Fourier transforms of $I(r)$and $f(r)$: $\tilde{f}(k) := (v_0)^{-1} \int f(r) d^3 r \exp(i\mathbf{k}r)$.

Thus, the order of the diagram is determined by the number of sums such as

$$\sum_l f(\mathbf{R}_l) \cdot J_{ll'}$$

contributing to this diagram. The self-consistent field approximation (or zeroth approximation) corresponds to the case $R_0 \to \infty$. It means that in this approximation only such diagrams in the series (4.2.15) are retained of which the corresponding analytical expressions do not contain sums like (4.2.17). These diagrams are called 'elementary one-tail' parts, or *tree diagrams.* They have no closed loops made by continuous, wavy or dashed lines, and it is always possible to make two one-tail diagrams from such a diagram by cutting any wavy line. One-tail parts are connected to the rest of the diagram through the vertices that correspond to $\hat{S}_l^z$ operators. That is why any one-tail part is a product of an exchange integral $J_{l,l'}$ and some diagram from the diagram series (4.2.15) for $\langle \hat{S}_{l'}^{\hat{z}} \rangle$ with summation over l'.

A complete set of elementary one-tail parts is called a *full one-tail part* with a corresponding diagram series

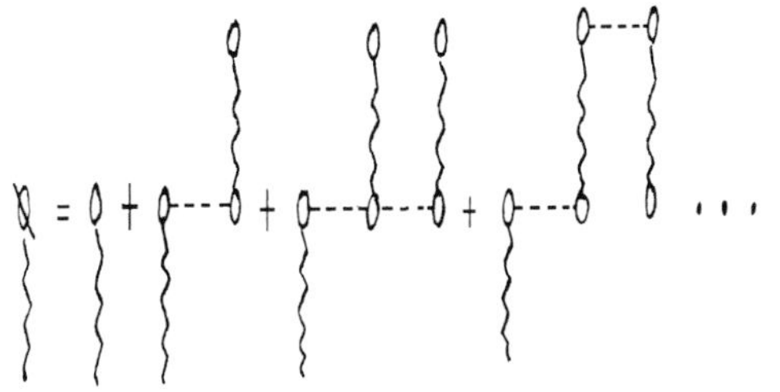

(4.2.18)

This series corresponds to the analytical expression

$$\sum_{l'} J_{ll'} \langle \hat{S}^z_{l'} \rangle^{(0)}$$

where $\langle \hat{S}^z_{l'} \rangle^{(0)}$ represents the diagram series (4.2.15) for $\langle \hat{S}^z_{l,l'} \rangle$ in the zeroth approximation, i.e. only zero-order terms with respect to the reciprocal radius of interaction are retained (that is the sum of those diagrams in (4.2.15) that 'survive' for $R_0 \to \infty$).

For the full one-tail part we use the graph ∅-----, where the symbol ∅ corresponds to the series $\langle \hat{S}^z_{l'} \rangle^{(0)}$. It is possible to carry out a partial summation in this series, which corresponds to a replacement of the 'bare' vertices O by 'dressed ones', i.e. by ∅:

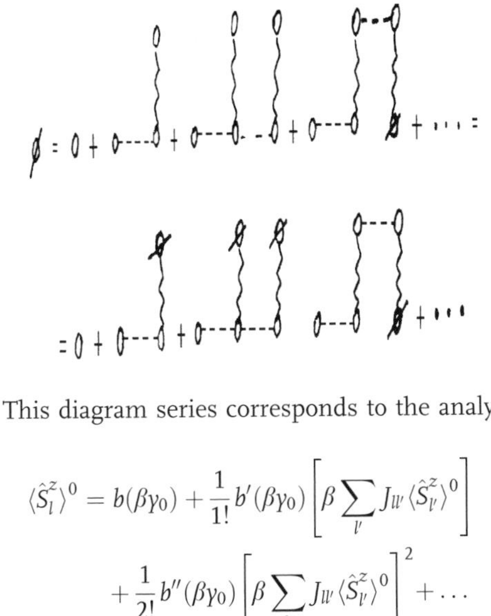

(4.2.19)

This diagram series corresponds to the analytical expression

$$\begin{aligned}\langle \hat{S}^z_l \rangle^0 &= b(\beta y_0) + \frac{1}{1!} b'(\beta y_0) \left[\beta \sum_{l'} J_{ll'} \langle \hat{S}^z_{l'} \rangle^0 \right] \\ &\quad + \frac{1}{2!} b''(\beta y_0) \left[\beta \sum_{l'} J_{ll'} \langle \hat{S}^z_{l'} \rangle^0 \right]^2 + \dots \\ &= b\left(\beta y_0 + \beta \sum_{l'} J_{ll'} \langle \hat{S}^z_{l'} \rangle^0 \right) \end{aligned} \qquad (4.2.20)$$

By substituting Eq. (4.2.20) into Eq. (4.2.6) we obtain the average spin value per atom in the mean-field approximation:

$$\bar{S}^{(0)} = \frac{1}{N}\sum_{l=1}^{N} \langle \hat{S}_l^z \rangle^{(0)}$$
$$= \frac{1}{N}\sum_{l=1}^{N} b\left(\beta y_0 + \beta \sum_{l'(l\neq l')} J_{ll'} \langle \hat{S}_{l'}^z \rangle^{(0)}\right) \tag{4.2.21}$$

If structural fluctuations can be neglected (they are small if we average over large regions), then in the r.h.s. of Eq. (4.2.21) we can substitute $\langle \hat{S}_l^z \rangle^{(0)}$ for $S^{(0)}$ and obtain the following equation for the mean spin in the self-consistent field approximation:

$$\bar{S}^{(0)} = \overline{\frac{1}{N}\sum_{l}^{N} b\left(\beta y_0 + \beta \sum_{l'\neq l} J_{ll'} \bar{S}^{(0)}\right)} \tag{4.2.22}$$

This equation can be easily solved for ferromagnets near the phase transition point to the paramagnetic phase (Curie point). Let us assume the external magnetic field to be zero and the temperature close to the critical value $T = T_C$. As the mean spontaneous spin per atom $\bar{S}_0 \equiv \bar{S}^{(0)}_{H=0}$ is equal to zero at $T = T_C$, we can expand the r.h.s. of Eq. (4.2.22) into a Taylor series. By preserving the first two terms of the expansion, we obtain

$$\bar{S}_0^{(0)} = a_0 \beta d_1 \bar{S}_0^{(0)} - \frac{c_0}{3}\beta^3 d_3 \overline{(S_0^{(0)})^3} \tag{4.2.23}$$

where

$$a_0 = \frac{S(S+1)}{3}, \quad c_0 = a_0 \frac{S^2 + S + 1}{5} \tag{4.2.24}$$

$$d_1 = \overline{\frac{1}{N}\sum_{l}\sum_{l'} J_{ll'}} = \frac{1}{v_0}\int d^3 r J(\mathbf{r}) g(\mathbf{r}) \tag{4.2.25}$$

$$d_3 = \overline{\frac{1}{N}\sum_{l}\left(\sum_{l'} J_{ll'}\right)^3} = \frac{1}{v_0}\int d^3 r J^3(\mathbf{r}) g(\mathbf{r})$$
$$+ \frac{3}{v_0^2}\int d^3 r_1 \int d^3 r_2 J^2(\mathbf{r}_1) J(\mathbf{r}_2) F_3(0, \mathbf{r}_1, \mathbf{r}_2)$$
$$+ \frac{1}{v_0^3}\int d^3 r_1 \int d^3 r_2 \int d^3 r_3 J(\mathbf{r}_1) J(\mathbf{r}_2) J(\mathbf{r}_3) F_4(0, \mathbf{r}_1, \mathbf{r}_2, \mathbf{r}_3) \tag{4.2.26}$$

From this equation we find the expression for spontaneous magnetization near the critical point:

$$S_{\mathrm{av}}^{0}=\left[\frac{3a_0^3}{c_0}\frac{d_1^3}{d_3}\left(1-\frac{T}{T_{\mathrm{C}}}\right)\right]^{1/2} \tag{4.2.27}$$

where

$$\begin{aligned}T_{\mathrm{C}}&=\frac{a_0 d_1}{k_{\mathrm{B}}}=\frac{S(S+1)}{3k_{\mathrm{B}}}\frac{1}{v_0}\int \mathrm{d}^3 r J(\mathbf{r})g(\mathbf{r})\\&=T_{\mathrm{C}}^{0}\left\{1+\frac{v_0}{(2\pi)^3}\int\frac{\tilde{J}(k)}{\tilde{J}(0)}[a(k)-1]\mathrm{d}^3 k\right\}\end{aligned} \tag{4.2.28}$$

is the Curie temperature in the molecular field approximation for amorphous ferromagnets, whereas

$$T_{\mathrm{C}}^{0}=S(S+1)\frac{\tilde{J}(0)}{3k_{\mathrm{B}}} \tag{4.2.29}$$

is the critical temperature for crystalline ferromagnets in the same approximation.
In Eq. (4.2.28) we have used the Fourier expansion

$$g(\vec{r}_1-\vec{r}_2)=F_2(\vec{r}_1-\vec{r}_2)=1+\frac{v_0}{(2\pi)^3}\int[a(k)-1]\mathrm{e}^{i\vec{k}\cdot(\vec{r}_1-\vec{r}_2)}\mathrm{d}^3 k$$

where $a(k)$ is the structure factor, as in the preceding chapters.

Equation (4.2.28) was obtained by Vakarchuk *et al.* (1981, 1985a,b) and Krasny *et al.* (1993). Numerous calculations show that structural fluctuations lower the critical temperature, when a transition from a crystalline to the amorphous state occurs. It is also obvious that T_{C} depends essentially on the range R_0 of the exchange interactions. That is easy to ensure if the functions $J(r)$ take the form (Landau and Lifschitz, 1963)

$$J_1(r)=\tilde{J}_1(0)\frac{v_0}{\pi r_0^3}\exp\left(-\frac{2r}{r_0}\right)\quad\text{with } r_0=R_0/\sqrt{3} \tag{4.2.30}$$

i.e. an *exponential* decay, respectively the somewhat different *Yukawa potential* decay

$$J_2(r)=\tilde{J}_2(0)\frac{v_0}{\pi}\frac{1}{r_0^2}\frac{1}{r}\exp\left(-\frac{2r}{r_0}\right)\quad\text{with } r_0=R_0/\sqrt{3} \tag{4.2.31}$$

and the structure factor corresponds to that of a hard-sphere system (Eq. (2.3.27)) (see Kcharkov *et al.*, 1979; Belashchenko, 1986). At the packing density $\eta=0.46$ the calculations show that $T_{\mathrm{C}}/T_{\mathrm{C}}^{0}$ is less than unity and reaches its limit at $R_0\to\infty$ (Fig. 4.2.1, Krasny *et al.*, 1993). Experimental values of $T_{\mathrm{C}}/T_{\mathrm{C}}^{0}$ in the range from 0.5 to 0.7 correspond to $R_0/a=1.5$–2.5, from which r_0 follows. Remember that the interaction range R_0 was already defined in Eq. (4.2.16).

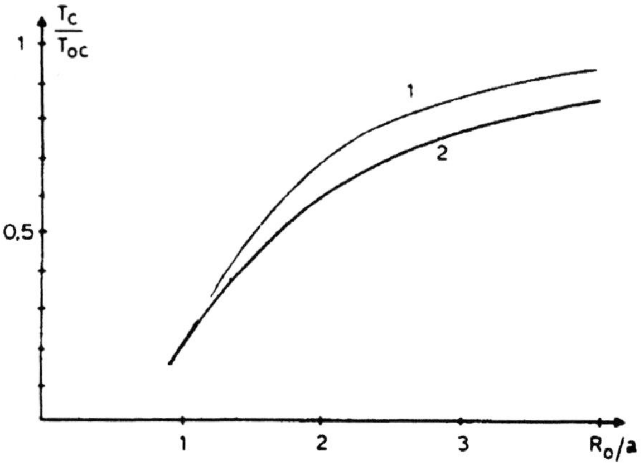

Fig. 4.2.1 Critical temperatures of amorphous magnetic alloys against the effective radius of the exchange interaction derived with Eq. (4.2.30) (case 1) and Eq. (4.2.31) (case 2), respectively

Equation (4.2.27) for the spontaneous magnetization takes a form known from the theory of crystalline ferromagnets; if $d_3 = d_1$, as we can see from Eqs (4.2.25) and (4.2.26), for amorphous ferromagnets we have $d_3 > d_1^3$. This is easy to prove, if we use a superposition approximation and neglect the correlations between particles 1, 2 and 3 (Tyablikov, 1965):

$$F_4(0, \mathbf{r}_1, \mathbf{r}_2, \mathbf{r}_3) \approx g(\mathbf{r}_1)g(\mathbf{r}_2)g(\mathbf{r}_3)g(\mathbf{r}_1 - \mathbf{r}_2)g(\mathbf{r}_1 - \mathbf{r}_3)g(\mathbf{r}_2 - \mathbf{r}_3) \approx g(\mathbf{r}_1)g(\mathbf{r}_2)g(\mathbf{r}_3) \tag{4.2.32}$$

$$F_3(0, \mathbf{r}_1, \mathbf{r}_2) \approx g(\mathbf{r}_1)g(\mathbf{r}_2)g(\mathbf{r}_1 - \mathbf{r}_2) \approx g(\mathbf{r}_1)g(\mathbf{r}_2) \tag{4.2.33}$$

Then

$$\frac{d_3}{d_1^3} = 1 + \frac{(3/v_0)\int d^3r\, J^2(\mathbf{r})g(\mathbf{r})}{[(1/v_0)\int d^3r\, J(\mathbf{r})g(\mathbf{r})]^2} + \frac{(1/v_0)\int d^3r\, J^3(\mathbf{r})g(\mathbf{r})}{[(1/v_0)\int d^3r\, J(\mathbf{r})g(\mathbf{r})]^3} > 1 \tag{4.2.34}$$

If we use Eqs (4.2.30) and (4.2.31) and the hard-sphere approximation for a numerical evaluation of this expression, then, according to the calculations (see Fig. 4.2.2), $d_1^3/d_3 \to 1$, when $R_0 \to \infty$ ·

So d_1^3/d_3 is always ≥1 when $R_0/a \sim 1.5$–2.5 then $d_1^3/d_3 = 0.55$–0.75 (Krasny *et al.*, 1993). This means that the spontaneous magnetization increases with the distance from the critical point (i.e. when T/T_C decreases) in a *slower* way for amorphous ferromagnets than for their crystalline analogues. This is precisely what was observed in experiment (see Fig. 4.1.2).

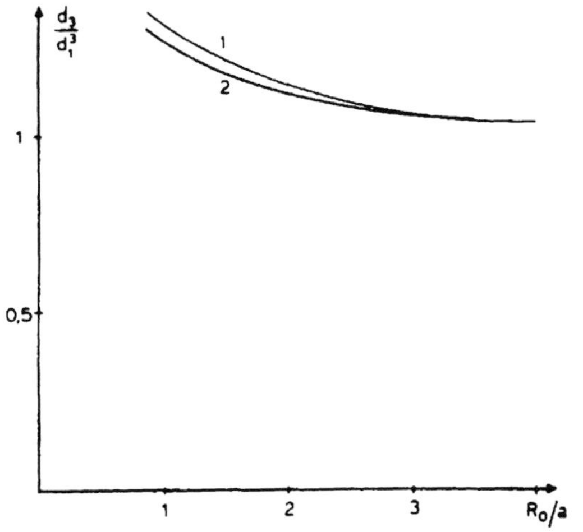

Fig. 4.2.2 Dependence of the ratio d_3/d_1^3 (see Eqs (4.2.25) and (4.2.26)) on the effective radius R_0 of the exchange interaction for the cases 1 and 2 defined in Eqs (4.2.30) and (4.2.32) for amorphous ferromagnets

The above results can be generalized for the situation of a two-component amorphous *alloy*, where spins are only localized on the atoms of the first component (e.g. on metal atoms in an amorphous *alloy of a transition metal with a non-metal*; Krasny *et al.*, 1993). We will further assume that there are N_1 atoms of the first component and N_2 of the second. Their corresponding concentrations are $X_1 = N_1/N$ and $X_2 = N_2/N$, where $N = N_1 + N_2$ is the total number of atoms of the system. Besides, we will assume that only the spins of the first component are connected by an exchange interaction. Then the magnetic part of the Hamiltonian of the system will be described by Eqs (4.2.1)–(4.2.3), in which the summation refers to the atoms of the first alloy component (Heisenberg model). Therefore, we can use our previously obtained results to find the average values of the characteristic magnetic properties of the system. For instance, the average value of the spin per atom of the first component in the self-consistent field approximation is defined by the expression

$$\bar{S} = \overline{\frac{1}{N_1}\sum_{l_1=1}^{N_1} b\left(\beta y_0 + \beta \sum_{l_1'=1}^{N_1} J_{l_1 l_1'} \langle \hat{S}^z_{l_1'} \rangle^0\right)} \tag{4.2.35}$$

Yet, for configurational averaging, this expression should take into consideration the non-magnetic interaction between all atoms of the system, i.e. between the atoms of the first and of the second components. To find the spontaneous magnetization near the critical point, as in the case of an one-component system, we assume $\langle \hat{S}^z_l \rangle^0|_{H=0} \approx S_0^{(0)}|_{H=0} = \bar{S}_0^{(0)}$ on the r.h.s. of Eq. (4.2.35) and expand it into a Taylor series, preserving the first two terms of the expansion. Equation (4.2.35) is then easily solved, and we obtain the following formula, which is similar to Eq. (4.2.27):

$$\bar{S}_0^{(0)} = \left[\frac{3a_0^3}{c_0}\frac{d_1^3}{d_3}\left(1 - \frac{T}{T_C}\right)\right]^{1/2} \tag{4.2.36}$$

However, the coefficients in this equation essentially depend on the concentration:

$$\begin{aligned} d_1 &= \frac{x_1}{v_0}\int d^3rJ(\mathbf{r})g_{11}(\mathbf{r}) \\ d_3 &= \frac{x_1}{v_0}\int d^3r\,J^3(\mathbf{r})g_{11}(\mathbf{r}) + \frac{3x_1^2}{v_0^2}\int d^3r_1\int d^3r_2J^2(\mathbf{r}_1)J(\mathbf{r}_2)F_3(0,\mathbf{r}_1,\mathbf{r}_2) \\ &\quad + \frac{x_1^3}{v_0^3}\int d^3r_1\int d^3r_2\int d^3r_3J(\mathbf{r}_1)J(\mathbf{r}_2)J(\mathbf{r}_3)F_4(0,\mathbf{r}_1,\mathbf{r}_2,\mathbf{r}_3) \end{aligned} \tag{4.2.37}$$

In Eq. (4.2.37), $F_S(\mathbf{r}_1,\mathbf{r}_2,\ldots\mathbf{r}_S)$ for $s=2, 3, 4$ is the correlation function of s particles of the *first* component. We have, as in the one-component case:

$$\begin{aligned} F_2(\mathbf{r}_1,\mathbf{r}_2) &= g_{11}(\mathbf{r}_1 - \mathbf{r}_2) \\ &= 1 + \frac{v_0}{(2\pi)^3}\int d^3k[a_{11}(\mathbf{k}) - 1]e^{\cdot i\mathbf{k}(\mathbf{r}_1-\mathbf{r}_2)} \end{aligned}$$

Then it is easy to show that

$$\begin{aligned} T_C &= \frac{a_0d_1}{k_B} \\ &= T_C^0x_1\left\{1 + \frac{v_0}{(2\pi)^3}\int\frac{\tilde{J}(\mathbf{k})}{\tilde{J}(0)}[a_{11}(\mathbf{k}) - 1]d^3k\right\} \\ T_C^0 &= S(S+1)\tilde{J}(0)/3k_B \end{aligned} \tag{4.2.38}$$

We can see from Eq. (4.2.38) that when the concentration x_1 of magnetically active atoms decreases, the critical temperature goes down. However, the change may be non-monotonic, because the concentration dependence of $a_{11}(\mathbf{k})$ has a complex character. This is easy to prove, if we use the hard-sphere model: let us assume that our alloy is a mixture of solid spheres with diameters D_1 and D_2, $(D_2/D_1{:=}\xi<1)$ while the magnetic interaction is described by Eqs (4.2.30) and (4.2.31). Then the structure factor has the form (see Kcharkov *et al.*, 1979):

$$a_{11}(k) = 1 + \frac{C_{11}(k) - x_2[C_{11}(k)C_{22}(k) - C_{12}^2(k)]}{1 - x_1C_{11}(k) - x_2C_2(k) + x_1x_2[C_{11}(k)C_{22}(k) - C_{12}^2(k)]} \tag{4.2.39}$$

where $C_{\alpha\beta}(k)$ have analytical expressions analogous to that of Eq. (2.3.27).

The expression for T_C/T_C^0 can be numerically evaluated if we use the well-known formulae for the structural factor of hard spheres. Figures 4.2.3 and 4.2.4 represent numerical calculations for the concentration dependence of T_C/T_C^0 at $\xi=0.6$, 0.8 and the packing density $\eta=\pi D^3/\sigma v_0=0.46$ (Krasny *et al.*, 1993).

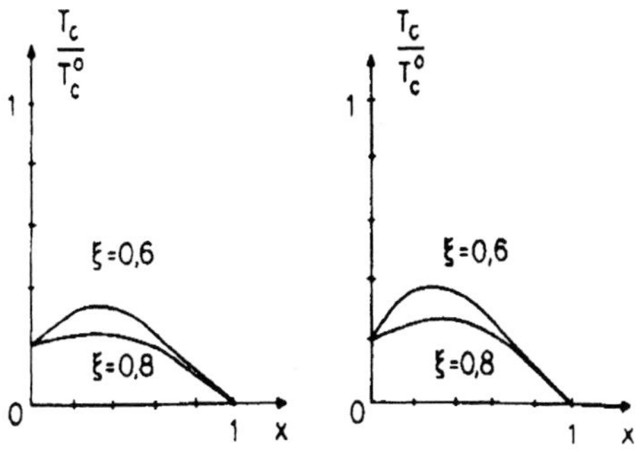

Fig. 4.2.3 The numerically determined values of the critical temperature are plotted over the concentration x of the magnetic component 1 of a two-component amorphous alloy, where ξ is the ratio D_2/D_1 of the hard-sphere diameters and $R_0/a=1$. a) Exponential decay (Eq. (4.2.30)) of the exchange; b) Yukawa decay (Eq. (4.2.31)) (taken from Krasny *et al.*, 1993)

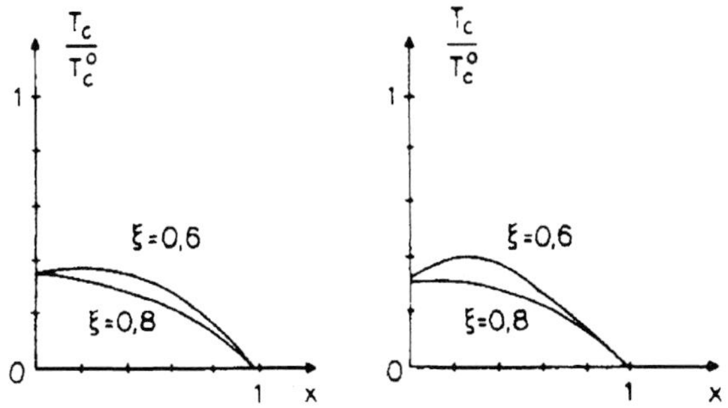

Fig. 4.2.4 The numerically determined values of the critical temperature are plotted over the concentration x of the magnetic component 1 of a two-component amorphous alloy, where ξ is the ratio D_2/D_1 of the hard-sphere diameters and $R_0/a=1.25$. a) Exponential decay (Eq. (4.2.30)) of the exchange; b) Yukawa decay (Eq. (4.2.31)) (taken from Krasny *et al.*, 1993)

We can see from these two figures that at $R_0/a=1$–1.25 and $\xi=0.6$, with x_1 decreasing, the critical temperature first goes up, then reaches its maximum at $x_1 \sim 0.3$–0.4 (which agrees with the experiments described in the books by Luborsky (1983) and Handrich and Kobe (1980)) and then begins to fall abruptly.

The greater the contrast in the size of the atoms and the smaller the magnetic interaction radius, the more evident is the non-monotonous character of the changes in the critical temperature.

The ratio $(d_1^3/d_3)^{1/2}$ is less than unity, both for pure magnetics and for amorphous metal alloys. Besides, as is evident from Eq. (4.2.38), the values of $(d_1^3/d_3)^{1/2}$ decrease as the concentration x_1 decreases. Because $T_C \sim d_1$, the concentration dependence of $S_0^{(0)}$at a given reduced temperature T/T_C will behave similarly to that of T_C, which is also confirmed by experiment.

4.3
The Spectrum of Quasi-Magnon Excitations in Amorphous Ferromagnets

In Section 4.1 we used the fact that the experimental investigation of the spin-wave dispersion law is based on the analysis of peaks in the double-differential cross-section of inelastic neutron scattering from the atomic magnetic moments. In the Born approximation, this cross-section is proportional to the Fourier transform of the double-time temperature *retarded* Green's function, which is constructed from the transversal components of the spin operators of the magnetic atom (Barjakhtar *et al.*, 1984). The Fourier transform of the retarded Green's function in turn can be obtained by an analytical continuation of the Fourier transform of the temperature *causal* Green's function (Barjakhtar *et al.*, 1984; Vakarchuk and Tkachuk, 1989), which is calculated diagrammatically for the so-called *Matsubara frequencies* defined on the imaginary axis (see below). The analytical continuation is then to frequencies on the real axis. In the following section we find the magnon spectrum from *poles* of the Green's functions.

To investigate a system of N identical atoms with magnetic moments randomly positioned at the points $\vec{R}_1, \dots, \vec{R}_N$ with spins interacting through exchange forces (the Heisenberg model), we employ the causal temperature Green's function (Krasny *et al.*, 1993c; Vakarchuk and Tkachuk, 1989):

$$K(\vec{r}_1 - \vec{r}_2; \tau_1 - \tau_2) = \frac{v_0^2}{N} \overline{\sum_{l_1, l_2} \delta(\vec{r}_1 - \vec{R}_{l_1}) K_{l_1 l_2}(\tau_1 - \tau_2) \delta(\vec{r}_2 - \vec{R}_{l_2})} \tag{4.3.1}$$

where

$$\begin{cases} K_{l,m}(\tau_1 - \tau_2) = -\frac{1}{2} \langle \hat{T}_\tau \{ \hat{\sigma}(\beta) \cdot \hat{S}_l^+(\tau_1) \cdot \hat{S}_m^-(\tau_2) \} \rangle_0 \\ \hat{S}_m^{+,-}(\tau) = \exp(+\hat{\mathrm{H}}_0 \tau) \cdot S_m^{+,-} \cdot \exp(-\hat{\mathrm{H}}_0 \tau) \end{cases} \tag{4.3.2}$$

The operators $\hat{S}_l^+, \hat{S}_l^-, \hat{\sigma}(\beta), \hat{\mathrm{H}} := \hat{\mathrm{H}}_0 + \hat{\mathrm{H}}_1$ are determined by Eqs (4.3.1) through (4.2.12); $\tau/\hbar$ is an imaginary time. As usual, the overbar in Eq. (4.3.1) denotes configuration averaging over all possible realizations of atomic random positions in our amorphous sample.

To obtain an equation for the Green's function $K_{l_1 l_2}(\tau)$, we use the diagrammatic method of Vaks *et al.* (1967); see Appendix B and the monographs by Barjakhtar *et al.* (1984) and Akhieser *et al.* (1967). The equation for the Fourier transform of the Green's function

$$\tilde{K}_{l_1 l_2}(i\omega_n) = \frac{1}{2} \int_{-\beta}^{+\beta} K_{l_1 l_2}(\tau) e^{i\omega_n \tau} d\tau \tag{4.3.3}$$

has the following form:

$$\begin{cases} \tilde{K}_{l_1 l_2}(i\omega_n) = \Pi_{l_1 l_2}(i\omega_n) - \Pi_{l_1 l_3}(i\omega_n) J_{l_3 l_4} \tilde{K}_{l_3 l_2}(i\omega_n) \\ \Pi_{l_1, l_2}(i\omega_n) = \tilde{K}_0 [M_{l_1, l_2}(i\omega_n) + E_{l_1, l_3}(i\omega_n) \Pi_{l_1, l_3}(i\omega_n)] \end{cases} \quad (4.3.4)$$

where the $\omega_n/\hbar$ are the above-mentioned Matsubara frequencies, i.e.

$$\begin{cases} \tilde{K}_0 \equiv \tilde{K}_0(i\omega_n) = \dfrac{1}{i\omega_n - \gamma_0} \\ \omega_n = n \cdot \pi \cdot (k_B T) \end{cases} \quad (4.3.5)$$

and the summation has to be taken over the repeated indices.

The functions $E_{l_1 l_2}(i\omega_n)$ and $M_{l_1 l_2}(i\omega_n)$ symbolize analytical expressions for the set of diagrams, which are *irreducible* after Larkin, i.e. indivisible by cutting only one interaction line: these functions correspond to diagrams joined to each other with two and one Green's function lines, respectively.

Solving the second part of Eq. (4.3.4) we find the matrix $\Pi_{l_1 l_2}$:

$$\Pi_{l_1 l_2} = \tilde{K}_0 [1 - \tilde{K}_0 \hat{E}]^{-1}_{l_1 l_3} M_{l_3 l_2}$$

and substituting it into the first part we find $\tilde{K}_{l_1 l_2}$:

$$\begin{aligned} \tilde{K}_0 &= \tilde{K}_0 [1 - \tilde{K}_0 (\hat{E} - \hat{M}\hat{J})]^{-1}_{l_1 l_3} M_{l_3 l_2} \\ &= \tilde{K}_0 \{ M_{l_1 l_2} + \tilde{K}_0 [E_{l_1 l_4} - M_{l_1 l_3} J_{l_3 l_4}] M_{l_4 l_2} + \ldots \} \end{aligned} \quad (4.3.6)$$

Here the operator symbols $\hat{B}$ correspond to the matrices $B_{ll'}$.

Let us represent the Green's function $K(\vec{r}_1 - \vec{r}_2, \tau_1 - \tau_2)$ from Eq. (4.3.1) by the Fourier series:

$$K(\vec{r}_1 - \vec{r}_2, \tau_1 - \tau_2) = \frac{\upsilon_0}{(2\pi)^3} \int d^3 q e^{i\vec{q}\cdot(\vec{r}_1 - \vec{r}_2)} \frac{1}{\beta} \sum_n e^{-i\omega_n(\tau_1 - \tau_2)} \tilde{K}(\vec{q}, i\omega_n) \quad (4.3.7)$$

where

$$\tilde{K}(\vec{q}, i\omega_n) = \frac{1}{N} \overline{\sum_{ll'} e^{-i\vec{q}\cdot\vec{R}_l} \tilde{K}_{ll'}(i\omega_n) e^{i\vec{q}\cdot\vec{R}_{l'}}} \quad (4.3.8)$$

Using now Eq. (4.3.6) we find for the Fourier transform of the Green's function:

$$\begin{aligned} \tilde{K}(\vec{q}, i\omega_n) = \tilde{K}_0(i\omega_n) \Big\{ & \frac{1}{N_0} \overline{e^{-i\vec{q}\cdot\vec{R}_1} M_{ll'}(i\omega_n) e^{i\vec{q}\cdot\vec{R}_{l'}}} \\ & + \overline{\tilde{K}_0(i\omega_n) \frac{1}{N} e^{-i\vec{q}\cdot\vec{R}_l} [E_{ll_2}(i\omega_n) - M_{ll_1}(i\omega_n) I_{l_1 l_2}] M_{l_2} l'(i\omega_n) e^{i\vec{q}\cdot\vec{R}_{l'}} + \ldots} \Big\} \end{aligned} \quad (4.3.9)$$

To sum up this infinite series we perform the following decoupling:

$$\frac{1}{N}\overline{e^{-i\vec{q}\cdot\vec{R}_l}(E_{ll_2}-M_{ll_1}I_{l_1l_2})(E_{l_2l_4}-M_{l_2l_3}I_{l_3l_4})\ldots(E_{l_{s-2}l_s}-M_{l_{s-2}l_{s-1}}I_{l_{s-1}l_s})M_{l_sl'}\cdot e^{i\vec{q}\cdot\vec{R}_{l'}}}$$
$$\approx[\tilde{E}(\vec{q},i\omega_n)-\tilde{f}(\vec{q},i\omega_n)]^s\cdot\tilde{M}(\vec{q},i\omega_n) \tag{4.3.10}$$

$$\begin{cases}\tilde{E}(\vec{q},i\omega_n)=\dfrac{1}{N}\overline{e^{-i\vec{q}\cdot\vec{R}_l}E_{ll'}(i\omega_n)e^{i\vec{q}\cdot\vec{R}_{l'}}}\\ \tilde{M}(\vec{q},i\omega_n)=\dfrac{1}{N}\overline{e^{-i\vec{q}\cdot\vec{R}_l}M_{ll'}(i\omega_n)e^{i\vec{q}\cdot\vec{R}_{l'}}}\\ \tilde{f}(\vec{q},i\omega_n)=\dfrac{1}{N}\overline{e^{-i\vec{q}\cdot\vec{R}_l}M_{ll_1}(i\omega_n)J_{l_1l'}e^{i\vec{q}\cdot\vec{R}_{l'}}}\end{cases} \tag{4.3.11}$$

In this approximation, the series (4.3.9) is summed and we obtain (Vakarchuk and Margolych, 1987):

$$\tilde{K}(\vec{q},i\omega_n)=\frac{\tilde{M}(\vec{q},i\omega_n)}{i\omega-[\gamma_0+\tilde{E}(\vec{q},i\omega_n)-\tilde{f}(\vec{q},i\omega_n)]} \tag{4.3.12}$$

The inelastic neutron scattering cross-section is described by the Fourier transform $\tilde{G}^R(\vec{q},\omega)$ of the *retarded* Green's function. To obtain this function, an analytic continuation of the *causal* Green's function $\tilde{K}(\vec{q},i\omega_n)$ has to be performed from the discrete set of Matsubara frequencies on the imaginary axis into the whole complex plane, namely $i\omega_n\to\hbar\omega+i\delta$, $\delta\to0^+$. In this way we find:

$$\tilde{G}^R(\vec{q},\omega)=\frac{\tilde{M}(\vec{q},\omega+i\delta)}{\hbar\omega-[g\mu_0H+\tilde{E}(\vec{q},\omega+i\delta)-\tilde{f}(\vec{q},\omega+i\delta)+i\delta]} \tag{4.3.13}$$

A pole of this function determines the spin-wave energy spectrum:

$$\hbar\omega=g\mu_0H+\tilde{E}(\vec{q},\omega+i\delta)-\tilde{f}(\vec{q},\omega+i\delta) \tag{4.3.14}$$

In the self-consistent field approach (i.e. a zero-order approximation in terms of reciprocal exchange-interaction radius), Eq. (4.3.11) takes the form (see Appendix B):

$$\begin{cases}\tilde{M}(\vec{q},\omega)\approx\dfrac{1}{N}\overline{\sum_{l=1}^{N}\langle\hat{S}_l^z\rangle}=\bar{S}\\ \tilde{E}(\vec{q},\omega)\approx\dfrac{1}{N}\overline{\sum_{l,l'}\langle\hat{S}_l^z\rangle J_{l,l'}}\approx S\left\{\tilde{J}(0)+\dfrac{v_0}{(2\pi)^3}\displaystyle\int\tilde{J}(\vec{k})[a(\vec{k})-1]d^3k\right\}\\ \tilde{f}(\vec{q},\omega)\approx\dfrac{1}{N}\overline{\sum_{l,l'}e^{-i\vec{q}\cdot\vec{R}_l}\langle\hat{S}_l^z\rangle J_{l,l'}e^{i\vec{q}\cdot\vec{R}_{l'}}}\approx S\left\{\tilde{J}(\vec{q})+\dfrac{v_0}{(2\pi)^3}\displaystyle\int\tilde{J}(\vec{q}+\vec{k})[a(\vec{k})-1]d^3k\right\}\end{cases} \tag{4.3.15}$$

Here it should be noted that in the present approximation $\tilde{E}(\vec{q},\omega)$ depends on $\vec{q}$ only through the magnitude q, because the same 'isotropy' refers to $a(k)$ and $\tilde{J}(k)$.

In the last two equations we have neglected the structural fluctuations in the average spin, assuming $\langle\hat{S}_l^z\rangle\approx S$, where S is the atomic spin. Then the energy spectrum of spin waves in the self-consistent field approximation becomes (Vakarchuk and Tkachuk, 1989; Krasny *et al.*, 1993c):

$$\hbar\omega_0(\vec{q}) = g\mu_0 H + S[\tilde{J}(0) - \tilde{J}(\vec{q})] + S\frac{v_0}{(2\pi)^3}\int [\tilde{J}(\vec{k}) - \tilde{J}(\vec{q}+\vec{k})]\cdot[a(\vec{k}) - 1]\mathrm{d}^3k \tag{4.3.16}$$

This equation allows us to determine the behavior of $\omega_0(\vec{q})$ at all values of the wavenumber q. To do this, we have to specify the structure of the amorphous material and to define explicitly the exchange integral. Here again we employ the model of the classical liquid of hard spheres with diameter ω and packing parameter $\eta = \pi\sigma^3/(6v_0)$ (see Eq. (1.2)), and exchange integrals in the form given above. The results for the above-mentioned cases 1 and 2 are presented in Figs 4.3.1 and 4.3.2 (Krasny *et al.*, 1993c).

As is evident from these figures, to the present accuracy the behavior of the dispersion curves is practically independent of the choice of $J(r)$, and the quasi-magnon dispersion curves themselves demonstrate a minimum at the q-value at which the structure factor has its maximum. This is similar as for the quasi-phonon dispersion and is in agreement with neutron scattering experiments, and with calculations of other authors.

In the long-wave limit ($q \to \infty$) we again get

$$\hbar\omega_0(q) = g\mu_0 H + D_0 \cdot q^2 \tag{4.3.17}$$

where

$$D_0 = D_{0_{\mathrm{cr}}}\left\{1 - \frac{4\pi}{3D_{0_{\mathrm{cr}}}} S\frac{v_0}{(2\pi)^3}\int_0^\infty \frac{\mathrm{d}}{\mathrm{d}q}\left(q^2\frac{\mathrm{d}\tilde{J}}{\mathrm{d}q}\right)[a(q) - 1]\mathrm{d}^3q\right\} \tag{4.3.18}$$

is the spin-wave stiffness in amorphous ferromagnets; $D_{0_{\mathrm{cr}}} = -\frac{1}{2}S\frac{\mathrm{d}^2\tilde{J}}{\mathrm{d}q^2}\Big|_{q=0}$ is the stiffness in the corresponding crystals (Akhieser *et al.*, 1967). For numerical calcu-

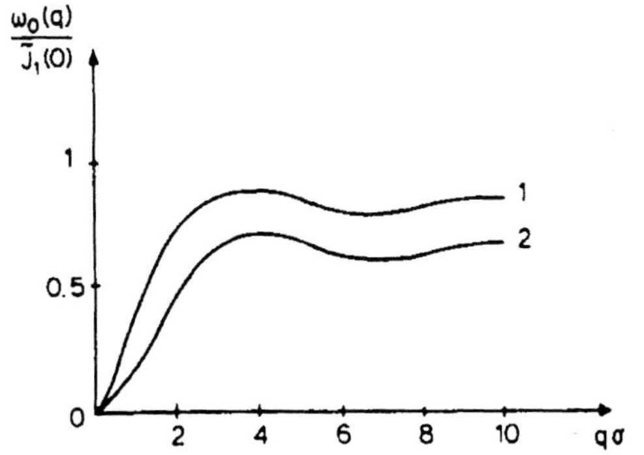

Fig. 4.3.1 Quasi-magnon dispersion curves for $R_0/a=1$ (further details in the caption of Fig. 4.3.3)

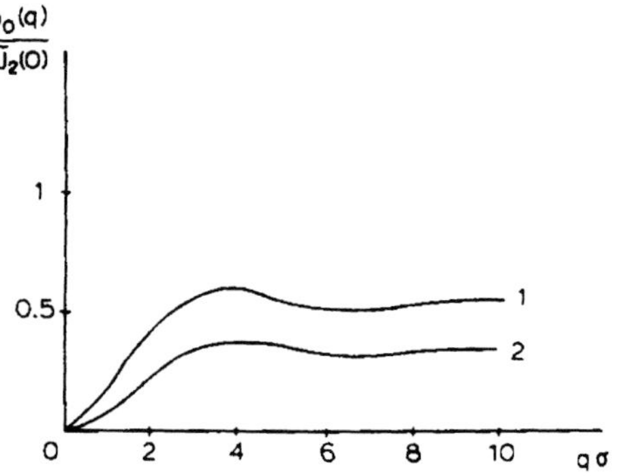

Fig. 4.3.2 Quasi-magnon dispersion curves for $R_0/a = 0.7$ (further details in the caption of Fig. 4.3.3)

lations we again used hard-sphere models and exchange integrals (4.2.30) and (4.2.31). Figure 4.3.3 shows D_0 as a function of R_0 (Krasny *et al.*, 1993c).

As follows from this figure, $D_0/D_{0_{cr}} \to 1$ when $R_0 \to \infty$, whereas for $D_0/D_{0_{cr}} = 0.7$ (determined by neutron scattering experiments) (Handrich and Kobe 1980, Hooper and de Graaf, 1973) we get $R_0/a \sim 1 - 1.15$. Nearly the same exchange interaction radius R_0 was obtained from the comparison of theoretical values of Curie temperature with the experimental ones (see Section 4.2).

Expressions (4.3.16) and (4.3.18) were derived by Vakurchuk *et al.* (1989), Krasny *et al.* (1993b) and Vakarchuk *et al.* (1987), by methods very different from the present ones.

To improve the results obtained it is necessary first to alter the decoupling procedure (4.3.10) and second to take into account corrections to the mean-field approximation. In the first case this means taking into account the scattering of magnons by the density fluctuations, and the appearance of *decay* (see Vakarchuk *et al.*, 1987). In the second case the temperature dependence of $\omega_0(q)$ (or D_0) is taken into consideration. To show this, it is necessary to find corrections of first order in the inverse interaction radius to the above-stated results (Krasny *et al.*, 1993b). For simplicity take H=0 and consider low temperatures such that

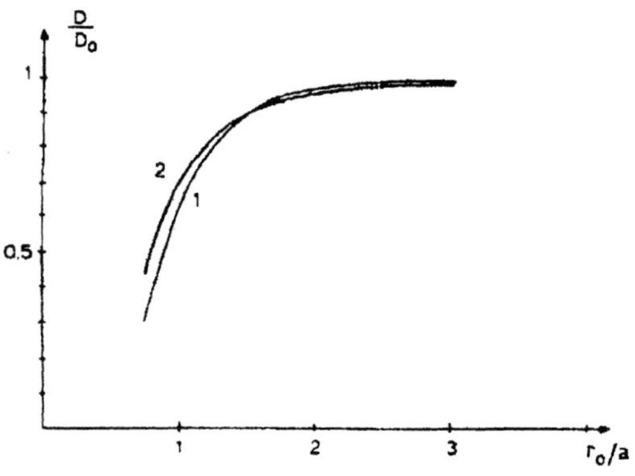

Fig. 4.3.3 Spin-wave stiffness D/D_0 ($=D_0/D_{0\sigma}$ in the text) versus the effective radius r_0/a ($=R_0/a$ in the text) of the exchange interaction. 1: Exponential decay (5.2.30) of the exchange; 2: Yukawa form (5.2.31) of the decay

$\exp\left(\frac{-T_C}{T}\right) \ll 1$. In this approximation the matrix elements $E_{ll'}$ and $M_{ll'}$ take the form (Barjakhtar *et al.*, 1984):

$$\begin{cases} M_{ll'} = \left[S - \frac{\upsilon_0}{(2\pi)^3}\int n_{\vec{k}} \cdot d^3k\right] \cdot \delta_{ll'} \\ E_{ll'} = \delta_{ll'}\left[S - \frac{\upsilon_0}{(2\pi)^3}\int n_{\vec{k}} \cdot d^3k\right] \cdot \sum_{l_1} J_{ll_1} \\ \quad +\delta_{ll'} \cdot \sum_{l_1} J_{ll_1} \cdot \frac{\upsilon_0}{(2\pi)^3}\int d^3k \cdot n_{\vec{k}} \cdot \exp[-i\vec{k} \cdot (\vec{R}_l - \vec{R}_{l_1})] \\ \quad -J_{ll'} \cdot \frac{\upsilon_0}{(2\pi)^3}\int d^3k \cdot n_{\vec{k}} \cdot \exp[-i\vec{k} \cdot (\vec{R}_l - \vec{R}_{l'})] \end{cases} \tag{4.3.19}$$

where

$$n_{\vec{k}} = \left[\exp\left(-\frac{\hbar\omega_0(k)}{k_B T}\right) - 1\right]^{-1}$$

After substituting the matrix elements into Eqs (4.3.11) and (4.3.13) we obtain the following expression for the retarded Green's function:

$$G^R(\vec{q},\omega) = \frac{S - \frac{\upsilon_0}{(2\pi)^3}\int d^3k \cdot n_{\vec{k}}}{\hbar\omega - \omega(\vec{q}) - i\delta} \tag{4.3.20}$$

where

$$\begin{aligned} \hbar\omega(\vec{q}) =& \left(S - \frac{\upsilon_0}{(2\pi)^3}\int n_{\vec{k}} \cdot d^3k\right) \\ & \times\left\{[\tilde{J}(0) - \tilde{J}(\vec{q})] + \frac{\upsilon_0}{(2\pi)^3}\int d^3k[\tilde{J}(\vec{k}) - \tilde{J}(\vec{q}+\vec{k})] \cdot [a(\vec{k}) - 1]\right\} \\ & + \frac{\upsilon_0}{(2\pi)^3}\int d^3k[\tilde{J}(\vec{k}) - \tilde{J}(\vec{q}+\vec{k})] \cdot n_{\vec{k}} \\ & + \left(\frac{\upsilon_0}{(2\pi)^3}\right)^2 \int d^3k \int d^3k'[\tilde{J}(\vec{q}+\vec{k}') - \tilde{J}(\vec{q}+\vec{k}+\vec{k}')] \cdot n_{\vec{k}}[a(\vec{k}') - 1] \end{aligned} \tag{4.3.21}$$

If we take into account that n_k has the maximal value at $k=0$ and approaches zero at $k \to \infty$, then it is possible to expand the integrand into a power series in k and keep the first three terms of the expansion. Then the function $\omega(\vec{q})$ takes the form:

$$\omega(\vec{q}) = \omega_0(\vec{q}) + \omega_1(\vec{q}) \tag{4.3.22}$$

where $\omega_0(\vec{q})$ is defined by Eq. (4.3.16) and

$$\hbar\omega_1(\vec{q}) = \frac{v_0}{(2\pi)^3 6}\int d^3k\cdot k^2\cdot n_{\vec{k}}\left\{[(\Delta_{\vec{q}}\cdot\tilde{J}(\vec{q}))|_{q=0} - \Delta_{\vec{q}}\tilde{J}(\vec{q})] + \frac{v_0}{(2\pi)^3}\int d^3k_1\cdot[a(\vec{k}_1)-1]\cdot\Delta_{\vec{k}_1}[\tilde{J}(\vec{k}_1) - \tilde{J}(\vec{k}_1+\vec{q})]\right\} \tag{4.3.23}$$

In the long-wave limit ($q \to 0$)

$$\hbar\omega(\vec{q}) = (D_0 + D_1)\cdot q^2 \tag{4.3.24}$$

where

$$D_1 = \frac{v_0\cdot 1.341\cdot\sqrt{\pi}}{8(2\pi)^2}\left(\frac{k_B T}{D_0}\right)^{5/2} + \left\{-\frac{3}{2}\frac{d^4\tilde{J}}{dq^4}\bigg|_{q=0} - \frac{v_0}{3(2\pi)^2}\int_0^\infty k^2 dk[\Delta_{\vec{k}}\cdot\Delta_{\vec{k}}\cdot\tilde{J}(\vec{k})]\cdot[a(k)-1]\right\} \tag{4.3.25}$$

and $\Delta_{\vec{k}}$ is the Laplace operator. Besides, we have taken into account that

$$\frac{v_0}{(2\pi)^3 6}\int d^3k\cdot k^2\cdot n_{\vec{k}} \approx \frac{v_0}{3(2\pi)^2}\left(\frac{k_B T}{D_0}\right)^{5/2}\int_0^\infty \frac{y^4 dy}{e^y - 1} = \frac{v_0\cdot 1.341\cdot\sqrt{\pi}}{8(2\pi)^2}\left(\frac{k_B T}{D_0}\right)^{5/2}$$

Numerical calculations show D_1 to be negative. Thus with increasing temperature the spin-wave stiffness decreases as $T^{5/2}$, and at $T=1/2\ T_C$, the stiffness becomes nearly one-half of its initial value, i.e. $\left(\frac{D_1}{D_0}\Big|_{T=\frac{T_C}{2}} \sim 0.5\right)$.

4.4 Low-temperature Magnetic Behavior of Amorphous Ferromagnets

If only the external magnetic field and the exchange interaction between the atomic magnetic moments is taken into account, the magnetic part of the free energy of an amorphous ferromagnet has the following form in the Heisenberg model (see Section 4.2):

$$F = -\frac{1}{\beta}\overline{\ln\{\mathrm{Tr}[\exp(-\beta\hat{H})]\}} \tag{4.4.1}$$

where

$$\hat{H} = -\gamma_0\sum_{l=1}^{N}\hat{S}_l^z - \frac{1}{2}\sum_{l,l'} J_{ll'}\left\{\hat{S}_l^z\hat{S}_{l'}^z + \frac{1}{2}[\hat{S}_l^-\hat{S}_{l'}^+ + \hat{S}_l^+\hat{S}_{l'}^-]\right\} \tag{4.4.2}$$

Here the term with $\gamma_0 := g\mu_B H$ represents the so-called *Zeeman energy* produced by the external magnetic field H, and the remaining terms represent the so-called exchange interaction of the spins, which can be written more simply as

$$-\frac{1}{2}\sum_{l,l'} J_{ll'}\{\hat{S}_l^z\hat{S}_{l'}^z + [\hat{S}_l^y\hat{S}_{l'}^y + \hat{S}_l^x\hat{S}_{l'}^x]\}\,.$$

In the ground state (i.e. at $T=0$) all atoms l have maximal z-projection, $\hat{S}_l^z = S$, of the atomic spin. In contrast, in the excited states $\hat{S}_l^z$ can become smaller than S. When the number of such excited atoms is much less than the total number of atoms, then the mean value of the operator $\hat{S}_l^z$ differs little from S. Using this fact, it is useful to introduce new operators $\hat{b}_l^+$, and, respectively, through the following equations, the *Holstein-Primakoff* representation (Barjakhtar *et al.*, 1984; Akhieser *et al.*, 1967):

$$\begin{cases} \hat{S}_l^z &= S - \hat{b}_l^+\hat{b}_l \\ \hat{S}_l^+ &= (2S)^{1/2}\sqrt{1-\dfrac{\hat{b}_l^+\hat{b}_l}{2S}}\cdot\hat{b}_l \\ &= \sqrt{2S}\left\{1-\dfrac{1}{4S}\hat{b}_l^+\hat{b}_l - \dfrac{1}{32\cdot S^2}\cdot\hat{b}_l^+\hat{b}_l\hat{b}_l^+\hat{b}_l + \dots\right\}\hat{b}_l \\ \hat{S}_l^- &= (2S)^{1/2}\cdot\hat{b}_l^+\cdot\sqrt{1-\dfrac{\hat{b}_l^+\hat{b}_l}{2S}} \\ &= \sqrt{2S}\cdot\hat{b}_l^+\cdot\left\{1-\dfrac{1}{4S}\hat{b}_l^+\hat{b}_l - \dfrac{1}{32\cdot S^2}\cdot\hat{b}_l^+\hat{b}_l\hat{b}_l^+\hat{b}_l + \dots\right\} \end{cases} \tag{4.4.3}$$

In these expressions $\hat{b}_l^+$ and $\hat{b}_l$ are the Bose creation and annihilation operators for spin excitations of the lth atom, obeying the commutation relations:

$$\hat{b}_l\hat{b}_{l'}^+ - \hat{b}_{l'}^+\hat{b}_l = [\hat{b}_l;\hat{b}_{l'}^+] = \delta_{ll'} \tag{4.4.4}$$

The operator $\hat{n}_l = \hat{b}_l^+\hat{b}_l$ represents the number of spin excitation for the atom l. It is easily verified that all commutation relations for the spin operators $\hat{S}_l^+, \hat{S}_l^-$ and $\hat{S}_l^z$ are fulfilled, if expressions (4.4.3) are substituted into them, with (4.4.4) taken into account. The use of Eq. (4.4.3) in the Hamiltonian (4.4.2) gives at low temperatures (i.e. when $\frac{1}{S}\langle\hat{n}_l\rangle \ll 1$) the following infinite series (Akhieser *et al.*, 1967):

$$\hat{\mathrm{H}} = E_0 + \hat{\mathrm{H}}_0 + \hat{\mathrm{H}}_{\mathrm{int}} \tag{4.4.5}$$

where

$$E_0 = -\gamma_0\cdot S\cdot N - \frac{S^2}{2}\sum_{\substack{l_1,l_2\\ l_1\neq l_2}} J_{l_1l_2} \tag{4.4.6}$$

$$\hat{\mathrm{H}}_0 = \gamma_0\cdot\sum_{l=1}^{N}\hat{b}_l^+\hat{b}_l \tag{4.4.7}$$

$$\hat{\mathrm{H}}_{\mathrm{int}} = \hat{\mathrm{H}}_2 + \hat{\mathrm{H}}_4 + \ldots \tag{4.4.8}$$

$$\hat{\mathrm{H}}_2 = -S \sum_{(l_1 \neq l_2)} J_{l_1 l_2} (\hat{b}_{l_1}^+ \hat{b}_{l_2} - \hat{b}_{l_1}^+ \hat{b}_{l_1}) \tag{4.4.9}$$

$$\hat{\mathrm{H}}_4 = \frac{1}{2} \sum_{(l_1 \neq l_2)} J_{l_1 l_2} \left\{ \frac{1}{2} \hat{b}_{l_2}^+ (\hat{b}_{l_2}^+ \hat{b}_{l_2} + \hat{b}_{l_1}^+ \hat{b}_{l_1}) \cdot \hat{b}_{l_1} - \hat{b}_{l_1}^+ \hat{b}_{l_1} \hat{b}_{l_2}^+ \hat{b}_{l_2} \right\} \tag{4.4.10}$$

and so on. Now the free energy of the amorphous magnet (4.4.1) takes the form (Krasny *et al.*, 1993c):

$$F = -\gamma_0 \cdot S \cdot N - \frac{S^2}{2} \overline{\sum_{(l_1 \neq l_2)} J_{l_1 l_2}} - \frac{1}{\beta} \overline{\ln\{\mathrm{Tr}[\exp(-\beta(\hat{\mathrm{H}}_0 + \hat{\mathrm{H}}_{\mathrm{int}}))]\}} \tag{4.4.11}$$

In cases where $\hat{\mathrm{H}}_{\mathrm{int}}$ can be considered as a small quantity, we can again use the thermodynamic perturbation theory, as in the preceding chapter, and represent F in the form (see Section 4.2.2):

$$F = -\gamma_0 \cdot S \cdot N - \frac{S^2}{2} \overline{\sum_{(l_1 \neq l_2)} J_{l_1 l_2}} + f_0 - \frac{1}{\beta} \{ \overline{\langle \hat{\sigma}(\beta) \rangle_{0c}} - 1 \} \tag{4.4.12}$$

where

$$\hat{\sigma}(\beta) = \sum_{n=0}^{\infty} \frac{(-1)^n}{n!} \int_0^\beta \mathrm{d}\tau_1 \ldots \int_0^\beta \mathrm{d}\tau_n \hat{T}_\tau \{ \hat{\mathrm{H}}_{\mathrm{int}}(\tau_1) \ldots \hat{\mathrm{H}}_{\mathrm{int}}(\tau_n) \} \tag{4.4.13}$$

$$\hat{\mathrm{H}}_{\mathrm{int}}(\tau) = \mathrm{e}^{\hat{\mathrm{H}}_0 \tau} \cdot \hat{\mathrm{H}}_{\mathrm{int}} \cdot \mathrm{e}^{-\hat{\mathrm{H}}_0 \tau}$$

The averaging in the angular brackets of Eq. (4.4.12) is performed with the eigenstates of the operator $\hat{\mathrm{H}}_0$, and

$$f_0 = -\frac{1}{\beta} \overline{\ln\{\mathrm{Tr}[\exp(-\beta \hat{\mathrm{H}}_0)]\}} \tag{4.4.14}$$

The last term in the brackets in Eq. (4.4.12) represents a sum of all *connected* diagrams entering the diagrammatic expansion of $\langle \sigma(\beta) \rangle_0$ (see below).

In the following we notice that the operator $\hat{\mathrm{H}}_0$, which only counts the z-component of the total spin, commutes with $\hat{\mathrm{H}}_{\mathrm{int}}$. So the eigenvalues of the operator $\hat{\mathrm{H}}_0$ are independent of the coordinates of the atoms, and also in the eigenfunctions we can change the positions of the atoms without changing the eigenvalues. As a consequence, in Eq. (4.4.11) the order of averaging can be inverted, i.e. we can *start* with averaging over the random positions of atoms, and afterwards over the spin variables (see Appendix B):

$$F = -\gamma_0 \cdot S \cdot N - \frac{S^2}{2}\overline{\sum_{l_1 \neq l_2} J_{l_1 l_2}} + f_0 - \frac{1}{\beta}\{\langle\overline{\hat{\sigma}(\beta)}\rangle_{0c} - 1\} \tag{4.4.15}$$

Let us make the following substitution in Eq. (4.4.13):

$$\hat{\mathrm{H}}_{\mathrm{int}}(\tau) = \hat{\hat{\mathrm{H}}}_{\mathrm{int}}(\tau) + \Delta\hat{\mathrm{H}}_{\mathrm{int}}(\tau) \tag{4.4.16}$$

where

$$\Delta\hat{\mathrm{H}}_{\mathrm{int}}(\tau) = \hat{\mathrm{H}}_{\mathrm{int}}(\tau) - \hat{\hat{\mathrm{H}}}_{\mathrm{int}}(\tau) \tag{4.4.17}$$

Then, after some regrouping of terms, the operator $\overline{\hat{\sigma}(\beta)}$ can be expressed in the form:

$$\begin{aligned}\overline{\hat{\sigma}(\beta)} &= \sum_{n=0}^{\infty}\frac{(-1)^n}{n!}\int_0^{\beta} \mathrm{d}\tau_1 \ldots \int_0^{\beta} \mathrm{d}\tau_n \hat{T}_{\tau}\{\hat{\mathrm{H}}_{\mathrm{int}}(\tau_1)\ldots\hat{\mathrm{H}}_{\mathrm{int}}(\tau_n)\} \\ &= \hat{T}_{\tau} \cdot \exp\left[-\int_0^{\infty} \hat{\mathrm{H}}_{\mathrm{int}}(\tau)\mathrm{d}\tau\right],\end{aligned} \tag{4.4.18}$$

where

$$\begin{aligned}\hat{\mathrm{H}}_{\mathrm{int}}(\tau_1) = \hat{\hat{\mathrm{H}}}_{\mathrm{int}}(\tau_1) &- \frac{(-1)^2}{2!}\int_0^{\beta}\mathrm{d}\tau_2 \cdot \hat{T}_{\tau}\overline{\{\Delta\hat{\mathrm{H}}_{\mathrm{int}}(\tau_1)\cdot\Delta\hat{\mathrm{H}}_{\mathrm{int}}(\tau_2)\}} \\ &- \frac{(-1)^3}{3!}\int_0^{\beta}\mathrm{d}\tau_2\int_0^{\beta}\mathrm{d}\tau_3 \cdot \hat{T}_{\tau}\overline{\{\Delta\hat{\mathrm{H}}_{\mathrm{int}}(\tau_1)\cdot\Delta\hat{\mathrm{H}}_{\mathrm{int}}(\tau_2)\cdot\Delta\hat{\mathrm{H}}_{\mathrm{int}}(\tau_3)\}} \\ &- \frac{(-1)^4}{4!}\int_0^{\beta}\mathrm{d}\tau_2\int_0^{\beta}\mathrm{d}\tau_3\int_0^{\beta}\mathrm{d}\tau_4 \cdot \hat{T}_{\tau}\{\overline{\Delta\hat{\mathrm{H}}_{\mathrm{int}}(\tau_1)\cdot\Delta\hat{\mathrm{H}}_{\mathrm{int}}(\tau_2)\cdot\Delta\hat{\mathrm{H}}_{\mathrm{int}}(\tau_3)\cdot\Delta\hat{\mathrm{H}}_{\mathrm{int}}(\tau_4)} \\ &- 3\overline{\Delta\hat{\mathrm{H}}_{\mathrm{int}}(\tau_1)\cdot\Delta\hat{\mathrm{H}}_{\mathrm{int}}(\tau_2)}\cdot\overline{\Delta\hat{\mathrm{H}}_{\mathrm{int}}(\tau_3)\cdot\Delta\hat{\mathrm{H}}_{\mathrm{int}}(\tau_4)}\} + \ldots\end{aligned} \tag{4.4.19}$$

Therefore

$$\begin{aligned}F &= -\gamma_0 \cdot S \cdot N - \frac{S^2}{2}\overline{\sum_{l_1 \neq l_2} J_{l_1 l_2}} + f_0 - \frac{1}{\beta}\left\{\left\langle \hat{T}_{\tau}\exp\left[-\int_0^{\beta}\hat{\mathrm{H}}_{\mathrm{int}}(\tau)\mathrm{d}\tau\right]\right\rangle_{0c} - 1\right\} \\ &= -\gamma_0 \cdot S \cdot N - \frac{S^2}{2}\overline{\sum_{l_1 \neq l_2} J_{l_1 l_2}} + f_0 - \frac{1}{\beta}\ln\left\{\left\langle \hat{T}_{\tau}\cdot\exp\left[-\int_0^{\beta}\hat{\mathrm{H}}_{\mathrm{int}}(\tau)\mathrm{d}\tau\right]\right\rangle_0\right\}\end{aligned} \tag{4.4.20}$$

Now the operator $\hat{\mathrm{H}}_{\mathrm{int}}$ commutes with $\hat{\mathrm{H}}_0$, because

$$\left[\left(\hat{b}_{l_1}^{+}\cdot\sqrt{1-\frac{\hat{b}_{l_1}^{+}\hat{b}_{l_1}}{2S}}\cdot\sqrt{1-\frac{\hat{b}_{l_2}^{+}\hat{b}_{l_2}}{2S}}\cdot\hat{b}_{l_2}\right);\ \sum_{l}\hat{b}_{l}^{+}\hat{b}_{l}\right]=0$$

This means that $\hat{\mathrm{H}}_{\mathrm{int}}(\tau)$ and, as a consequence, $\hat{\hat{\mathrm{H}}}_{\mathrm{int}}(\tau)$ and $\Delta\hat{\mathrm{H}}_{\mathrm{int}}(\tau)$ also do not depend on τ. Therefore

$$\hat{\mathrm{H}}_{\mathrm{int}}(\tau)\equiv\hat{\mathrm{H}}_{\mathrm{int}}=\hat{\hat{\mathrm{H}}}_{\mathrm{int}}-\frac{(-1)^2}{2!}\beta\overline{\Delta\hat{\mathrm{H}}_{\mathrm{int}}^{2}}$$
$$-\frac{(-1)^3}{3!}\beta^2\overline{\Delta\hat{\mathrm{H}}_{\mathrm{int}}^{3}}-\frac{(-1)^4}{4!}\beta^3\{\overline{\Delta\hat{\mathrm{H}}_{\mathrm{int}}^{4}}-3(\overline{\Delta\hat{\mathrm{H}}_{\mathrm{int}}^{2}})^2\}+\ldots \tag{4.4.21}$$

and the free energy of amorphous magnets takes the following form:

$$F=-\gamma_0\cdot S\cdot N-\frac{S^2}{2}\overline{\sum_{l_1\neq l_2}J_{l_1l_2}}-\frac{1}{\beta}\ln\{\mathrm{Tr}[\exp(-\beta(\hat{\mathrm{H}}_0+\hat{\mathrm{H}}_{\mathrm{int}}))]\} \tag{4.4.22}$$

Because

$$\ln\left\{\mathrm{Tr}\left[\exp\left(-\beta\gamma_0\sum_{l=1}^{N}\hat{b}_{l}^{+}\hat{b}_{l}\right)\right]\right\}=\ln\left[\frac{1}{1-\exp(-\gamma_0\beta)}\right] \tag{4.4.23}$$

does not depend on the atomic displacements (see above) we can proceed from discrete to continuous variables, that is we can change $\hat{b}^{+}(\vec{R}_l)$ and $\hat{b}(\vec{R}_l)$ for $\hat{b}_l^{+}$ and $\hat{b}_l$, and consider the former as continuous functions. To do this we divide the volume V into the N elementary sub-volumes ΔV_i and bring each one into a correspondence with Bose particle operator $\Delta\hat{N}_i$. Then Eq. (4.4.23) can be rewritten in the following form:

$$\ln\left\{\mathrm{Tr}\left[\exp\left(-\beta\gamma_0\sum_{l}\hat{b}_{l}^{+}\hat{b}_{l}\right)\right]\right\}=\ln\left[\frac{1}{1-\exp(-\gamma_0\beta)}\right]^{N}$$
$$=\ln\left\{\prod_{i=1}^{N}\mathrm{Tr}[\exp(-\beta\gamma_0\Delta\hat{N}_i)]\right\}=\ln\left\{\mathrm{Tr}\left[\exp\left(-\beta\gamma_0\sum_{i=1}^{N}\Delta\hat{N}_i\right)\right]\right\}$$

Let us define

$$\Delta\hat{N}_i=\frac{1}{v_0}\hat{b}^{+}(\vec{r}_i)\hat{b}(\vec{r}_i)\cdot\Delta V_i$$

where operators $\hat{b}^{+}(\vec{r}_i)$ and $\hat{b}(\vec{r}_i)$ obey the commutation relation

$$\hat{b}(\vec{r}_i)\hat{b}^{+}(\vec{r}_j)-\hat{b}^{+}(\vec{r}_j)\hat{b}(\vec{r}_i)=v_0\frac{\delta_{ij}}{\Delta V_i}$$

Now let us proceed to the continuum limit $\Delta V_i\to 0$:

$$\ln\left\{\mathrm{Tr}\left[\exp\left(-\beta y_0 \sum_l \hat{b}_l^+ \hat{b}_l\right)\right]\right\} = \ln\left\{\mathrm{Tr}\left[\exp\left(-\beta y_0 \int_V \hat{b}^+(\vec{r})\hat{b}(\vec{r})\frac{d^3 r}{v_0}\right)\right]\right\}$$

where

$$\hat{b}(\vec{r})\hat{b}^+(\vec{r}\,') - \hat{b}^+(\vec{r}\,')\hat{b}(\vec{r}) = v_0 \cdot \delta(\vec{r} - \vec{r}\,')$$

Therefore we can take in Eq. (4.4.22)

$$\hat{H}_0 = \frac{y_0}{v_0}\int d^3 r \hat{b}^+(\vec{r})\hat{b}(\vec{r}) \tag{4.4.24}$$

Now we perform a configurational average over the positions of the atoms and use the frozen-liquid model. Then

$$\begin{aligned}\overline{\sum_{l_1 \neq l_2} J_{l_1 l_2}} &= \frac{1}{v_0^2}\int d^3 r_1 \int d^3 r_2 J(\vec{r}_1 - \vec{r}_2)\cdot g(\vec{r}_1 - \vec{r}_2) \\ &= \frac{N}{v_0}\int d^3 r \cdot J(r)\cdot g(r)\end{aligned} \tag{4.4.25}$$

and the Hamiltonian takes the form:

$$\hat{H}_0 + \hat{H}_{\mathrm{int}} = \hat{H}_2 + \hat{H}_4 + \ldots \tag{4.4.26}$$

where

$$\begin{aligned}\hat{H}_2 = \hat{H}_0 + \hat{\hat{H}}_2 &= \frac{y_0}{v_0}\int d^3 r \cdot \hat{b}^+(\vec{r})\hat{b}(\vec{r}) \\ &- \frac{S}{v_0^2}\int d^3 r_1 \int d^3 r_2 \cdot J(\vec{r}_1 - \vec{r}_2)\cdot g(\vec{r}_1 - \vec{r}_2)[\hat{b}^+(\vec{r}_1)\cdot\hat{b}(\vec{r}_2) - \hat{b}^+(\vec{r}_1)\cdot\hat{b}(\vec{r}_1)]\end{aligned} \tag{4.4.27}$$

$$\begin{aligned}\hat{H}_4 = \hat{\hat{H}}_4 - \frac{\beta}{2}\overline{\Delta\hat{H}_2^2} = \hat{\hat{H}}_4 - \frac{\beta}{2}\left\{\overline{\hat{H}_2^2} - (\overline{\hat{H}_2})^2\right\} &= \frac{1}{2v_0^2}\int d^3 r_1 \int d^3 r_2 \cdot J(\vec{r}_1 - \vec{r}_2)\cdot g(\vec{r}_1 - \vec{r}_2) \\ &\times \left\{\frac{1}{2}\hat{b}^+(\vec{r}_2)[\hat{b}^+(\vec{r}_2)\hat{b}(\vec{r}_2) + \hat{b}^+(\vec{r}_1)\hat{b}(\vec{r}_1)]\hat{b}(\vec{r}_1) - \hat{b}^+(\vec{r}_1)\hat{b}(\vec{r}_1)\hat{b}^+(\vec{r}_2)\hat{b}(\vec{r}_2)\right\} \\ &- \frac{S^2\beta}{2v_0^4}\int d^3 r_1 \int d^3 r_2 \int d^3 r_3 \int d^3 r_4 \cdot J(\vec{r}_1 - \vec{r}_2)\cdot J(\vec{r}_3 - \vec{r}_4) G_4(\vec{r}_1, \vec{r}_2, \vec{r}_3, \vec{r}_4) \\ &\times \hat{b}^+(\vec{r}_1)\hat{b}^+(\vec{r}_3)[\hat{b}(\vec{r}_2) - \hat{b}(\vec{r}_1)][\hat{b}(\vec{r}_4) - \hat{b}(\vec{r}_3)] \\ &- \frac{S^2\beta}{2v_0^3}\int d^3 r_1 \int d^3 r_2 \int d^3 r_3 \cdot J(\vec{r}_1 - \vec{r}_2)\cdot J(\vec{r}_2 - \vec{r}_3)\cdot G_3(\vec{r}_1, \vec{r}_2, \vec{r}_3) \\ &\times [\hat{b}^+(\vec{r}_1) - \hat{b}^+(\vec{r}_2)][\hat{b}(\vec{r}_3) - \hat{b}(\vec{r}_2)]\end{aligned} \tag{4.4.28}$$

and

$$g(\vec{r}_1 - \vec{r}_2) = F_2(\vec{r}_1, \vec{r}_2) = \upsilon_0^2 \overline{\sum_{l_1 \neq l_2} \delta(\vec{r}_1 - \vec{R}_{l_1}) \cdot \delta(\vec{r}_2 - \vec{R}_{l_2})} \tag{4.4.29}$$

$$\begin{aligned} G_3(\vec{r}_1, \vec{r}_2, \vec{r}_3) &= F_3(\vec{r}_1, \vec{r}_2, \vec{r}_3) - g(\vec{r}_1 - \vec{r}_2) g(\vec{r}_2 - \vec{r}_3) \\ &= \upsilon_0^3 \overline{\sum_{l_1 \neq l_2 \neq l_3} \delta(\vec{r}_1 - \vec{R}_{l_1}) \cdot \delta(\vec{r}_2 - \vec{R}_{l_2}) \cdot \delta(\vec{r}_3 - \vec{R}_{l_3})} - g(\vec{r}_1 - \vec{r}_2) \cdot g(\vec{r}_2 - \vec{r}_3) \end{aligned} \tag{4.4.30}$$

$$\begin{aligned} &G_4(\vec{r}_1, \vec{r}_2, \vec{r}_3, \vec{r}_4) \\ &= \upsilon_0^4 \overline{\sum_{l_1 \neq l_2} \delta(\vec{r}_1 - \vec{R}_{l_1}) \cdot \delta(\vec{r}_2 - \vec{R}_{l_2}) \sum_{l_3 \neq l_4} \delta(\vec{r}_3 - \vec{R}_{l_3}) \cdot \delta(\vec{r}_4 - \vec{R}_{l_4})} - g(\vec{r}_1 - \vec{r}_2) \cdot g(\vec{r}_3 - \vec{r}_4) \end{aligned} \tag{4.4.31}$$

and so on.

The operator $\hat{H}_2$ is an effective Hamiltonian of non-interacting bosons (the quasi-magnon gas, see below). The first term in the operator $\hat{H}_4$ presents a two-magnon interaction and is of purely kinematic origin. It remains upon the transition to a crystalline magnet. The second and the third terms in $\hat{H}_4$ come out of the effective quasi-magnon interactions with micro-inhomogeneities of amorphous magnets. These terms vanish upon transition to crystals because in this case the functions G_3 and G_4 of Eqs (4.4.30) and (4.4.30) tend to zero.

In fact, at very low temperatures ($T \ll T_k$) magnons form an ideal gas of quasi-particles obeying Bose-Einstein statistics, i.e. anharmonicities do not play a role; but due to the inhomogeneous structure of the amorphous system the **k**-vector is not a good quantum number. As in Chapter 3 with the phonon case, this can be seen in inelastic neutron scattering experiments (see below), where we again find, as in the previous case of harmonic vibrational excitations, that in an amorphous sample the inelastic scattering cross-section ($d^2\sigma/dEd\Omega$) as a function of the energy $dE = \hbar d\omega$ and the momentum $dp = \hbar dq$ transferred to the neutron *cannot* simply be defined by a simple dispersion curve $\omega(q)$ and the usual relation $(d^2\sigma/dEd\Omega) \propto \delta(\omega - \omega(q))$ as in a crystal, except at low wavenumbers and perhaps again approximately in the 'roton region' $|q| \approx 2\pi/r_0$, where r_0 is the nearest-neighbor distance. However, in the thermodynamic perturbation theory presented above, if we retain only the term $\propto \hat{H}_2$, we get not only a harmonic approximation, but a special harmonic approximation *without* lifetime broadening, which we call the *'quasi-magnon' approximation*. The free energy in this approximation is

$$F_0 = -\gamma_0 \cdot S \cdot N - \frac{N}{2\upsilon_0} S^2 \int d^3r \cdot J(r) \cdot g(r) - \frac{1}{\beta} \ln\{\mathrm{Tr}[\exp(-\beta \hat{H}_2)]\} \tag{4.4.32}$$

Let us introduce the Fourier representation:

$$\begin{cases} \hat{b}(\vec{r}) = \sqrt{\frac{\upsilon_0}{(2\pi)^3}} \int \hat{b}(\vec{k}) \cdot e^{i\vec{k}\cdot\vec{r}} d^3k \\ \hat{b}^+(\vec{r}) = \sqrt{\frac{\upsilon_0}{(2\pi)^3}} \int \hat{b}^+(\vec{k}) \cdot e^{i\vec{k}\cdot\vec{r}} d^3k \\ J(\vec{r}) = \frac{\upsilon_0}{(2\pi)^3} \int \tilde{J}(\vec{k}) e^{i\vec{k}\cdot\vec{r}} d^3k \\ g(\vec{r}) = 1 + \frac{\upsilon_0}{(2\pi)^3} \int [a(\vec{k}) - 1] e^{i\vec{k}\cdot\vec{r}} d^3k \end{cases} \quad .. \tag{4.4.33}$$

It is easy to verify that operators $\hat{b}^+(\vec{k})$ and $\hat{b}(\vec{k})$ obey the following commutation relations:

$$[\hat{b}(\vec{k});\ \hat{b}^+(\vec{k}')] = \delta(k - k')$$

Then the Hamiltonian $\hat{H}_2$ and free energy F_0 are given by

$$\hat{H}_2 = \int d^3k \cdot \varepsilon(\vec{k}) \hat{b}^+(\vec{k}) \hat{b}(\vec{k}) \tag{4.4.34}$$

$$F_0 = -\gamma_0 \cdot S \cdot N - \frac{N}{2\upsilon_0^2} S^2 \int d^3r \cdot J(r) \cdot g(r) + \frac{1}{\beta} N \frac{\upsilon_0}{(2\pi)^3} \int d^3k \ln\{1 - \exp[-\beta\varepsilon(\vec{k})]\} \tag{4.4.35}$$

where

$$\varepsilon_0(\vec{k}) = g\mu_0 H + S[\tilde{J}(0) - \tilde{J}(\vec{k})] + S \frac{\upsilon_0}{(2\pi)^3} \int d^3k_1 [\tilde{J}(\vec{k}_1) - \tilde{J}(\vec{k} - \vec{k}_1)] \cdot [a(\vec{k}_1) - 1] \tag{4.4.36}$$

is the energy spectrum of magnon excitations in amorphous ferromagnets in our approach.

An analogous expression for $\varepsilon(\vec{k})$ was obtained in Section 4.3.16 with the Green's function method.

At low temperatures the main contribution to the third-term integral in Eq. (4.4.35) gives the region of small wavevectors k where the dispersion law is given by

$$\varepsilon_0(\vec{k}) = g\mu_0 H + D_0 k^2 \tag{4.4.37}$$

where

$$\begin{cases} D = D_{0_{cr}} \left\{ 1 - \frac{4\pi}{3 D_{0_{cr}}} S \frac{\upsilon_0}{(2\pi)^3} \int_0^\infty dk_1 \frac{\partial}{\partial k_1} \left(k_1^2 \frac{\partial \tilde{J}}{\partial k_1} \right) [a(k_1) - 1] \right\} \\ D_{0_{cr}} = -\frac{1}{2} S \left. \frac{\partial^2 \tilde{J}}{\partial k^2} \right|_{k=0} \end{cases} \tag{4.4.38}$$

From $\varepsilon_0(k)$ it is possible to calculate the free energy F_0 and then the spontaneous magnetic moment density and the magnetic contribution to the heat capacity of the amorphous magnet at $H=0$:

$$M_z = -\frac{1}{V}\frac{\partial F_0}{\partial H}\bigg|_{H=0} = M_0(1 - B_0 T^{3/2}) \tag{4.4.39}$$

$$C_s = -\frac{1}{V}T\frac{\partial^2 F_0}{\partial T^2}\bigg|_{H=0} = A_0 T^{3/2} \tag{4.4.40}$$

where

$$M_0 = \frac{3\mu_0 S}{v_0}$$

$$B_0 = 0.116\frac{\upsilon_0}{2S}\left(\frac{k_B}{D_0}\right)^{3/2} \tag{4.4.41}$$

$$A_0 = 0.113 k_B\left(\frac{k_B}{D_0}\right)^{3/2} \tag{4.4.42}$$

So at low temperatures the *magnetic moment* M_z of amorphous (and crystalline) ferromagnets decreases, and the *heat capacity* increases, with increasing temperature according to Bloch's famous $T^{3/2}$ law, $M_z(T) = M_z(0) - \text{const.}\ T^{3/2}$. But at the same time in amorphous magnets M_z decreases more quickly than in the crystalline case, because for the coefficient B_0 we have $B_0 \sim 1/D_0^{3/2}$, and the spin-wave stiffness D_0 for amorphous magnets is less than $D_{0\mathrm{cr}}$ in crystals. This fact is confirmed by experiments (Section 4.1). Experiments also show that D_{exp}, as obtained from experiments on the temperature dependence of M_z and C_s, is *less* than D_0. This means that B is influenced by quasi-magnon scattering on density fluctuations (Krasny *et al.*, 1993).

To take into account this effect we keep the two terms $\hat{H}_2$ and $\hat{H}_4$ in the Hamiltonian (4.4.26). Now, setting $\hat{H}_4$ infinitesimally small we employ again the thermodynamic perturbation theory. Then a correction to the free energy is

$$\Delta F = \mathrm{Tr}\{\hat{H}_4 \exp[F_0 - \hat{H}_2]\} \tag{4.4.43}$$

If Eq. (4.4.28) is converted to its Fourier transform, the correction will be

$$\begin{aligned}\Delta F = &-\frac{\upsilon_0}{(2\pi)^3}\frac{1}{2}\frac{V}{(2\pi)^3}\int d^3k_1 \int d^3k_2 \cdot n_{k_1}\cdot n_{k_2}\\ &\times\{-\tilde{I}(\vec{k}_2-\vec{k}_1)+\tilde{I}(0)-\tilde{I}(\vec{k}_1)-\tilde{I}(\vec{k}_2)+S^2\beta[\varphi_4(\vec{k}_1,\vec{k}_2;\vec{k}_1,\vec{k}_2)\\ &+\varphi_4(\vec{k}_1,\vec{k}_2;\vec{k}_2,\vec{k}_1)]\} - \frac{S^2\beta}{2}\frac{V}{(2\pi)^3}\int d^3k\, n_{\vec{k}}\,\varphi_3(\vec{k})\end{aligned} \tag{4.4.44}$$

where

$$n_{\vec{k}} = \{\exp[\beta\varepsilon_0(\vec{k})] - 1\}^{-1}$$

$$\tilde{I}(\vec{k}) = \tilde{J}(\vec{k}) + \frac{v_0}{(2\pi)^3}\int d^3k_1 \tilde{J}(\vec{k}+\vec{k}_1)[a(\vec{k}_1) - 1]$$

$$\varphi_3(\vec{k}) = \frac{v_0^2}{(2\pi)^6}\int d^3k_1 \int d^3k_2\, \tilde{G}_3(\vec{k}_1, \vec{k}_2, \vec{k}_1 - \vec{k}_2) \times [\tilde{J}(\vec{k}+\vec{k}_1) - \tilde{J}(\vec{k}_1)] \cdot [\tilde{J}(\vec{k}+\vec{k}_2) - \tilde{J}(\vec{k}_2)]$$

$$\varphi_4(\vec{k}_1, \vec{k}_2; \vec{k}_3, \vec{k}_4) = \frac{v_0^2}{(2\pi)^6}\int d^3k_5 \int d^3k_6\, \tilde{G}_4(\vec{k}_5, \vec{k}_1 - \vec{k}_3 - \vec{k}_5, \vec{k}_6, \vec{k}_2 - \vec{k}_4 - \vec{k}_6) \times [\tilde{J}(\vec{k}_1 - \vec{k}_5 - \vec{k}_3) - \tilde{J}(\vec{k}_1 - \vec{k}_5)] \cdot [\tilde{J}(\vec{k}_2 - \vec{k}_6 - \vec{k}_4) - \tilde{J}(\vec{k}_2 - \vec{k}_6)]$$

Here $\tilde{G}_3$ and $\tilde{G}_4$ are the Fourier transforms of the functions G_3 and G_4. Because the function $n_{\vec{k}}$ falls off steeply when $|\vec{k}| \to \infty$, we approximate $\varepsilon_0(\vec{k})$ by

$$\varepsilon_0(\vec{k}) = g\mu_0 H + D_0 k^2 + D_1 k^4 + D_2 k^6 + \ldots$$

(D_1 and D_2 will be neglected below) and expand the functions under the integrals together with $n_{\vec{k}}$ into a Taylor series, retaining only the first non-vanishing terms. It can readily be shown that this leads to corrections for the magnetic moment and the heat capacity, which are of the order $D_1 b_{5/2} T^{5/2} + b_3 T^3 + D_2 b_{7/2} T^{7/2} + b_4 T^4$ and only slightly influence moment and specific heat at low temperatures. Moreover the term $b_3 T^3$ vanishes rigorously (at least in crystals) according to a famous paper by F. Dyson (1956); so if the corrections, of which only the term $\propto D_1$ is sometimes measured, are neglected altogether in the following, we get

$$\Delta F \approx -\frac{S^2\beta}{2}\frac{V}{(2\pi)^3}\int d^3k \cdot n_{\vec{k}} \cdot \varphi(\vec{k}) \approx -\gamma \cdot \frac{V}{(2\pi)^3}\int d^3k \cdot n_{\vec{k}} \cdot k^2 \qquad (4.4.45)$$

where

$$\gamma = \frac{S^2\beta}{6}\frac{v_0^2}{(2\pi)^6}\int d^3k_1 \int d^3k_2 \tilde{G}_3(-\vec{k}_1, \vec{k}_2, \vec{k}_1 - \vec{k}_2)\frac{\partial\tilde{J}(\vec{k}_1)}{\partial\vec{k}_1} \cdot \frac{\partial\tilde{J}(\vec{k}_2)}{\partial\vec{k}_2}$$

The coefficient γ is evaluated in superposition approximation. Then,

$$F_3(\vec{r}_1, \vec{r}_2, \vec{r}_1 - \vec{r}_2) = g(\vec{r}_1) \cdot g(\vec{r}_2) \cdot g(\vec{r}_1 - \vec{r}_2)$$

$$\tilde{G}(-\vec{k}_1, \vec{k}_2, \vec{k}_1 - \vec{k}_2) = \frac{v_0}{(2\pi)^3}\int d^3k_3\, \tilde{g}(\vec{k}_1 + \vec{k}_3) \cdot \tilde{g}(\vec{k}_2 + \vec{k}_3)[a(\vec{k}_3) - 1]$$

and γ takes the form

$$\gamma = \frac{S^2\beta}{6}\frac{v_0^3}{(2\pi)^9}\int d^3k_1\,\tilde{g}(\vec{k}_1)\int d^3k_2\,\tilde{g}(\vec{k}_2)\int d^3k_3[a(\vec{k}_3)-1] \times \frac{\partial\tilde{J}(\vec{k}_1-\vec{k}_3)}{\partial\vec{k}_1}\cdot\frac{\partial\tilde{J}(\vec{k}_2-\vec{k}_3)}{\partial\vec{k}_2} \tag{4.4.46}$$

Let us define the *direct correlation function* $c(\vec{r})$ of which the Fourier transform $\tilde{c}(\vec{k})$ is related to the structure factor $a(\vec{k})$ as follows (Kcharkov *et al.*, 1979; Belashchenko, 1986):

$$a(\vec{k}) - 1 = \frac{\tilde{c}(\vec{k})}{1-\tilde{c}(\vec{k})}$$

Substituting this expression into Eq. (4.4.44), and taking into account that the functions $\tilde{J}(\vec{k}-\vec{k}_3)$ and $\tilde{c}(\vec{k}_3)$ tend steeply to zero when $|k_3| \to \infty$, we can put

$$1-\tilde{c}(\vec{k}) \approx 1 - c(0) = \left(\frac{\chi_T}{v_0\beta}\right)^{-1}$$

where χ_T is the isothermal compressibility of the material.

Expanding $\tilde{J}(\vec{k}_1-\vec{k}_3)$ and $\tilde{J}(\vec{k}_2-\vec{k}_3)$ into a Taylor series in terms of the variable $\vec{k}_3$, and taking into account a power-law decrease in $\tilde{J}(\vec{k})$, we can now demonstrate that the coefficient γ is, with a high accuracy,

$$\gamma \approx 2\frac{\chi_T}{v_0}D_0^2\left(-\left.\frac{d^2c(r)}{dr^2}\right|_{r=r_0}\right) \tag{4.4.47}$$

Hence, corrections to the magnetic moment and heat capacity are defined by the expressions

$$\begin{cases} \Delta M = -\left.\dfrac{1}{V}\dfrac{\partial\Delta F}{\partial H}\right|_{H=0} = -M_0\cdot B_1\cdot T^{3/2} \\ \Delta C_s = -\left.\dfrac{1}{V}T\dfrac{\partial^2\Delta F}{\partial T^2}\right|_{H=0} = A_1\cdot T^{3/2} \end{cases} \tag{4.4.48}$$

where

$$\begin{cases} B_1 = B_0\left(-3\dfrac{\chi_T}{v_0}D_0\left.\dfrac{d^2c}{dr^2}\right|_{r=r_0}\right) \\ A_1 = A_0\left(-3\dfrac{\chi_T}{v_0}D_0\left.\dfrac{d^2c}{dr^2}\right|_{r=r_0}\right) \end{cases} \tag{4.4.49}$$

The expressions show *fluctuation corrections* to the thermodynamic functions of amorphous magnets at low temperatures, behaving as $T^{3/2}$, and in view of these corrections the stiffness coefficient D is renormalized also, i.e. we get

$$\begin{cases} M_z = M_0[1 - B \cdot T^{3/2}] \\ C_s = A \cdot T^{3/2} \end{cases} \tag{4.4.50}$$

with

$$\begin{cases} B = B_0 + B_1 = 0.116 \dfrac{v_0}{2S} \left(\dfrac{k_B}{D}\right)^{3/2} \\ A = A_0 + A_1 = 0.113\, k_B \left(\dfrac{k_B}{D}\right)^{3/2} \\ D = \dfrac{D_0}{\left[1 - 3D_0 \dfrac{\chi_T}{v_0} \dfrac{d^2 c}{dr^2}\bigg|_{r=r_0}\right]^{2/3}} \end{cases} \tag{4.4.51}$$

As a rule,

$$\frac{d^2 c}{dr^2}\bigg|_{r=r_0} < 0$$

and therefore D is usually *significantly smaller* than D_0 (see below), in agreement with experiments.

To evaluate D_0/D quantitatively, we again take advantage of the hard-sphere model of frozen liquid metals, where expressions for $c(r)$ and χ_T are known (see Kcharkov *et al.*, 1979). In terms of this model,

$$\frac{D_0}{D} = \left[1 + \frac{27}{32(S+1)} \left(\frac{r_0}{a}\right)^3\right]^{2/3}$$

Passing in this equation from r_0 to the exchange interaction radius R_0 (see Eq. (4.2.16)) and putting $S = \frac{1}{2}$, we have $D_0/D = 1.1$ for the exchange integral $J_1(r)$ when $R_0/a \sim 1.1$, while $D_0/D = 1.2$ for $J_2(r)$ and $R_0/a \sim 1$.

The above evaluations show that D, which enters into the thermodynamical formulae (4.4.50), is 10–20% smaller than the spin-wave stiffness D_0 at $T=0$ determined from neutron scattering experiments and theoretically with the Green's function method.

4.5
Beyond the Quasi-Magnon Approach: Computer Simulations

One of the present authors (Krey, 1978) has performed computer simulations for a soft-sphere model of an amorphous ferromagnet, where the magnetism has also been described by a Heisenberg model. Again the exchange integral $J_{l,m}$ ($=J(r)$) depended in a similar way as above on the distance $r = r_{lm}$ between the sites l and m, namely as

$$J(r) = J_0 \exp\{a[(r/r_0) - 1]\} \tag{4.5.1}$$

where the interaction was cut off for $r>1.25r_0$; r_0 was again the average nearest-neighbor distance of the computer model, which was the same soft-sphere model generated by L. von Heimendahl (1979) for his study of phonon excitations in amorphous metals (see Section 3.4); a was a variational parameter ranging between –8 and +8.

The simulations mentioned above go beyond the analytical quasi-magnon approach, because the one-magnon excitations of the present model of an amorphous Heisenberg ferromagnet were principally calculated *exactly* by a Green's function formalism evaluated numerically with a continued-fraction technique, using the *realistic* non-homogeneous 'scattering potential' where $\mathbf{k}$ is not a good quantum number and the inelastic neutron scattering cross-section $(\mathrm{d}^2\sigma/\mathrm{d}E\mathrm{d}\Omega)(\mathbf{k},\omega)$ depends in a non-trivial way on both arguments (see Section 3.4).

In this way, numerical results for the energy density $g(E)$ of one-magnon excitations have been obtained, which then were used in the Tyablikov decoupling approximation (Tyablikov, 1965) of the thermodynamic Green's function, to calculate the temperature dependence of the quantity $\sigma(T) := M_s(T)/M_s(0)$ for the whole temperature range between $T=0$ and the critical temperature T_C. At low temperatures the approach is asymptotically exact, i.e. it yields correctly the decrease due to non-interacting magnons. At higher temperatures it corresponds to an approximation that is better than the usual molecular field approach; namely altogether the calculation corresponds to a non-interacting Boson approximation of the magnons, but with a temperature-dependent *renormalization of the exchange*, namely $J(r) \to J(r)\sigma(T)$ (see, e.g., Tyablikov, 1965).

This leads for $s=1/2$ to the formula

$$\sigma(T) = \left\{1 + 2\int g(E)\mathrm{d}E/(\exp[\sigma(T)E/(k_B T)] - 1)\right\}^{-1} \tag{4.5.2}$$

Numerical results are presented in Fig. 4.5.1, taken from Krey (1978).

The behavior up to roughly $T_C/3$ can well be described by an extended Bloch's law, namely for $s=1/2$

$$M_s(T)/M_s(0) = 1 - 0.587(T/T_C(a))^{3/2} - 0.156(T/T_C(0))^{5/2} + \dots \tag{4.5.3}$$

This result displays a remarkable amount of universality: the prefactor in front of the $T^{5/2}$ term does not depend at all on a according to this approximation, whereas the prefactor in front of $T^{3/2}$ is 0.587 $T_C(a)^{-3/2}$. For $s \neq 1/2$ we have found similar results, e.g. for $s=1$ the numbers 0.587 and 0.156 in Eq. (4.5.3) are replaced by 0.512 and 0.195.

As already mentioned, methods and results for the *magnon* excitations of the amorphous ferromagnet in the above-mentioned simulation (Krey, 1978) are similar to those obtained by L. von Heimendahl (1979) for the corresponding *phonon* case (see Section 3.5). Again, as in the case of the quasi-phonon approach, the analytical

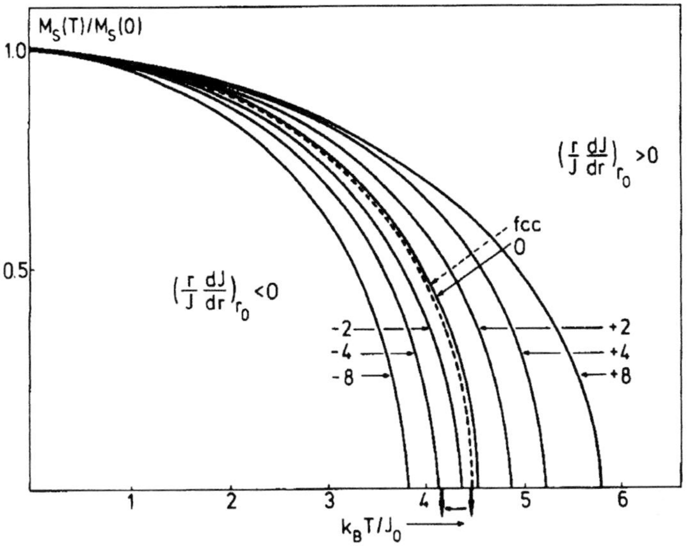

Fig. 4.5.1 For $s=1/2$ the reduced magnetization $\sigma(T):=M_s(T)/M_s(0)$ is presented over the reduced temperature $k_B T/J_0$. The curve parameter α and the exchange unit J_0 have been defined in Eq. (4.5.1). For $\alpha=0$ the result is hardly distinguishable from the result obtained for the fcc lattice, the dashed line, which leads to $k_B T_C = 4.45 J_0$. This corresponds to the upper arrow on the abscissa. The lower arrow at $4.15 J_0$ corresponds to the *exact* value of $k_B T_C$ for the fcc lattice (Rushbrook and Wood, 1958), which gives an impression of the quality of the results

results concerning the thermodynamics of the system, as obtained by the quasi-magnon approach described in the preceding sections, are essentially confirmed by the simulation results. In particular, the simulation yielded not only Bloch's famous $T^{3/2}$ law, but also the $T^{5/2}$ correction in good agreement with experiments, e.g. Chien and Hasegawa (1977). Thereby it was observed – as already mentioned above – that not only was the prefactor in front of the $(T/T_C)^{3/2}$ term four to five times larger in the amorphous system than in corresponding crystalline ones, but also the range of validity of the behavior $M(T) = M(0) - b_{3/2} T^{3/2} - b_{5/2} T^{5/2} + \dots$ was unusually large.

4.6
The Thermodynamics of Amorphous Ferrimagnets

In a recent publication (Krasny *et al.*, 1997), we have also treated the magnetism of *amorphous ferrimagnets*. Such systems, as for example the amorphous alloy $Gd_{1-x}Co_x$, are quite interesting for applications in magnetic storage media. The important point is that in such systems, i.e. amorphous alloys of the type $A_x B_{1-x}$, the exchange interaction between neighbors of the same type is usually ferromagnetic, i.e. $J_{AA}(\mathbf{r}_l, \mathbf{r}_m) \geq 0$ and $J_{BB}(\mathbf{r}_l, \mathbf{r}_m) \geq 0$, whereas the interaction between *different* compounds is *antiferromagnetic*, i.e. $J_{AB}(\mathbf{r}_l, \mathbf{r}_m) = J_{BA}(\mathbf{r}_l, \mathbf{r}_m) \leq 0$. In this case the total magnetization is the *difference* of the two contributions of the A-compound and the B-compound, respectively, and usually the system behaves as a *ferrimagnetic*, i.e. with a positive (or negative) difference $M=|M_A|-|M_B|$. For given concentration x, the total magnetization M vanishes usually only for one *particular* temperature, the so-called *compensation temperature* T_K, which is typically well below the critical temperature T_C: at T_C, both 'sublattice magnetizations' vanish, and the system is no longer ferrimagnetic, but paramagnetic above T_C, i.e. $M_A=M_B=0$ for $T \geq T_C$ in the case of vanishing external magnetic fields, whereas at the compensation temperature T_K both M_A and M_B are finite, but identical in magnitude.

The vicinity of the compensation point is usually the most interesting temperature region for the applications. Therefore it is of particular interest to study the question of compensation points for *amorphous* ferrimagnets systematically as a function of x. In fact, in the above-mentioned paper we have performed such a systematic study, by generalization of the techniques developed in the Sections 4.1–4.4. Based on these techniques, we have found that generally not one, but *two* compensation temperatures may exist. To draw such conclusions, only the low-temperature behavior and the behavior near T_C had to be carefully analyzed, while the rest follows from continuity arguments.

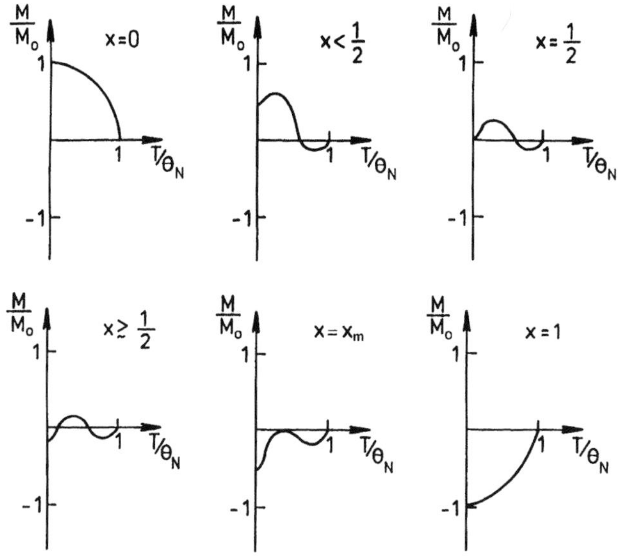

Fig. 4.6.1 Typical 'scenarios' for the magnetization curve of an amorphous two-component ferrimagnetic alloy with ferromagnetic rsp. anti-ferromagnetic exchange interaction between spins belonging to the same rsp. different alloy components. θ_N is the ordering temperature. Particular attention should be given to the *compensation points*, where the magnetization M/M_0 changes sign. For example in Fig. 4 there are two such points: x is the concentration of one of the two alloy components. In Fig. 5, the two compensation points have just merged together at the particular concentration $x=x_m$

Figure 4.6.1 presents typical 'scenarios' for the magnetization curve of an amorphous two-component magnetic alloy with exchange interactions of the above-mentioned type. Here the sequence with x increasing from 0 to 1 is from the upper left to the lower right segment of the figure (parts Fig. 1 to Fig. 6). The appearance of the *two* compensation points is visible in the lower left segment, i.e. Fig. 4. In Fig. 5, for a particular concentration $x = x_m$, the two compensation points merge together, and there we observe a *slow* quadratic increase, i.e. $M(T) \propto |T - T_K|^2$, instead of the usual rapid increase $M(T) \propto |T - T_K|^1$.

4.7 Itinerant Magnetism and Itinerant Spin-glass Behavior in Amorphous Alloys

In principle, the magnetism of amorphous metals should not be described by the Heisenberg model, which is known to correspond to *localized* spins, but instead by a theory of *itinerant magnetism*, where the *magnetic* properties of the system are described on the same level as, for example, the *conductance* properties of the metal. Thus we should use, in principle, a band-theoretic approach that would yield an accurate description, for example, of the behavior of the 3d, 4s and 4p electrons in crystalline and amorphous 3d metallic alloys.

The main signature of metallic ferromagnetism is of course the fact that the atomic magnetic moments, $\mu = 2s\mu_B$, do not correspond to integer or half-integer values of the effective spin quantum number s. Instead, for bcc Fe, hcp Co and fcc Ni, we have $\mu = 2.2\mu_B$, $1.7\mu_B$ and $0.6\mu_B$, respectively, and in *alloys* we often observe results depending continuously on the concentration x of the compounds involved.

However, this does not imply that the description of metallic magnetism by a Heisenberg model is simply wrong; rather this description should be considered as a *phenomenological* one, covering the most important collective itinerant excitations, the spin-wave excitations ('magnons') correctly, if the parameter s is chosen as a 'real fit-parameter', and if the spin-wave stiffness D, relating the excitation energy $\varepsilon(\mathbf{k})$ of a magnon excitation in the ferromagnetic metal to its wavenumber $\mathbf{k}$, is taken from the experiment. For Bloch's $T^{3/2}$ law, for example, it does not matter that the magnon excitations in a metal are well-defined quasi-particles only up to a typical wavenumber k_c that may be significantly smaller than the value $k_{max} = \pi/a$ corresponding to the edge of the Brillouin zone of a crystal. There is, however, one caveat to this simple correspondence.

Namely, in disordered crystalline alloys, and in amorphous metallic ferromagnets, it turns out that not only the *local directions* of the atomic moments, but also the *local magnitudes* μ_l of the atomic magnetic moment for the atom l depend strongly on the position, i.e. the surrounding of that site, and the electronic structure of the disordered metallic ferromagnet is not of secondary importance for the magnetism, but is actually the dominating entity. So it makes no sense to think of the different 'atomic spins', which can also be defined in an amorphous metal, just as different vectors of fixed length, which only change their direction. Rather, changes in the local spin directions may change the electronic structure in an essential way, which can lead to important changes in the effective exchange integrals describing the mutual coupling of neighbored spins. So these effective exchange integrals are secondary quantities, not the primary ones (Paintner *et al.*, 1996).

Concerning the local properties, the *volume* of the *Voronoi* polyhedron surrounding the atom considered seems to play an important role.[1)] This is particularly true for iron-rich amorphous alloys such as $Fe_{1-x}B_x$ or $Fe_{1-x}Zr_x$. For example, the magnetic properties of the last-mentioned compound can strongly be influenced by *hydrogenation*, i.e. for iron-rich amorphous $Fe_{1-x}Zr_x$ we can introduce two H atoms for every Zr, leading to an amorphous ternary compound $Fe_{1-x}Zr_xH_y$, with $0 \le y \le 2x$. In numerical calculations we have found that the drastic enhancement of the atomic moment induced by such a hydrogenation is actually due to the significant increase in the volume of the system, whereas actually the alloying compound H itself plays a rather passive role (Krompiewski *et al.*, 1989 a, b).

In Fig. 4.7.1 the calculated behavior of the magnetization m^{Fe} per Fe atom of the amorphous alloys $Fe_{1-x}Zr_xH_y$ with $y=0$ and $y=2x$ is presented over x, as calculated for computer samples of 54 Fe and/or Zr atoms plus the necessary H atoms for $T=0$ K. The computer models have the correct density; measurements were performed at He temperatures (i.e. $T=4$ K).

The amorphous alloy exists at room temperature for $x \ge 0.07$ and remains magnetic up to $x=0.6$. There is at first a linear *increase* in m^{Fe} with x decreasing from $x=0.6$ down to $x=0.13$, followed by a drastic *decrease* with further decrease of x down to $x=0.07$. This difference corresponds to a transition from *strong itinerant ferromagnetism* for $x>0.13$ to *weak itinerant ferromagnetism* for $x<0.13$. For $x>0.13$, the *majority-spin* d-band is full while with decreasing x there is an increasing number of holes in the *minority* d-bands; whereas for $x<0.13$ there will be holes in both bands, i.e. also in the *majority-spin* d-band. However, this transition to weak itinerant ferromagnetism is suppressed by the hydrogenation and, as already mentioned above, the calculation shows through the results presented by the crosses that the reason for this effect is not the hydrogen itself, but the *volume enhancement* $\Delta V/V=0.315y$ induced by the hydrogenation.

So in these systems, and generally in iron-rich amorphous alloys, there are strong magneto-volume effects.

In the following we describe (i) how these calculations are performed, and (ii) how they have been extended to take into account *non-collinear itinerant magnetism*, i.e. where the atomic spins are not simply parallel or anti-parallel to each other, but point in various individual directions.

(i) Concerning *collinear* itinerant magnetism, we simply describe it in a semi-empirical way by a symmetrical Hamiltonian matrix acting on vectors composed out of real probability amplitudes $c(l,\alpha,s)$, describing the probability $|c(l,\alpha,s)|^2$ that the electron is found at the atom l in the orbital α with spin s ($=\uparrow$ for the 'major-

1) The *Voronoi* polyhedron surrounding a given atom is constructed by a *bisecting-plane* algorithm: the planes considered bisect the different vectors joining the atom considered with its irregularly spaced neighbors. This construction is similar to that of the so-called Wigner-Seitz cell in a crystal.

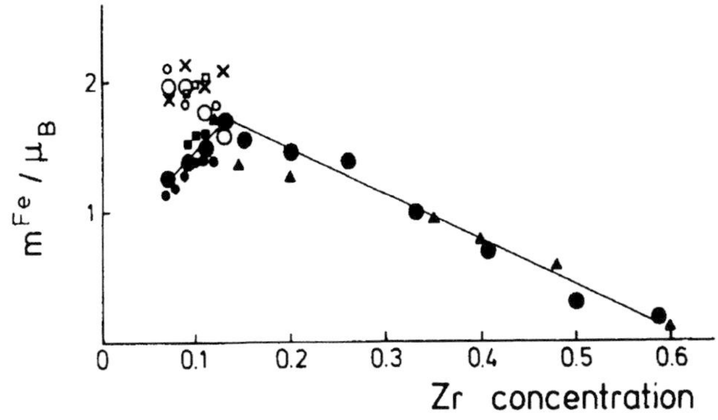

Fig. 4.7.1 The magnetization per Fe atom is given in units of μ_B for amorphous $Fe_{1-x}Zr_xH_y$ alloys. The filled symbols represent hydrogen-free samples ($y=0$) and the open samples hydrogenated ones ($y=2x$). The large circles represent results from the numerical calculation, the small circles and triangles represent experimental results. Finally, the crosses represent results for models of fictitious $Fe_{1-x}Zr_x$, without H, but with the same volume expansion $\Delta V/V=0.315\,y$, as would result with H for $y=2x$ (after Krompiewski *et al.*, 1989b)

ity-spin' electrons, and ↓ for the 'minority-spin'). We assume that the corresponding states $|l,a,s\rangle$ are already orthogonalized. If we describe the *paramagnetic* state by a *Hamiltonian matrix* $H^{para}(l,a;l',a')$, the single-particle electronic states and the corresponding single-particle energies $\varepsilon_\nu(s)$ for given spin direction are calculated from the set of eigenvalue equations

$$\varepsilon_\nu(s)c^\nu(l,a,s) = H^{para}(l,a;l',a')c^\nu(l',a',s) + U(l,a)[\langle n(l,a,-s)\rangle - \langle n^{para}(l,a)\rangle]c^\nu(l,a,s) \tag{4.7.1}$$

Here a summation over the indices l' and a', which appear twice on the r.h.s. of Eq. (4.7.1), is understood as usual (Einstein's summing convention). The important quantity $U(l,a)$ is an effective Coulomb energy, which punishes energetically that an electron with spin s is added to an orbital $|l,a\rangle$ already occupied by another electron, which according to the Pauli principle must then have spin $-s$. From our studies of disordered and amorphous compounds with Fe, Ni and Mn (Paintner *et al.*, 1996) we have obtained good results by assuming $U(l,a) \neq 0$ only if a is a d-orbital, namely $U(l,d)=5.3\,\text{eV}$, if the site l is occupied by Fe or Ni, and $2.9\,\text{eV}$ for Mn.

The quantities $\langle n(l,a,-s)\rangle$ must be calculated *self-consistently* in the ferromagnetic state – and the corresponding quantity $\langle n^{para}(l,a)\rangle$ in the paramagnetic state – from the equation

$$\langle n(l,a,-s)\rangle = \sum_{\nu=1}^{\nu f} |c^\nu(l,a,-s)|^2 \tag{4.7.2}$$

where the sum is over the *occupied* lowest energy states of spin $-s$. This self-consistency is obtained by up to 50 to 100 iterations.

That the terms $\propto U(l,a)$ induce ferromagnetism can be seen by the following decomposition, which shows that it induces *a self-consistent effective magnetic field*, which is similar to the *molecular field* introduced by P. Weiss into the 'mean field theory' of magnetism. Namely, if we add an external magnetic field H^{ext} with z-direction, we are led to the following energy shift:

$$\begin{aligned}\Delta\varepsilon(s) &= U(l,a)\{\langle n(l,a,-s)\rangle - \langle n^{para}(l,a)\rangle\} - s\mu_B H^{ext} \\ &= U(l,a)(1/2)\{[\langle n(l,a,\uparrow) + n(l,a,\downarrow) - 2n^{para}(l,a)\rangle] \\ &\quad - s[\langle n(l,a,\uparrow) - n(l,a,\downarrow)\rangle]\} - s\mu_B H^{ext}\end{aligned} \tag{4.7.3}$$

So the term in the second line punishes strong changes of the charge density, when the system changes from paramagnetic to a magnetic state, whereas the third line corresponds to the above-mentioned effective 'molecular field' H^{MF}, as given by $\mu_B H^{MF} = \mu_B H^{ext} + 1/2U(l,a)\langle\sigma_z\rangle(l,a)$, where $\langle\sigma_z\rangle(l,a) = \langle n(l,a,\uparrow) - n(l,a,\downarrow)\rangle$ is the expectation value of the *local spin polarization* for electrons occupying the orbital a at the site l considered.

Finally, the matrix elements $H^{para}(l,a;l',a')$ describing the paramagnetic state of the amorphous system in this formalism can be derived in a Slater-Koster representation (Slater and Koster, 1954) based on the results presented in the comprehensive book by Papaconstantopoulos (1986).

In fact, one of the present authors has also studied *amorphous Fe* itself in this formalism (Krauss and Krey, 1991), although at room temperature amorphous Fe apparently does not exist in nature, except as a very thin film of thickness <0.24 nm on amorphous Y or Gd substrate (Handschuh *et al.*, 1993). However, because computer models of purely amorphous Fe can of course be prepared without difficulty, we compare our computer results with the above-mentioned experiments of Handschuh *et al.* (1993). These authors found by Mössbauer spectroscopy a significant reduction in the Fe moment of the amorphous alloy, namely values of roughly 1.2 μ_B, compared with 2.2 μ_B for bcc Fe. This reduction is in agreement with the results for a-$Fe_{1-x}Zr_x$ presented above in Fig. 4.7.1, extrapolated to $x \to 0$, but actually it is drastically dependent on the specific volume of the sample, as suggested in this figure by the different results for the hydrogenated system. In the above-mentioned paper (Krauss and Krey, 1991) we generated 17 different samples of amorphous Fe, with 54 atoms in each sample, and with densities of the samples ranging from ρ=7.39 to 9.19 g cm^{-3} (the density of bcc Fe would be 7.87 g cm^{-3}). Furthermore, for every site l of the 918 sites in these samples we have calculated the volume V_l of the corresponding *Voronoi* polyhedron.

Figure 4.7.2 is a plot of the moment distribution obtained in this way, i.e. the 'cloud' of 918 pairs (μ_l, V_l).

These results should be compared to similar results obtained for (at T=0 K fictitious) fcc iron obtained by Moruzzi *et al.* (1989), and contrasted to the results for bcc Fe (Moruzzi *et al.*, 1988). This comparison is shown in Fig. 4.7.3.

For *non-collinear spin configurations*, the self-consistent eigenvalue quations correspond to the natural extension of Eq. (4.7.1) in view of Eq. (4.7.3) and $s\delta_{s,s'} = (\sigma_z)_{s,s'}$, namely

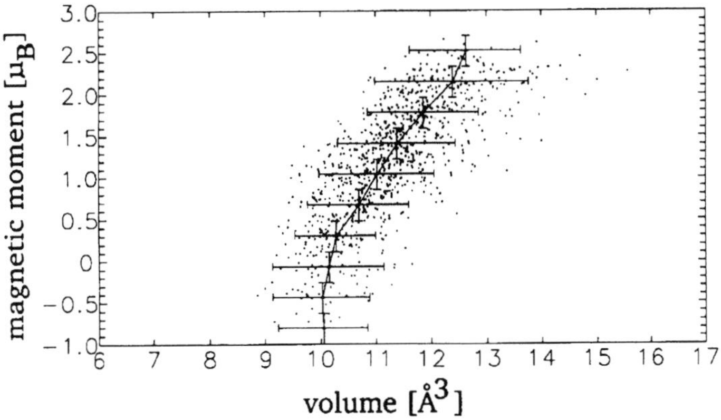

Fig. 4.7.2 All 918 points (μ_l, V_l) drawn from all atoms l of the 17 different computer samples of amorphous Fe. V_l is the volume of the *Voronoi* polyhedron surrounding atom l. Note that some negative moments appear. For ten different μ-intervals, 'horizontal averages' have been performed, which yield the solid line joining the averages, and the 'error bars' corresponding to the standard deviation

$$\begin{aligned}\varepsilon_\nu c^\nu(l,a,s) &= H^{\text{para}}(l,a;l',a')c^\nu(l',a',s)\\ &\quad + 0.5\mathrm{U}(l,a)[\langle \mathrm{n}(l,a,\uparrow) + \mathrm{n}(l,a,\downarrow) - 2\mathrm{n}^{\text{para}}(l,a)\rangle]c^\nu(l,a,s)\\ &\quad - \{0.5[\mathrm{U}(l,a)\delta_{aa'} + \mathrm{I}(l,a,a')(1-\delta_{aa'})]\langle\sigma\rangle_{l,a'} + \mu_\mathrm{B}\mathbf{H}_l^{\text{ext}}\}\cdot\sigma_{s,s'}c^\nu(l,a,s')\end{aligned} \quad (4.7.4)$$

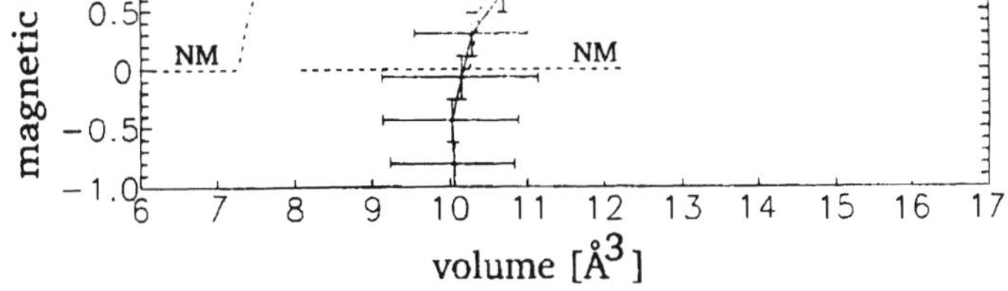

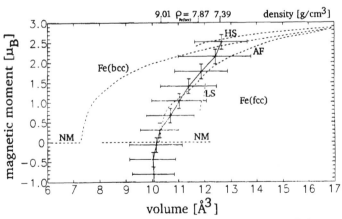

Fig. 4.7.3 The averages and error bars from Fig. 4.7.2 are plotted together with results for crystalline (fictitious at T=0 K) fcc Fe and for bcc Fe. Fe|μ|-V relation for the type-I-antiferromagnetic state (AF) and for the high-spin (HS) and low-spin (LS) states are taken from Moruzzi *et al.* (1989), and the results for bcc Fe from Moruzzi *et al.* (1988). The densities of bcc Fe, ρ=7.87 g cm^{-3} and the densities corresponding to ρ=9.01 and 7.39 g cm^{-3} in Fig. 7.5.2 are marked by arrows

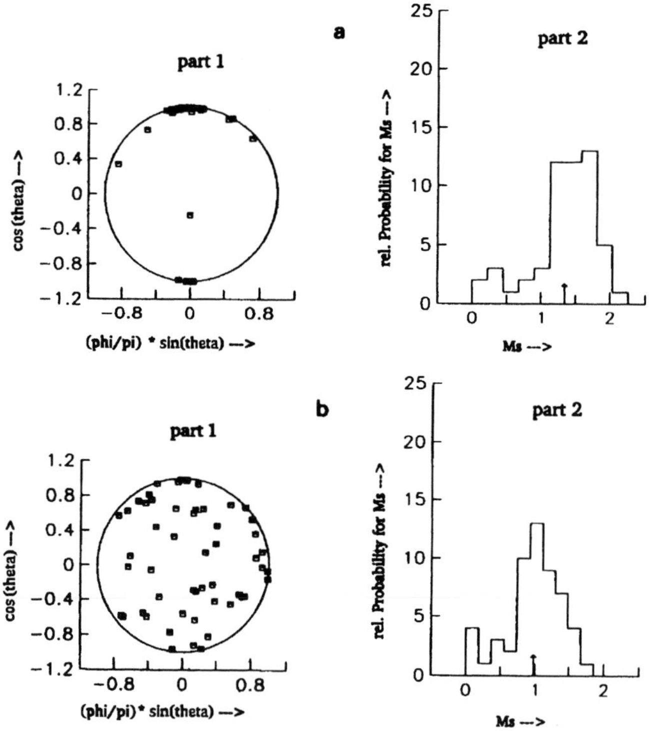

Fig. 4.7.4 Part 1: the directions of the atomic moments as calculated by Eq. (4.7.6) for a small computer model of 54 atoms (periodically continued) representing amorphous $Fe_{0.93}Zr_{0.07}$. It turns out that the results are *preparation dependent*: in the case of the upper results, 'a', the distribution of the localized initializing fields, which are switched off after the first iteration, had a preference for the z-direction, leading to a converged spin configuration presented in part 'a' of the figure, which corresponds more to an *asperomagnet* (Moorjani and Coey, 1984) than to a spin-glass (the moments corresponding to $\theta=\pi$ belong to the Zr atoms). Instead, by an essentially isotropic arrangement of the initializing random fields, the *isotropic spin-glass* state is generated in part 'b' of the figure. The distribution of the *magnitudes* of the moments, plotted on the r.h.s., part 2, is quite different in both cases

Here again we use Einstein's summing convention, summing over the primed indices; as usual, $\delta_{aa'}$ is Kronecker's symbol, $=1$ if $a=a'$, $=0$ otherwise; $\sigma_{s,s'}$ is the *vector* of the 2×2 Pauli matrices, namely

$$\sigma_{s,s'} = \left\{ \begin{pmatrix} 0 & 1 \\ 1 & 0 \end{pmatrix}, \begin{pmatrix} 0 & -i \\ i & 0 \end{pmatrix}, \begin{pmatrix} 1 & 0 \\ 0 & -1 \end{pmatrix} \right\}_{s,s'} \tag{4.7.5}$$

$\mathbf{H}_l^{\text{ext}}$ is the external magnetic field, which can depend greatly on the position l and on the time, i.e. after initializing a non-trivial magnetic state, the magnetic fields $\mathbf{H}_l^{\text{ext}}$ can be switched off in course of the iterations leading to self-consistency; furthermore, $\mathbf{H}_l^{\text{ext}} \cdot \sigma_{s,s'}$ and $\langle\sigma\rangle_{l,a'} \cdot \sigma_{s,s'}$ are the usual scalar products, e.g. $\mathbf{a}\cdot\mathbf{b}=a_x b_x + a_y b_y + a_z b_z$. In particular, the self-consistent value of the *local polarization vectors* are

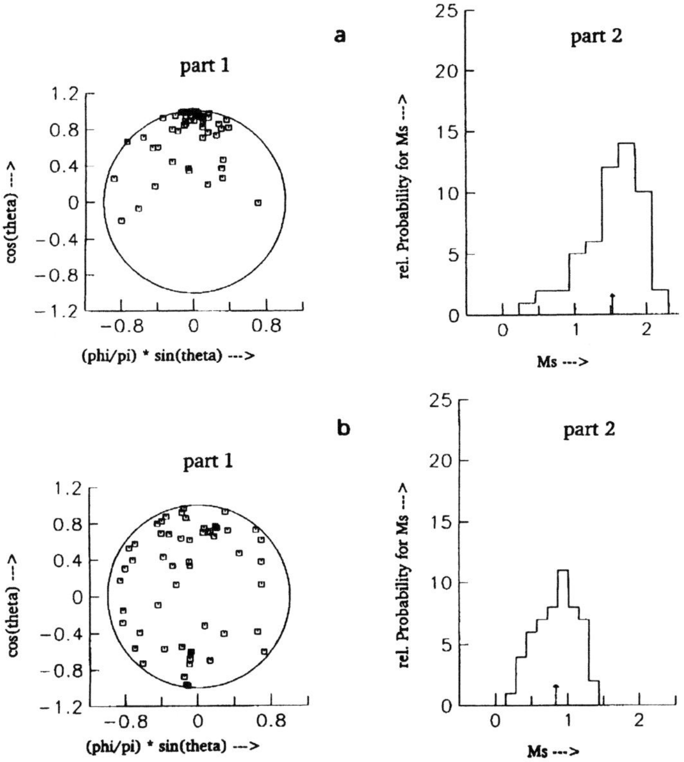

Fig. 4.7.5 Similar results as for Fig. 4.7.4, but for two models of amorphous Fe with different densities, namely ρ=7.39 g cm^{-3} ('a') and ρ=7.87 g cm^{-3} ('b'). In both cases the same roughly homogeneous distribution of the initializing random fields was used, which led to the results 'a' of Fig. 4.7.4. The arrows in part 2 denote the 'median' of the moment distribution. The distribution itself is rather broad

$$\langle\sigma\rangle_{l,a'} = \sum_{\nu=1}^{\nu f}\sum_{s,s'} c^{\nu}(l,a',s)^{*}\sigma_{s,s'}c^{\nu}(l,a',s) \tag{4.7.6}$$

Finally, in addition to Hubbard's Coulomb repulsive energy U(l,a), which energetically *disfavors* the occupation of one orbital by two electrons,[2] we have also taken into account in Eq. (4.7.4) the *Hund's rule* exchange parameters I(l,a,a'): these parameters are positive (a simple proof is given, for example, on p. 392 of the book on Quantum Mechanics by W. Döring, 1962) and favor the occupation of different atomic d-orbitals with *parallel spin*. In our calculations we have neglected the I(l,a,a') with respect to U(l,a), assuming $I \equiv 0$, but U=5.3 eV, for the five different d-orbitals of iron, whereas other authors (e.g. Lorenz and Hafner, 1995) assume

2) Two electrons occupying the *same* orbital must of course have opposite spin because of the Pauli principle.

$U \equiv I$ (=0.95 eV for Fe) and obtain results that are not much different. Probably the reality is somewhere in between, i.e. I and U should be treated as two independent parameters instead of fixing their ration.

There are hints that amorphous $Fe_{1-x}Zr_x$ alloys in the region of $x<0.13$, where there is only *weak* itinerant magnetism in the above-mentioned sense, i.e. holes not only in the minority d-band, may acquire *spin-glass properties,* e.g. long relaxation times, strong hysteresis, metastable non-collinear and apparently disordered paramagnetic states; also amorphous Fe itself, which at least can be prepared as a computer model, shows similar behavior (Uchida and Kakehashi, 1997, 1998; Kakehashi *et al.*, 1997; Yu and Kakehashi, 1996) in the high density region, where fcc iron is non-magnetic or weakly magnetic (see Fig. 4.7.3). We have performed numerical studies on the non-collinear magnetic behavior of amorphous Fe (Krey *et al.*, 1990) and of amorphous $Fe_{1-x}Zr_xH_y$ alloys (Krey *et al.*, 1992).

Figures 4.7.4 and 4.7.5 visualize the non-collinear spin configurations obtained for amorphous $Fe_{1-x}Zr_x$ alloys and amorphous Fe, respectively; on the l.h.s. these figures represent the distribution of the spin *directions*, and on the r.h.s. the distribution of the *magnitudes* of the atomic moments.

Note that, by hydrogenation, leading to the large volume expansion mentioned above, we would get for amorphous $Fe_{0.93}Zr_{0.07}H_{0.14}$ a conventional collinear ferromagnetic state (see, e.g., Krey *et al.*, 1992).

Results for similar computer models of 'amorphous Fe' are presented in Fig. 4.7.5.

Somewhat different results were obtained by Lorenz and Hafner (1995) and Kakehashi *et al.* (1997), but we do not go into details here.

In conclusion we can say that the magnetic properties of amorphous alloys, particularly of the iron-rich alloys, are sometimes more complicated than we would expect from a Heisenberg model and need a complicated itinerant description, taking into account strong deviations from collinearity.

5
Superconductivity of Glassy Metals

5.1
The Eliashberg Equations for Amorphous Metals

In all known superconducting metals,[1] including amorphous ones, the electron–phonon interaction is a determining mechanism responsible for the superconductive condition. This mechanism and its implication for various properties of metals has been extensively investigated by many authors (Vonsovskij *et al.*, 1977; Ginzburg and Kirzhnitz, 1972, 1977; Dolgov and Maximov, 1983; Maximov, 1969). However, a number of parameters of a superconductor change as the system transforms from the crystalline to the amorphous state. The experiment shows that for simple metals and their compounds the characteristic temperature T_s for the onset of superconductivity in amorphous metals differs significantly from that in crystalline ones (Nemoshkalenko *et al.*, 1987; Poon, 1987; Bergmann, 1976). This change can be attributed to the increase in the electron-phonon coupling constant, due to diffuse electron scattering in amorphous metals, which arises because the phonon spectrum is *softened* by the disturbed long-range order. To prove this we have to generalize the *Eliashberg superconductivity equations* to amorphous metals and to analyze the equations obtained in the neighborhood of T_C, the transition temperature to superconductivity (Krasny *et al.*, 1990b, 1993, 1995, 1996; Krasny and Kovalenko, 1990).

If a simple metal is considered as a mixture of two types of charged particles – the conduction electrons and ions – the Hamiltonian of such a system has the form (see Section 5.3):

$$\hat{H} - N\mu_i - z_0 N\mu_e = \sum_{\nu=1}^{N}\left(\frac{\hat{\vec{P}}_\nu^2}{2M} - \mu_i\right) + \frac{1}{2}\sum_{\substack{\nu,\nu' \\ (\nu\neq\nu')}}\frac{z_0^2 e^2}{|\vec{R}_\nu - \vec{R}_{\nu'}|} + \sum_{j=1}^{z_0 N}\left(\frac{\hat{\vec{P}}_j^2}{2m} - \mu_e\right)$$
$$+\frac{1}{2}\sum_{\substack{j,j' \\ (j\neq j')}}\frac{e^2}{|\vec{r}_j - \vec{r}_{j'}|} + \sum_{\nu}\sum_{j} w_0(\vec{R}_\nu - \vec{r}_j) = \hat{H}_i + \hat{H}_e + \hat{H}_{ie} \qquad (5.1.1)$$

[1] Here we exclude of course high-temperature superconducting materials, which are – by the way – based on non-metallic systems, where CuO_2 groups play a fundamental role, and where apparently the phonons are *not* responsible for the superconductivity.

Here $w_0(\vec{r})$ is the local pseudopotential of the electron-ion $\hat{\vec{P}}$ coupling; $\hat{\vec{P}}_\nu$ and $\hat{\vec{P}}_j$, $\vec{R}_\nu$ and $\vec{r}_j$, M and m, μ_i and μ_e are the momentum operators, the radii vectors, masses, and chemical potentials of an *ion* and an *electron* respectively; e is the electronic charge; $z_0 e$ is the ionic charge; N is the total number of ions and z_0 is the valency of the ion;

$$\hat{H}_i = \sum_{\nu=1}^{N}\left(\frac{\hat{\vec{P}}_\nu^2}{2M} - \mu_i\right) + \frac{1}{2}\sum_{\substack{\nu,\nu' \\ (\nu\neq\nu')}}\frac{z_0^2 e^2}{|\vec{R}_\nu - \vec{R}_{\nu'}|} + \sum_{\nu}\frac{z_0 N}{V}\int_V w_0(\vec{R}_\nu - \vec{r})d^3r$$
$$+\frac{1}{2}\frac{N^2 z_0^2}{V^2}\int\frac{e^2}{|\vec{r}-\vec{r}\,'|}d^3r\cdot d^3r' = \sum_{\nu=1}^{N}\left(\frac{\hat{\vec{P}}_\nu^2}{2M} - \mu_i\right) + \frac{1}{2}\sum_{\substack{\nu,\nu' \\ (\nu\neq\nu')}} z_0^2\cdot V_e(\vec{R}_\nu - \vec{R}_{\nu'}) \tag{5.1.2}$$

is the Hamiltonian of the ions inserted into the negative charge smeared uniformly through the volume V of the body;

$$\hat{H}_e = \sum_{j=1}^{z_0 N}\left(\frac{\hat{\vec{P}}_j^2}{2m} - \mu_e\right) + \frac{1}{2}\sum_{\substack{j,j' \\ (j\neq j')}}\frac{e^2}{|\vec{r}_j - \vec{r}_{j'}|} - \sum_{j=1}^{z_0 N}\frac{N}{V}\int_V\frac{z_0 e^2 d^3r}{|\vec{r}_j - \vec{r}|} + \frac{1}{2}\frac{N^2 z_0^2}{V^2}\int\frac{e^2\cdot d^3r\cdot d^3r'}{|\vec{r}-\vec{r}\,'|}$$
$$= \sum_{j=1}^{zN}\left(\frac{\hat{\vec{P}}_j^2}{2m} - \mu_e\right) + \frac{1}{2}\sum_{j,j'} V_e(\vec{r}_j - \vec{r}_{j'}) \tag{5.1.3}$$

is the Hamiltonian of the interacting electron gas embedded within the positive background;

$$\hat{H}_{ie} = \sum_{\nu=1}^{N}\sum_{j=1}^{z_0 N}\left\{w_0(\vec{R}_\nu - \vec{r}_j) - \frac{1}{V}\int_V w_0(\vec{R}_\nu - \vec{r})d^3r\right\} = \sum_{\nu,j} w(\vec{R}_\nu - \vec{r}_j) \tag{5.1.4}$$

is the Hamiltonian describing the electron-ion interaction.

In Eqs (5.1.1)–(5.1.4) we assume as usual the thermodynamic limit, i.e., $V, N \to \infty (V/N = v_0 = \text{const.})$ and

$$\begin{cases} \int_V w_0(\vec{R}-\vec{r})d^3r = \int_V w_0(\vec{r})d^3r \\ \int_V \frac{d^3r}{|\vec{R}-\vec{r}|} = \int_V \frac{d^3r}{r} = \frac{1}{V}\int d^3r\int\frac{d^3r'}{|\vec{r}-\vec{r}\,'|} \end{cases} \tag{5.1.5}$$

and we introduce the following abbreviations

$$\begin{cases} V_e(\vec{r}-\vec{r}\,') = \dfrac{e^2}{|\vec{r}-\vec{r}\,'|} - \dfrac{e^2}{V}\displaystyle\int \dfrac{d^3r}{r} & (5.1.6) \\ w(\vec{R}-\vec{r}) = w_0(\vec{R}-\vec{r}) - \dfrac{1}{V}\displaystyle\int w_0(r)\cdot d^3r & (5.1.7) \\ b = \displaystyle\int_V \left[w(r) + \dfrac{z_0e^2}{r}\right] d^3r & (5.1.8) \end{cases}$$

Later on it will be more convenient to write the Hamiltonian $\hat{H}$ in the *second-quantized representation.* In this representation we introduce the following field operators for the electron gas:

$$\hat{\Psi}_\sigma(\vec{r}) = \frac{1}{\sqrt{V}}\sum_{\vec{k}} \hat{a}_{\vec{k}\sigma}\cdot\exp(i\vec{k}\cdot\vec{r});\ \hat{\Psi}^+_\sigma(\vec{r}) = \frac{1}{\sqrt{V}}\sum_{\vec{k}} \hat{a}^+_{\vec{k}\sigma}\cdot\exp(-i\vec{k}\cdot\vec{r}) \tag{5.1.9}$$

where the *Fermionic* creation and annihilation operators $\hat{a}^+_{\vec{k}\sigma}$ and $\hat{a}_{\vec{k}\sigma}$ for the electron with momentum $\hbar\vec{k}$ and spin $\sigma = \left(+\frac{1}{2};\ -\frac{1}{2}\right) = (\uparrow;\downarrow)$ obey the usual *anti-commutation* rules:

$$\left[\hat{a}_{\vec{k}\sigma};\hat{a}^+_{\vec{k}'\sigma}\right]_+ = \delta_{\vec{k}\vec{k}'}\cdot\delta_{\sigma\sigma'};\ \left[\hat{a}_{\vec{k}\sigma};\hat{a}_{\vec{k}'\sigma'}\right]_+ = \left[\hat{a}^+_{\vec{k}\sigma};\hat{a}^+_{\vec{k}'\sigma'}\right]_+ = 0 \tag{5.1.10}$$

To take into account the superconductive correlations, spinor operators (the so-called *Nambu* operators) are introduced (Vonsovskij *et al.*, 1977):

$$\hat{\Psi}(\vec{r}) = \begin{pmatrix}\hat{\Psi}_\uparrow(\vec{r}) \\ \hat{\Psi}^+_\downarrow(\vec{r})\end{pmatrix};\ \hat{\Psi}^+(\vec{r}) = \left(\hat{\Psi}^+_\uparrow(\vec{r});\ \hat{\Psi}_\downarrow(\vec{r})\right) \tag{5.1.11}$$

Then the electronic part of the Hamiltonian $\hat{H}$ in the second quantized representation has the form:

$$\begin{aligned}\hat{H}_e + \hat{H}_{ie} = &\int d^3r\hat{\Psi}^+(\vec{r})\cdot\hat{\tau}_3\left(-\frac{\hbar^2}{2m}\Delta - \mu_e\right)\hat{\Psi}(\vec{r}) \\ &+\frac{1}{2}\int d^3r\int d^3r'\hat{\Psi}^+(\vec{r}\,')\hat{\tau}_3\hat{\Psi}(\vec{r}\,')V_e(\vec{r}\,'-\vec{r})\hat{\Psi}^+(\vec{r})\hat{\tau}_3\hat{\Psi}(\vec{r}) \\ &+\int d^3r\hat{\Psi}^+(\vec{r})\hat{\tau}_3\hat{\Psi}(\vec{r})\sum_\nu w(\vec{r}-\vec{R}_\nu)\end{aligned} \tag{5.1.12}$$

Here τ_i $(i = 0, 1, 2, 3)$ are the Pauli matrices:

$$\tau_0 = \begin{pmatrix}1 & 0\\ 0 & 1\end{pmatrix};\ \tau_1 = \begin{pmatrix}0 & 1\\ 1 & 0\end{pmatrix};\ \tau_2 = \begin{pmatrix}0 & -i\\ i & 0\end{pmatrix};\ \tau_3 = \begin{pmatrix}1 & 0\\ 0 & -1\end{pmatrix}. \tag{5.1.13}$$

If the metal is amorphous, for every randomly realized mth spatial configuration of ions, the position of the νth ion is described by the vector

$$\vec{R}_{\nu}^{(m)} = \vec{R}_{0\nu}^{(m)} + \vec{\xi}_{\nu}^{(m)} \tag{5.1.14}$$

where $\vec{R}_{0\nu}^{(m)}$ is the *quasi-equilibrium position* of atom ν within the mth configuration of ions, while $\vec{\xi}_{\nu}^{(m)} = \vec{\xi}\left(\vec{R}_{0\nu}^{(m)}\right)$ describes the displacement of the atom ν from its quasi-equilibrium position.

The Hamiltonian corresponding to the mth ion configuration is represented by

$$\hat{\mathrm{H}}^{(m)} - \hat{N}\mu_{\mathrm{i}} - \hat{N}_{\mathrm{e}}\mu_{\mathrm{e}} = \hat{\mathrm{H}}_{\mathrm{i}}^{(m)} + \hat{\mathrm{H}}_{\mathrm{e}} + \hat{\mathrm{H}}_{\mathrm{ie}}^{(m)} \tag{5.1.15}$$

To describe the superconducting state of amorphous metals we introduce an averaged single-particle electronic Green's function $G(x, x')$ [2] that is related to single-particle experimental quantities (such as the specific heat, etc.). Hence, to obtain this function the Green's function $G^{(m)}(x, x')$ for a particular quasi-equilibrium configuration of ions needs first to be calculated by averaging over the quickly relaxing subsystems (the electronic and phonon ones), and by subsequent averaging over all quasi-equilibrium spatial arrangements of ions $\left(\vec{R}_{01}^{(m)}, \vec{R}_{02}^{(m)}, \ldots, \vec{R}_{0N}^{(m)}\right)$. Such averaging is carried out with the normalized probability density for realizations of the given configuration.

Hereafter we will use the frozen-liquid model applied to correlation functions or structure factors. Thus the single-electron Green's function in the amorphous metal is

$$\begin{aligned} G(x - x') &= \overline{G^{(m)}(x, x')} = -\overline{\langle T_{\tau}\{\hat{\Psi}(x)\hat{\Psi}^{+}(x')\}\rangle^{(m)}} \\ &= \begin{pmatrix} g(x - x'), & f(x - x') \\ f^{+}(x - x'), & g^{+}(x - x') \end{pmatrix} \end{aligned} \tag{5.1.16}$$

where

$$\begin{cases} g(x - x') = -\overline{\left\langle T_{\tau}\left\{\hat{\Psi}_{\uparrow}(x) \cdot \hat{\Psi}_{\uparrow}^{+}(x')\right\}\right\rangle^{(m)}} = \overline{g^{(m)}(x, x')} \\ f(x - x') = -\overline{\left\langle T_{\tau}\left\{\hat{\Psi}_{\uparrow}(x) \cdot \hat{\Psi}_{\downarrow}(x')\right\}\right\rangle^{(m)}} = \overline{f^{(m)}(x, x')} \end{cases} \tag{5.1.17}$$

Here the operators $\hat{\Psi}(x)$ and $\hat{\Psi}^{+}(x)$ are the *Heisenberg operators* corresponding to the Schrödinger field operators $\hat{\Psi}(\vec{r})$ and $\hat{\Psi}^{+}(\vec{r})$, obtained with the Hamiltonian $\hat{\mathrm{H}}^{(m)}$ as follows:

$$\begin{cases} \hat{\Psi}(x) := \mathrm{e}^{(\hat{\mathrm{H}}^{(m)} - \hat{N}\mu_{\mathrm{i}} - \hat{N}_{\mathrm{e}}\mu_{\mathrm{e}})\tau} \cdot \hat{\Psi}(\vec{r}) \cdot \mathrm{e}^{-(\hat{\mathrm{H}}^{(m)} - \hat{N}\mu_{\mathrm{i}} - \hat{N}_{\mathrm{e}}\mu_{\mathrm{e}})\tau} \\ \hat{\Psi}^{+}(x) := \mathrm{e}^{(\hat{\mathrm{H}}^{(m)} - \hat{N}\mu_{\mathrm{i}} - \hat{N}_{\mathrm{e}}\mu_{\mathrm{e}})\tau} \cdot \hat{\Psi}^{+}(\vec{r}) \cdot \mathrm{e}^{-(\hat{\mathrm{H}}^{(m)} - \hat{N}\mu_{\mathrm{i}} - \hat{N}_{\mathrm{e}}\mu_{\mathrm{e}})\tau} \end{cases} \tag{5.1.18}$$

$x = \{\vec{r}, \tau\}$ is an abbreviation for the set of coordinates and the 'thermodynamic imaginary time' τ, which is specified within the interval $0 \leq \tau \leq \beta = \dfrac{1}{k_{\mathrm{B}}T}$. Further-

2) Due to the averaging, $G(x, x')$ is translation invariant and can thus be written as $G(x - x')$.

more, as usual, T_τ is the well-known symbol of *ordering* of these 'time variables' τ, e.g. $T_\tau\hat{\Psi}(x)\hat{\Psi}(x') = \hat{\Psi}(x)\hat{\Psi}(x')$, if $\tau > \tau'$, but $= -\hat{\Psi}(x')\hat{\Psi}(x)$ if $\tau' > \tau$; $\langle\ldots\rangle^{(m)}$ symbolizes statistical averaging with the Hamiltonian $\hat{\mathrm{H}}^{(m)}$ over the electron and phonon variables:

$$\langle\ldots\rangle^{(m)} = \frac{\mathrm{Tr}\left\{\ldots\exp\left[-\beta\left(\hat{\mathrm{H}}^{(m)} - \hat{N}\mu_{\mathrm{i}} - \hat{N}_{\mathrm{e}}\mu_{\mathrm{e}}\right)\right]\right\}}{\mathrm{Tr}\left\{\exp\left[-\beta\left(\hat{\mathrm{H}}^{(m)} - \hat{N}\mu_{\mathrm{i}} - \hat{N}_{\mathrm{e}}\mu_{\mathrm{e}}\right)\right]\right\}} \tag{5.1.19}$$

We split the Hamiltonian $\hat{\mathrm{H}}^{(m)}$ into two parts:

$$\hat{\mathrm{H}}^{(m)} - \hat{N}\cdot\mu_{\mathrm{i}} - \hat{N}_{\mathrm{e}}\mu_{\mathrm{e}} = \hat{\mathrm{H}}_0^{(m)} + \hat{\mathrm{H}}_{\mathrm{int}}^{(m)} \tag{5.1.20}$$

where

$$\hat{\mathrm{H}}_0^{(m)} = \hat{\mathrm{H}}_{\mathrm{i}}^{(m)} + \hat{\mathrm{H}}_{0\mathrm{e}} \tag{5.1.21}$$

$$\hat{\mathrm{H}}_{\mathrm{int}}^{(m)} = \hat{\mathrm{H}}_{\mathrm{ee}} + \hat{\mathrm{H}}_{\mathrm{ie}}^{(m)} \tag{5.1.22}$$

and

$$\begin{cases} \hat{\mathrm{H}}_{\mathrm{i}}^{(m)} = \sum\limits_{\nu}\left(\dfrac{\vec{P}_\nu^2}{2M} + \dfrac{z\beta}{v_0} - \mu_{\mathrm{i}}\right) + \frac{1}{2}\sum\limits_{\nu\neq\nu'} z^2 V_{\mathrm{e}}\left(\vec{R}_\nu^{(m)} - \vec{R}_{\nu'}^{(m)}\right) \\ \hat{\mathrm{H}}_{0\mathrm{e}} = \displaystyle\int \mathrm{d}^3 r\,\hat{\Psi}^+(\vec{r})\tau_3\left(-\frac{\hbar^2}{2m}\Delta - \mu_{\mathrm{e}}\right)\hat{\Psi}(\vec{r}) \\ \hat{\mathrm{H}}_{\mathrm{ee}} = \dfrac{1}{2}\displaystyle\int \mathrm{d}^3 r \int \mathrm{d}^3 r'\,\hat{\Psi}^+(\vec{r})\tau_3\hat{\Psi}^+(\vec{r}')V_{\mathrm{e}}(\vec{r}' - \vec{r})\hat{\Psi}(\vec{r}')\tau_3\hat{\Psi}(\vec{r}) \\ \hat{\mathrm{H}}_{\mathrm{ie}}^{(m)} = \displaystyle\int \mathrm{d}^3 r\,\hat{\Psi}^+(\vec{r})\tau_3\hat{\Psi}(\vec{r})\sum_\nu w(\vec{r} - \vec{R}_\nu^{(m)}) \end{cases}$$

Considering $\hat{\mathrm{H}}_{\mathrm{int}}^{(m)}$ as a *perturbation* of $\hat{\mathrm{H}}_0^{(m)}$ we can employ perturbation theory to calculate the Green's function corresponding to the mth configuration of ions. To do this we introduce the operator $\hat{\Psi}_0(x)$ in the interaction representation, again formally with imaginary times $\tau(= \mathrm{i}t/\hbar)$:

$$\hat{\psi}_0(x) = \mathrm{e}^{\hat{\mathrm{H}}_0^{(m)}\tau}\hat{\Psi}(\vec{r})\mathrm{e}^{-\hat{\mathrm{H}}_0^{(m)}\tau} = \mathrm{e}^{\hat{\mathrm{H}}_{0\mathrm{e}}^{(m)}\tau}\hat{\Psi}(\vec{r})\mathrm{e}^{-\hat{\mathrm{H}}_{0\mathrm{e}}^{(m)}\tau} \tag{5.1.23}$$

because $\left[\hat{\mathrm{H}}_{\mathrm{i}}^{(m)}; \hat{\mathrm{H}}_{0\mathrm{e}}\right] = 0$ and $\left[\hat{\mathrm{H}}_{\mathrm{i}}^{(m)}; \hat{\Psi}(\vec{r})\right] = 0$. Then the Green's function $G^{(m)}(x, x')$ can be written in the form (see Appendix B):

$$G^{(m)}(x, x') = -\frac{1}{Z^{(m)}}\mathrm{Tr}\left\{\mathrm{e}^{-\beta\hat{\mathrm{H}}_0^{(m)}}\,T_\tau\left[\hat{\Psi}_0(x)\hat{\Psi}_0^+(x')\,\hat{S}^{(m)}(\beta)\right]\right\} \tag{5.1.24}$$

where

$$Z^{(m)} = \mathrm{Tr}\left\{ e^{-\beta \hat{H}_0^{(m)}} \hat{S}^{(m)}(\beta) \right\} \tag{5.1.25}$$

and $\hat{S}^{(m)}(\beta)$ is the 'temperature scattering matrix':

$$\begin{aligned} \hat{S}^{(m)}(\beta) &= T_\tau \exp\left\{ -\int_0^\beta \hat{H}_{\mathrm{int}}^{(m)}(\tau) d\tau \right\} \\ &= \sum_{n=0}^{\infty} \frac{(-1)^n}{n!} \int_0^\beta d\tau_1 \ldots \int_0^\beta d\tau_n T_\tau \left\{ \hat{H}_{\mathrm{int}}^{(m)}(\tau_1) \ldots \hat{H}_{\mathrm{int}}^{(m)}(\tau_n) \right\} \end{aligned} \tag{5.1.26}$$

The operator $\hat{H}_{\mathrm{int}}^{(m)}(\tau)$ is defined as follows:

$$\hat{H}_{\mathrm{int}}^{(m)}(\tau) = e^{\hat{H}_0^{(m)}\tau} \hat{H}_{\mathrm{int}}^{(m)} e^{-\hat{H}_0^{(m)}\tau} = \hat{H}_{\mathrm{ee}}(\tau) + \hat{H}_{\mathrm{ie}}^{(m)}(\tau) \tag{5.1.27}$$

$$\hat{H}_{\mathrm{ee}}(\tau) = e^{\hat{H}_{0e}\tau} \hat{H}_{\mathrm{ee}} e^{-\hat{H}_{0e}\tau} = \frac{1}{2} \int d^3r \int d^3r' \hat{\Psi}_0^+(x') \tau_3 \hat{\Psi}_0^+(x) V_e(\vec{r}' - \vec{r}) \hat{\Psi}_0(x) \tau_3 \hat{\Psi}_0(x') \tag{5.1.28}$$

$$\begin{aligned} \hat{H}_{\mathrm{ie}}^{(m)}(\tau) &= e^{\hat{H}_0^{(m)}\tau} \hat{H}_{\mathrm{ie}}^{(m)} e^{-\hat{H}_0^{(m)}\tau} = \int d^3r \hat{\Psi}_0^+(x) \tau_3 \hat{\Psi}_0(x) \hat{W}^{(m)}(x) \\ \hat{W}^{(m)}(x) &= \hat{W}^{(m)}(\vec{r}, \tau) = e^{\hat{H}_i^{(m)}\tau} \sum_\nu w\left(\vec{r} - \vec{R}_\nu^{(m)}\right) e^{-\hat{H}_i^{(m)}\tau} \end{aligned} \tag{5.1.29}$$

As in the usual case of normal-conducting Fermi systems, for a superconductor, the calculation of the Green's function reduces to the calculation of the averaged chronological field-operator products in the interaction representation. This can be implemented by a Wick's theorem generalized to include the *anomalous Green's functions* involved in Eq. (5.1.16). Hence, when finding the average of the T_τ product of the Nambu operators we should formally proceed as usual: take into account the pairing between operators $\hat{\Psi}_0(x)$ and $\hat{\Psi}_0^+(x)$. By this is meant that the perturbative series will be depicted with the same diagrams as in normal metals; the difference lies only in the matrix nature of the Nambu Green's functions, and in the matrix character of the corresponding vertices. The structure of the latter is readily available from the interaction Hamiltonian, and the zeroth approximation Green's function, $G_0(x, x')$, being independent of ion positions, has the form:

$$\begin{aligned} G_0(x, x') &= -\left\langle T_\tau \left[\hat{\Psi}_0(x) \hat{\Psi}_0^+(x') \right] \right\rangle_0 = \frac{-1}{\mathrm{Tr}\, e^{-\beta \hat{H}_0^{(m)}}} \mathrm{Tr}\left\{ T_\tau \left[\hat{\Psi}_0(x) \hat{\Psi}_0^+(x') \right] e^{-\beta \hat{H}_0^{(m)}} \right\} \\ &= -\frac{1}{\mathrm{Tr}\, e^{-\beta \hat{H}_{0e}}} \mathrm{Tr}\left\{ T_\tau \left[\hat{\Psi}_0(x) \hat{\Psi}_0^+(x') \right] e^{-\beta \hat{H}_{0e}} \right\} \equiv -\left\langle T_\tau \left[\hat{\Psi}_0(x) \hat{\Psi}_0^+(x') \right] \right\rangle_{0e} \end{aligned} \tag{5.1.30}$$

Differentiating $G_0(x, x')$ with respect to τ

$$-\frac{\partial G_0}{\partial \tau} + \left\langle T\left[\frac{\partial \hat{\Psi}_0(x)}{\partial \tau}\hat{\Psi}_0^+(x')\right]\right\rangle_{0e} = \delta(x - x')$$

and taking into account that

$$\frac{\partial \hat{\Psi}_0}{\partial \tau} = [\hat{\mathrm{H}}_{0e}; \hat{\Psi}_0] = -\tau_3\left(-\frac{\hbar^2}{2m}\Delta - \mu_e\right)\hat{\Psi}_0(x)$$

we can readily obtain the equation for $G_0(x, x')$:

$$\left\{-\tau_0\frac{\partial}{\partial \tau} - \tau_3\left(-\frac{\hbar^2}{2m}\Delta - \mu_e\right)\right\}G_0 = \delta(x - x') \cdot \tau_0 \tag{5.1.31}$$

If we move to the momentum representation:

$$G_0(x, x') = \frac{1}{\beta}\sum_n \int \frac{d^3k}{(2\pi)^3} e^{i\vec{k}\cdot(\vec{r}-\vec{r}') - i\hbar\omega_n\cdot(\tau-\tau')}\tilde{G}_0(\vec{k}, \omega_n)$$

where $\hbar\omega_n = (2n+1)\frac{\pi}{\beta}$ refer to the *Fermion Matsubara frequencies*, and if we take into consideration

$$\delta(x - x') = \delta(\vec{r} - \vec{r}')\delta(\tau - \tau') = \frac{1}{\beta}\sum_n \int \frac{d^3k}{(2\pi)^3} e^{i\vec{k}\cdot(\vec{r}-\vec{r}') - i\hbar\omega_n(\tau-\tau')}$$

then we have the following equation for $G_0(\vec{k}, \omega_n)$:

$$\left[i\hbar\omega_n\tau_0 - \left(\frac{\hbar^2k^2}{2m} - \mu_e\right)\tau_3\right]G_0(\vec{k}, \omega_n) = \tau_0$$

or

$$G_0^{-1}(\vec{k}, \omega_n) = i\hbar\omega_n\tau_0 - \left(\frac{\hbar^2k^2}{2m} - \mu_e\right)\tau_3 \tag{5.1.32}$$

To calculate the Green's function $G^{(m)}(x, x')$ we employ the adiabatic approximation, i.e. first we average over the electronic variables, the ions considered to be fixed, and then over the phonon ones:

$$G^{(m)}(x, x') = -\frac{1}{Z^{(m)}}\mathrm{Tr}_{\mathrm{Ph}}\left\{e^{-\beta\hat{\mathrm{H}}_i^{(m)}} \cdot Z_e^{(m)} \cdot \frac{1}{Z_e^{(m)}}\mathrm{Tr}_e\left[e^{-\beta\hat{\mathrm{H}}_{0e}} \cdot T_\tau\left[\hat{\Psi}_0(x)\hat{\Psi}_0^+(x')S^{(m)}(\beta)\right]\right]\right\}$$

and

$$Z^{(m)} = \mathrm{Tr}_{\mathrm{Ph}}\left\{e^{-\beta\hat{\mathrm{H}}_i^{(m)}} \cdot Z_e^{(m)}\right\}$$

Here

$$Z_{\mathrm{e}}^{(m)} = \mathrm{Tr}_{\mathrm{e}}\left\{ \mathrm{e}^{-\beta \hat{\mathrm{H}}_{0\mathrm{e}}} S^{(m)}(\beta) \right\}$$

It has been shown (see Section 5.3) that in the adiabatic approximation (i.e. $\left[\hat{\mathrm{H}}_{\mathrm{i}}^{(m)};\ Z_{\mathrm{e}}^{(m)}\right] \approx 0$):

$$\mathrm{e}^{-\beta \hat{\mathrm{H}}_{\mathrm{i}}^{(m)}} \cdot Z_{\mathrm{e}}^{(m)} = \mathrm{e}^{-\beta \hat{\mathrm{H}}_{\mathrm{i}}^{*(m)}} \cdot Z_{\mathrm{e}}$$

where

$$\hat{\mathrm{H}}_{\mathrm{i}}^{*(m)} = \sum_{\nu} \left(\frac{\vec{P}_{\nu}^{2}}{2M} + \frac{Z\beta}{\nu_0} - \mu_{\mathrm{i}} \right) + \mathrm{U}\left(\vec{R}_1^{(m)} \dots \vec{R}_{\nu}^{(m)} \right) \tag{5.1.33}$$

is an effective Hamiltonian describing the behavior of the ions with the effective coupling $\mathrm{U}\left(\vec{R}_1^{(m)} \dots \vec{R}_n^{(m)} \right)$; in the quadratic approximation in the pseudopotential

$$\mathrm{U}\left(\vec{R}_1^{(m)} \dots \vec{R}_n^{(m)} \right) = \frac{1}{2} \sum_{\nu_1 \neq \nu_2} \Phi\left(\vec{R}_{\nu_1}^{(m)} - \vec{R}_{\nu_2}^{(m)} \right)$$

Here $\Phi(\vec{R}_1 - \vec{R}_2)$ is the binary interionic coupling including the Coulomb repulsion and the ion-electron-ion attraction; $Z_{\mathrm{e}} = \mathrm{Tr}_{\mathrm{e}}\{\exp \beta(\hat{\mathrm{H}}_{0\mathrm{e}} + \hat{\mathrm{H}}_{\mathrm{ee}})\}$ is the statistical sum of the interacting electron gas embedded into a neutralizing background. Then $G^{(m)}(x, x')$ can be represented as

$$G^{(m)}(x, x') = \left\langle G_{\mathrm{e}}^{(m)}(x, x') \right\rangle_{\mathrm{Ph}}^{(m)} \tag{5.1.34}$$

where

$$G_{\mathrm{e}}^{(m)}(x, x') = -\frac{\left\langle T_{\tau}\left[\hat{\Psi}_0(x) \cdot \hat{\Psi}_0^{+}(x') S^{(m)}(\beta) \right] \right\rangle_{0\mathrm{e}}}{\langle S^{(m)}(\beta) \rangle_{0\mathrm{e}}} \tag{5.1.35}$$

$$\langle \dots \rangle_{\mathrm{Ph}}^{(m)} = \frac{1}{\mathrm{Tr}_{\mathrm{Ph}}\left\{ \exp\left(-\beta \hat{\mathrm{H}}_{\mathrm{i}}^{*(m)} \right) \right\}} \mathrm{Tr}_{\mathrm{Ph}}\left\{ \mathrm{e}^{-\beta \hat{\mathrm{H}}_{\mathrm{i}}^{*(m)}} \dots \right\} \tag{5.1.36}$$

$$\langle \dots \rangle_{0\mathrm{e}} = \frac{1}{\mathrm{Tr}_{\mathrm{e}}\{ \exp(-\beta \hat{\mathrm{H}}_{0\mathrm{e}}) \}} \mathrm{Tr}_{\mathrm{e}}\left\{ \mathrm{e}^{-\beta \hat{\mathrm{H}}_{0\mathrm{e}}} \dots \right\} \tag{5.1.37}$$

The function $G_{\mathrm{e}}^{(m)}(x, x')$ is the Green's function of the electron gas in the field of stationary ions. When every term in Eq. (5.1.26) of the scattering matrix $\hat{S}^{(m)}(\beta)$ is matched up by a Feynman diagram, there are connected and disconnected ones among them; the latter consist of two or more closed diagrams without any linking lines. Let $\langle \dots \rangle_{0\mathrm{ec}}$ symbolize the sum of all *connected* diagrams. Then it can readily be shown that

$$G_{\mathrm{e}}^{(m)}(x, x') = -\left\langle T_\tau \left[\hat{\Psi}_0(x) \cdot \hat{\Psi}_0^+(x') \hat{S}^{(m)}(\beta)\right]\right\rangle_{0\mathrm{ec}} \tag{5.1.38}$$

The full electron *causal Green's function* can be found by averaging Eq. (5.1.38) over phonon variables and all quasi-equilibrium configurations of ions:

$$G(x, x') = \frac{1}{\beta}\sum_n \int \frac{\mathrm{d}^3 k}{(2\pi)^3} \mathrm{e}^{i\vec{k}\cdot(\vec{r}-\vec{r}') - i\hbar\omega_n(\tau-\tau')} \tilde{G}(\vec{k}, \omega_n) = \overline{\left\langle G_l^{(m)}(x, x')\right\rangle_{\mathrm{Ph}}} \tag{5.1.39}$$

The Green's function can be evaluated by expanding the S-matrix into a series in powers of the pseudopotential. The diagrammatic analysis of this expansion shows that the Fourier transform of the Green's function has the following form:

$$\begin{cases} G^{-1}(\vec{k}, \omega_n) = i\hbar\omega_n\tau_0 - \left(\dfrac{\hbar^2 k^2}{2m} - \mu_{\mathrm{e}}\right)\tau_3 - \sum(\vec{k}, \omega_n) \\ \sum(\vec{k}, \omega_n) = \left[1 - Z(\vec{k}, \omega_n)\right] i\hbar\omega_n\tau_0 + \chi(\vec{k}, \omega_n)\tau_3 + \varphi(\vec{k}, \omega_n)\tau_1 \end{cases} \tag{5.1.40}$$

Then we take into account that m/M and a/l, where, as before, a is the mean interatomic distance in the amorphous metal and l is the mean-free path, are small quantities. In this case the self-energy $\sum(\vec{k}, \omega_n)$ is a linear functional of $G^{-1}(\vec{k}, \omega_n)$ (see Appendix D). Finally, if we make an analytical continuation from the imaginary axis to the real one by the substitution

$$i\omega_n \to \omega + i\delta$$

and if we take into account that the connection between sum and integral is as follows

$$\frac{1}{\beta}\sum_n \ldots \to \frac{1}{2\pi}\int \mathrm{d}z \ldots$$

and that by this continuation the causal Green's function $G(\vec{k}, \omega_n)$ changes to the *retarded* Green's function $G_R(\vec{k}, \omega)$, we obtain the following system of integral equations (see Appendix D):

$$\sum_{\mathrm{Ph}}(\vec{k}, \omega) = -\int \frac{\mathrm{d}^3 k'}{(2\pi)^3} \int_{-\infty}^{+\infty} \frac{\mathrm{d}z}{2\pi} \int_{-\infty}^{+\infty} \frac{\mathrm{d}z'}{2\pi} b(\vec{k} - \vec{k}', z)\tau_3 \,\mathrm{Im}\, G_R(\vec{k}', z')\tau_3$$
$$\times \frac{\tanh\left(\dfrac{\beta z'}{2}\right) + \coth\left(\dfrac{\beta z}{2}\right)}{\hbar\omega - z - z' + i\delta} \tag{5.1.41}$$

$$\sum_{\mathrm{C}}(\vec{k},\omega) = -\int \frac{d^3k'}{(2\pi)^3} V_e(\vec{k}-\vec{k}') \int_{-\infty}^{+\infty} \frac{dz'}{2\pi} \tau_3 \operatorname{Im} G_R(\vec{k}',z')\tau_3 \tanh\left(\frac{\beta z'}{2}\right) \tag{5.1.42}$$

$$\sum_{\mathrm{Scat}}(\vec{k},\omega) = \int \frac{d^3k'}{(2\pi)^3} \frac{|\tilde{w}(\vec{k}-\vec{k}_1)|^2}{v_0} S_0(\vec{k}-\vec{k}')\tau_3 G_R(\vec{k}',\omega)\tau_3 \tag{5.1.43}$$

where in Eq. (5.1.41)

$$b(\vec{q},z) = \frac{|\tilde{w}(\vec{q})|^2}{V} \sum_{\vec{k}\lambda} \frac{\hbar[\vec{q}\cdot\vec{\varepsilon}(\vec{k}\lambda)]^2 S_0(\vec{k}+\vec{q})}{2M\omega_\lambda(\vec{k})} [\delta(z-\hbar\omega_\lambda(\vec{k}))-\delta(z+\hbar\omega_\lambda(\vec{k}))]\cdot 2\pi$$

Then we obtain

$$\tau_3 G_R(\vec{k},\omega)\tau_3 = \frac{Z(\vec{k},\omega)\hbar\omega\tau_0 + \varepsilon(\vec{k},\omega)\tau_3 - \varphi(\vec{k},\omega)\tau_1}{|Z(\vec{k},\omega)\hbar\omega|^2 - \varphi^2(\vec{k},\omega) - \bar{\varepsilon}^2(\vec{k},\omega)} \tag{5.1.44}$$

$$\begin{aligned}\sum(\vec{k},\omega) &= \sum_{\mathrm{Ph}}(\vec{k},\omega) + \sum_{\mathrm{C}}(\vec{k},\omega) + \sum_{\mathrm{Scat}}(\vec{k},\omega)\\ &= [1 - Z(\vec{k},\omega)]\hbar\omega\tau_0 + \chi(\vec{k},\omega)\tau_3 + \varphi(\vec{k},\omega)\tau_1\end{aligned} \tag{5.1.45}$$

Equations (5.1.41)–(5.1.45) constitute one matrix equation determining three quantities: $Z(\vec{k},\omega)$, $\varphi(\vec{k},\omega)$ and $\chi(\vec{k},\omega)$. The quantity $\chi(\vec{k},\omega)$, as is evident from Eq. (5.1.40), is responsible for a shift in the chemical potential μ_e, i.e. for an effect that is almost insignificant for the development of the superconducting state. Thus in Eqs (5.1.44) and (5.1.45) we immediately put $\chi(\vec{k},\omega) \approx 0$, then

$$\varepsilon(\vec{k},\omega) = \frac{\hbar^2k^2}{2m} - \mu_e + \chi(\vec{k},\omega) \approx \frac{\hbar^2k^2}{2m} - \mu_e = \xi_{\vec{k}} \tag{5.1.46}$$

which is equal to the energy of free electrons measured from the Fermi level. So we have one matrix equation for $Z(\vec{k},\omega)$ and $\varphi(\vec{k},\omega)$.

Equations (5.1.41)–(5.1.45) can be simplified by averaging over the Fermi sphere and changing to integral equations of one variable $z = \hbar\omega$. Finally, if we introduce the new unknown functions $\eta(\omega)$ and $Z'(\omega)$ by

$$\frac{Z_{\mathrm{Scat}}(\omega)}{Z(\omega)} = \frac{\varphi_{\mathrm{Scat}}(\omega)}{\varphi(\omega)} = -\eta(\omega) + 1 \tag{5.1.47}$$

$$Z'(\omega) = \eta(\omega)Z(\omega) = Z(\omega) - Z_{\mathrm{Scat}}(\omega) \tag{5.1.48}$$

while $\Delta'(\omega)$ is equal to $\Delta(\omega)$, the *gap function*, namely

$$\Delta(\omega) = \frac{\varphi(\omega)}{Z(\omega)} = \frac{\eta(\omega)\cdot\varphi(\omega)}{\eta(\omega)\cdot Z(\omega)} = \frac{\varphi_{\mathrm{Ph}}(\omega) + \varphi_{\mathrm{C}}(\omega)}{Z_{\mathrm{Ph}}(\omega) + Z_{\mathrm{C}}(\omega)} = \Delta'(\omega) \tag{5.1.49}$$

then we obtain the following system of integral equations (see Appendix D):

$$[1 - Z'(\omega)]\omega\hbar = -\int_{-\infty}^{+\infty} dz' K_{Ph}(z', \omega) \mathrm{Re}\frac{z'}{\sqrt{z'^2 - \Delta^2(z')}} \mathrm{sign}(z') \tag{5.1.50}$$

$$Z'(\omega)\Delta(\omega) = \int_{-\infty}^{+\infty} dz' K_{Ph}(z', \omega) \cdot \mathrm{Re}\frac{\Delta(z')}{\sqrt{z'^2 - \Delta^2(z')}} \mathrm{sign}(z') - U_C N(0) \int_0^{\hbar\omega_c} dz' \tanh\frac{\beta z'}{2} \cdot \mathrm{Re}\frac{\Delta(z')}{\sqrt{z'^2 - \Delta^2(z')}} \tag{5.1.51}$$

where

$$K_{Ph}(z', \omega) = \int_0^{\infty} dz \cdot a^2(z) F(z) \times \frac{1}{2}\left\{ \frac{\tanh\left(\frac{\beta z'}{2}\right) + \coth\left(\frac{\beta z}{2}\right)}{z + z' - \hbar\omega - i\delta} - \frac{\tanh\left(\frac{\beta z'}{2}\right) - \coth\left(\frac{\beta z}{2}\right)}{z' - z - \hbar\omega - i\delta} \right\} \tag{5.1.52}$$

Here the following abbreviation is introduced:

$$a^2(z)F(z) = \left(\int_{S_F} \frac{d^2 k}{\hbar v_{\vec{k}}} \right)^{-1} \int_{S_F} \frac{d^2 k}{\hbar v_{\vec{k}}} \int \frac{d^2 k'}{\hbar v_{k'}} \frac{|\tilde{w}(\vec{k} - \vec{k}')|^2}{v_0} \times \sum_{\vec{k}_1 \lambda} \frac{\hbar[(\vec{k} - \vec{k}')\vec{\varepsilon}(\vec{k}_1 \lambda)]^2}{2NM\omega_\lambda(\vec{k}_1)} S_0(\vec{k}_1 + \vec{k} - \vec{k}') \cdot \delta[z - \hbar\omega_\lambda(\vec{k}_1)] \tag{5.1.53}$$

This function is referred to as the *spectral density of the electron-phonon interaction* (the so-called *Eliashberg function*). Therewith $F(\omega)$ is the phonon density of states; $a^2(\omega)$ is the constant of the interaction between electrons and phonons of a given energy $\hbar\omega$.

Sometimes another form of these equations is convenient – the one where the integral with respect to z' is taken only over a positive region. Such a form can readily be obtained by substituting Eq. (5.1.52) for the kernel of the phonon interaction into Eqs (5.1.50) and (5.1.51):

$$[1 - Z'(\omega)]\hbar\omega = \frac{1}{2}\int_0^{\infty} dz a^2(z) F(z) \int_0^{\infty} dz' \cdot \mathrm{Re}\frac{z'}{\sqrt{z'^2 - \Delta^2(z')}}$$

$$\times\left\{\left(\tanh\frac{\beta z'}{2}+\coth\frac{\beta z'}{2}\right)\cdot\left(\frac{1}{z'+z+\hbar\omega+i\delta}-\frac{1}{z'+z-\hbar\omega-i\delta}\right)\right.$$
$$\left.-\left(\tanh\frac{\beta z'}{2}-\coth\frac{\beta z}{2}\right)\cdot\left(\frac{1}{z'-z+\hbar\omega+i\delta}-\frac{1}{z'-z-\hbar\omega-i\delta}\right)\right\} \tag{5.1.54}$$

$$Z'(\omega)\Delta(\omega)=\frac{1}{2}\int_0^\infty dz\alpha^2(z)F(z)\int_0^\infty dz'\cdot\mathrm{Re}\frac{\Delta(z')}{\sqrt{z'^2-\Delta^2(z')}}$$
$$\times\left\{\left(\tanh\frac{\beta z'}{2}+\coth\frac{\beta z}{2}\right)\cdot\left(\frac{1}{z'+z+\hbar\omega+i\delta}+\frac{1}{z'+z-\hbar\omega-i\delta}\right)\right.$$
$$\left.-\left(\tanh\frac{\beta z'}{2}-\coth\frac{\beta z}{2}\right)\cdot\left(\frac{1}{z'-z+\hbar\omega+i\delta}+\frac{1}{z'-z-\hbar\omega-i\delta}\right)\right\}$$
$$-U_C N(0)\int_0^{\hbar\omega_c} dz'\tanh\frac{\beta z'}{2}\cdot\mathrm{Re}\frac{\Delta(z')}{\sqrt{z'^2-\Delta^2(z')}} \tag{5.1.55}$$

Thus we have a pair of connected nonlinear integral equations for $Z'(\omega)$ and $\Delta(\omega)$. The former describes the renormalization of the electron spectrum through the electron-phonon interaction, the latter represents the development of an anomalous part of the Green's function through the rearrangement of the normal metal ground state.

Initially these equations were obtained (without the Coulomb term in the energy) by Eliashberg (1960), and later (including the Coulomb term) by Morel and Anderson (1962). As a matter of established convention Eqs (5.1.50) and (5.1.51) are referred to as the *Eliashberg equations.*

Let us also write down the Green's function $G(\vec{k},\omega)$ of a superconductor for a momentum $\hbar\vec{k}$ at the Fermi surface: using Eqs (5.1.47) and (5.1.48) it follows that

$$G(\vec{k},\omega)=\frac{[Z'(\omega)+Z_{\mathrm{Scat}}(\omega)]\hbar\omega\tau_0+\xi_{\vec{k}}\tau_3+[\varphi'(\omega)+\varphi_{\mathrm{Scat}}(\omega)]\tau_1}{[Z'(\omega)+Z_{\mathrm{Scat}}(\omega)]^2\hbar^2\omega^2-[\varphi'(\omega)+\varphi_{\mathrm{Scat}}(\omega)]^2-\xi_{\vec{k}}^2}$$
$$=\frac{\eta^{-1}(\omega)Z'(\omega)\hbar\omega\tau_0+\xi_{\vec{k}}\tau_3+\eta^{-1}(\omega)\varphi'(\omega)\tau_1}{\eta^{-2}(\omega)Z'^2(\omega)\hbar^2\omega^2-\eta^{-2}(\omega)\varphi'^2(\omega)-\xi_{\vec{k}}^2}$$

Thus the Green's function for amorphous metal in the superconducting state can be derived from the Green's function of the crystalline superconductor by the simple substitution

$$\{Z'(\omega);\ \varphi'(\omega)\}\to\{\eta^{-1}(\omega)Z'(\omega);\ \eta^{-1}(\omega)\varphi'(\omega)\}$$

Let us demonstrate that in coordinate space this operation means a multiplication of the Green's function of the crystalline superconductor by $\exp(-r/2l)$. Indeed,

$$G(x) = \int \frac{d^3k}{(2\pi)^3} \int \frac{d\omega}{2\pi} e^{i\vec{k}\cdot\vec{r} - i\omega t} G(\vec{k}, \omega) \approx \frac{4\pi}{(2\pi)^4} \int d\omega \cdot e^{i\omega t} \int k^2 dk \frac{\sin kr}{r}$$
$$\times \frac{\eta^{-1}(\omega) Z'(\omega)\hbar\omega\tau_0 + \eta^{-1}(\omega)\varphi'(\omega)\tau_1 + \xi_k\tau_3}{\left[\eta^{-1}\sqrt{Z'^2(\omega)\hbar^2\omega^2 - \varphi'^2(\omega)} - \xi_{\vec{k}}\right]\left[\eta^{-1}\sqrt{Z'^2\hbar^2\omega^2 - \varphi'^2} + \xi_{\vec{k}}\right]}$$

Assuming that (Abrikosov *et al.*, 1965)

$$\eta^{-1}(\omega)\sqrt{Z'^2(\omega)\hbar^2\omega^2 - \varphi'^2(\omega)} \approx \sqrt{Z'^2(\omega)\hbar^2\omega^2 - \varphi'^2(\omega)} + \frac{1}{2} i \frac{\omega}{|\omega|} \frac{\hbar}{\tau}$$

we change to a new integration variable $\xi_k^r = \dfrac{\hbar^2 k^2}{2m} - \varepsilon_F$ instead of k:

$$k = k_F + (k - k_F) \approx k_F + \frac{\xi}{\hbar v_F}$$
$$k^2 dk \approx k_F \frac{m}{\hbar^2} d\xi_k$$

and let the lower limit of integration with respect to ξ_k tend to $-\infty$ (because the integrand decreases rapidly when $\xi_k^r \to \pm\infty$). Then, integrating over ξ_k^r and taking the residuum, we get

$$G(x) = G'(x) e^{-r/2l}$$

with $l = v_F \cdot \tau$, and $G'(x)$ being the (fictitious) Green's function of the superconductor where the electron-electron and electron–phonon interactions are as in a crystal.

Taking advantage of the Eliashberg equations (5.1.50) and (5.1.51), let us now find the *transition temperature* $T_s = 1/(\beta_c k_B)$, which is defined as the temperature when a non-vanishing solution $\Delta(\omega)$ of Eqs (5.1.50) and (5.1.51) just comes to existence, i.e. the onset temperature of superconductivity. Thus to obtain $1/(k_B T_s = \beta_c$, we need only the *linearized* equations, linearized with respect to $\Delta(\omega)$:

$$[1 - Z'(\omega)]\omega\hbar = -\int_{-\infty}^{+\infty} dz' K_{Ph}(z', \omega) \tag{5.1.57}$$

$$Z'(\omega)\Delta = \int_{-\infty}^{+\infty} dz' K_{Ph}(z', \omega) \cdot \mathrm{Re}\frac{\Delta(z')}{z'} - U_C N(0) \int_0^{\hbar\omega_c} dz' \tanh\frac{z'\beta_c}{2} \cdot \mathrm{Re}\frac{\Delta(z')}{z'}$$

The former equation gives an explicit expression for $Z'(\omega)$, and the latter determines $\Delta(\omega)$. It is just the solvability condition of the second equation that specifies the transition temperature T_s – which is a functional of $\alpha^2(z)F(z)$.

However, an exact solution of Eq. (5.1.57) is, unfortunately, impossible. Therefore, we will try to consider a number of approximate solutions, making a distinction between *weak, intermediate* and *strong coupling* according to the strength of the inequality $\hbar\omega_D/(k_B T_c) \gg 1$.

Weak Coupling In the weak-coupling limit of the electron-phonon interaction the fundamental equations of the *BCS theory* should be derived from the Eliashberg equations. This conversion is possible upon some approximation of the interaction kernel $K_{Ph}(z', \omega)$. As is obvious from Eq. (5.1.134), $K_{Ph}(z', \omega)$ decreases steeply when $|z'|$, $|\omega| \gtrsim \omega_D$. By ω_D we denote the Debye frequency $\omega_D = S_{av} k_D$, where the *Debye wavenumber* k_D and the average sound velocity S_{av} are defined with the longitudinal and transversal sound velocities S_l and S_t:

$$\begin{cases} \dfrac{1}{S_{av}^3} = \dfrac{1}{3}\left(\dfrac{2}{S_t^3} + \dfrac{1}{S_l^3}\right) \\ \dfrac{4}{3}\pi k_D^3 = \dfrac{(2\pi)^3}{L^3} N, \quad \text{or} \quad k_D^3 = \dfrac{6\pi^2 N}{L^3} = \dfrac{6\pi^2}{v_0} \end{cases}$$

We therefore accept the following approximation: let the kernel $K_{Ph}(z', \omega)$ be constant with respect to ω, when the arguments z' and ω lie between 0 and the frequency ω_D, and let $K_{Ph}(z', \omega)$ vanish outside this interval. Notice that for $z' \to 0$, $K_{Ph}(z', 0)$ may be written as

$$K_{Ph}(z', 0) = \frac{\lambda}{2} \tanh \frac{z'\beta}{2} \quad (z' \to 0) \tag{5.1.58}$$

where

$$\lambda = 2 \int_0^\infty dz \frac{\alpha^2(z) F(z)}{z} \tag{5.1.59}$$

λ is known as the *electron-phonon coupling constant*. Thus our approximation for the interaction kernel can be specified by the following analytical expression:

$$K_{Ph}(z', \omega) \approx \frac{\lambda}{2} \tanh \frac{\beta z'}{2} \theta(\omega_D - |z'|)\theta(\omega_D - |\omega|) \tag{5.1.60}$$

Then Eq. (5.1.57) takes the form

$$[1 - Z'(\omega)]\omega = \frac{\lambda}{2} \int_{-\omega_D}^{\omega_D} \tanh \frac{\beta z'}{2} dz' = 0$$

or

$$Z'(\omega) = 1 \tag{5.1.61}$$

Because, according to Eq. (5.1.60), the kernel factorizes, the integral equation (5.1.57) can be solved exactly. First of all, $\Delta(\omega)$ is obviously real. Considering the evenness of $\Delta(\omega)$ we rewrite Eq. (5.1.57), with allowance made for Eqs (5.1.60) and (5.1.61), as

$$\Delta(\omega) = \lambda\theta(\omega_D - |\omega|) \int_0^{\hbar\omega_D} \frac{dz'}{z'} \tanh\frac{\beta_C z'}{2} \Delta(z') - U_C N(0) \int_0^{\hbar\omega_D} \frac{dz'}{z'} \tanh\frac{\beta_c z'}{2} \Delta(z') \tag{5.1.62}$$

The solution of this equation can be found with the *Ansatz*

$$\Delta(\omega) = \begin{cases} \Delta_{Ph} & |\omega| < \omega_D \\ \Delta_C & \omega_D < |\omega| < \omega_C \end{cases} \tag{5.1.63}$$

where Δ_{Ph} and Δ_C are certain constants. They are defined by the following pair of homogeneous algebraic equations:

$$\begin{cases} \left\{1 - [\lambda - U_C N(0)] K\left(\frac{\beta_C \hbar\omega_D}{2}\right)\right\} \Delta_{Ph} + U_C N(0) \ln\frac{\omega_C}{\omega_D} \Delta_C = 0 \\ U_C N(0) K\left(\frac{\beta_C \hbar\omega_D}{2}\right) \cdot \Delta_{Ph} + \left\{1 + U_C N(0) \ln\frac{\omega_C}{\omega_D}\right\} \Delta_C = 0 \end{cases} \tag{5.1.64}$$

with the abbreviation

$$K(y) = \int_0^y dx \frac{\tanh x}{x} \tag{5.1.65}$$

Here the following approximate result is used

$$\int_{\hbar\omega_D}^{\hbar\omega_C} \frac{dz'}{z'} \tanh\frac{\beta_C z'}{2} \cong \int_{\hbar\omega_D}^{\hbar\omega_C} \frac{dz'}{z'} = \ln\frac{\omega_C}{\omega_D}, \quad \left(\frac{\beta_C \hbar\omega_D}{2} \gg 1\right)$$

Let us formulate the solvability condition for the homogeneous set of equations (5.1.64), namely the vanishing of the corresponding determinant

$$\begin{vmatrix} 1 - [\lambda - U_C N(0)] K\left(\frac{\beta_C \hbar\omega_D}{2}\right); & U_C N(0) \ln\frac{\omega_C}{\omega_D} \\ U_C N(0) K\left(\frac{\beta_C \hbar\omega_D}{2}\right); & 1 + U_C N(0) \ln\frac{\omega_C}{\omega_D} \end{vmatrix} = 0$$

or

$$K\left(\frac{\beta_C \hbar\omega_D}{2}\right) = \frac{1}{\lambda - \mu^*} \tag{5.1.66}$$

where

$$\mu^* = \frac{U_C N(0)}{1 + U_C N(0) \ln \dfrac{\omega_C}{\omega_D}} = \frac{V_e N(0)}{1 + V_e N(0) \ln \dfrac{E_F}{\hbar\omega_D}} \tag{5.1.67}$$

In writing Eq. (5.1.67) we have used the expression for the intermediate pseudopotential U_C.

Because for all known superconductors we have $\hbar\omega_D \gg k_B T_C$, the integral (5.1.65) should be calculated under the condition that $y \gg 1$, which leads to

$$K(y) = \ln\left(\frac{4\gamma}{\pi} y\right) \tag{5.1.68}$$

where γ is expressed through the Euler constant: $\ln\gamma = C = 0.577\ldots$. Thus Eq. (5.1.66) yields the following expression for T_C:

$$\frac{1}{\beta_C} = k_B T_C = \frac{2\gamma}{\pi} \hbar\omega_D \exp\left(-\frac{1}{\lambda - \mu^*}\right) \tag{5.1.69}$$

which coincides with the formula for T_C in the BCS theory supplemented with the Coulomb electron-electron interaction: the parameter μ^* represents the effective interelectronic Coulomb repulsion at the Fermi surface. As is evident from Eq. (5.1.67) – which was originally derived by Bogoliubov *et al.* (1958) – the Coulomb repulsion parameter μ^* is essentially weakened owing to a large logarithmic term in the denominator.

Intermediate Coupling In the above section we converted the strong-coupling equations of the superconductivity theory being discussed into the weak-coupling expressions of the BCS theory simply by formally approximating the genuine kernel of the electron-phonon interaction with Eq. (5.1.60). Let us now consider a more consistent solution of the linearized equations of the superconductivity theory that gives certain corrections to the BCS formula for T_C, thus corresponding to an intermediate-coupling case in the electron-phonon interaction. This enables us to obtain a weak-coupling criterion, much as later on we can deduce an intermediate-coupling criterion by considering the strong-coupling case. Now it is convenient to proceed not from Eq. (5.1.57), but from the linearized equations (5.1.54) and (5.1.55) in which the kernel of electron-phonon interaction is substituted explicitly. The characteristic frequencies z of the phonon spectral function $F(z)$ are in agreement with the main peaks of the distribution $F(z)$ and will be considered to be much higher than the characteristic transition temperature ($z \gg k_B T$). This permits us to neglect the distribution function $N(z) = (\exp\beta_C z - 1)^{-1}$ of the thermal phonons in Eqs (5.1.54) and (5.1.55). Because

$$\coth\frac{\beta_C z}{2} = 1 + 2N(z)$$

the above approximation corresponds to $\coth\frac{\beta_C z}{2}$ being replaced by unity. The combinations of fractions proportional to $\tanh\frac{\beta_C z'}{2}$ in Eqs (5.1.54) and (5.1.55) fall into two categories, one of which gives finite expressions when the arguments ω, z and z' tend to zero – here we can put $\tanh\frac{\beta_C z'}{2} \to 1$ in view of the same inequality constraint $z' \gg \frac{1}{\beta_C}$. Another category, when $\omega, z \to 0$, develops under the integral over z' a singularity in z' of the $1/z'$ type – to compensate for it the term $\tanh\frac{\beta_C z'}{2}$ should be retained. Upon such approximation and linearization of Eqs (5.1.54) and (5.1.55) we obtain the following homogeneous set of equations to calculate T_C:

$$\begin{cases} Z(\omega)\Delta(\omega) = \displaystyle\int_0^{+\infty} \frac{dz'}{z'} \tanh\frac{\beta_C z'}{2} \operatorname{Re}\Delta(z') \cdot \left[\int_0^{+\infty} dz\alpha^2(z)F(z)K_+(z', z, \omega) - \mu(\omega, z') \right] \\ [1 - Z(\omega)]\omega = \displaystyle\int_{-\infty}^{+\infty} dz' \int_0^{+\infty} dz\alpha^2(z)F(z)f(-z')K_-(z', z, \omega) \end{cases} \tag{5.1.70}$$

with the abbreviations

$$K_\pm(z', z, \omega) = \frac{1}{z' + z + \hbar\omega + i\delta} \pm \frac{1}{z' + z - \hbar\omega - i\delta} \tag{5.1.71}$$

$$\mu(\omega, z') = U_C N(0)\theta(\omega_C - \omega) \cdot \theta(\hbar\omega_C - z') \tag{5.1.72}$$

In the second part of Eq. (5.1.70), $f(z)$ is the Fermi distribution function $f(z) = (\exp\beta z + 1)^{-1}$, which arises from the identities

$$\tanh\frac{\beta z'}{2} = 1 - 2f(z') = -1 + 2f(-z')$$

We call the equations in Eq. (5.1.70), which are derived from the general superconductivity equations (5.1.54) and (5.1.55) by neglecting the thermal phonon contribution, the *intermediate-coupling equations.* These equations were solved by Karakozov, Maksimov and Mashkov in 1975 by an iteration procedure in the parameter $\frac{\lambda}{1+\lambda}$. Here only the end result for the transition temperature T_s is presented:

$$k_B T_s = \frac{2\gamma}{\pi} \hbar\omega_0 \exp\left(- \frac{1+\lambda}{\lambda - \mu^*\left(1 + \frac{\lambda A}{1+\lambda}\right)} \right) \tag{5.1.73}$$

Thus T_s is determined by four variables: λ, A, ω_0 and μ^*. The first three of these depend on the shape of the phonon spectrum $F(z)$, namely λ through Eq. (5.1.59), and A and ω_0 according to the following definitions:

$$A = -\frac{4}{\lambda^2} \int_0^\infty dz_1 a^2(z_1)F(z_1) \int_0^\infty dz_2 a^2(z_2)F(z_2) \cdot \left(\ln\frac{z_1}{z_2}\right) \cdot \frac{z_2}{z_1} \frac{1}{z_2^2 - z_1^2} \tag{5.1.74}$$

$$\omega_0 = \omega_{\log} \cdot e^{-A} \tag{5.1.75}$$

where

$$\ln(\omega_{\log}/\omega_e) = \frac{2}{\lambda} \int_0^\infty dz \frac{a^2(z) \cdot F(z)}{z} \cdot \ln(z/\omega_e) \tag{5.1.76}$$

is some mean logarithmic frequency.[3)]

It follows from Eq. (5.1.73) that the transition to the weak-coupling theory, which gives the BCS formula in Eq. (5.1.69) for T_C, succeeds provided that $\lambda \ll 1$ – it is this condition that determines the weak-coupling criterion. In contrast, for the strong-coupling case ($\lambda \sim 1$), it is rather difficult to formulate a particular criterion of the intermediate coupling. We should merely remember the approximations having been assumed upon derivation of the main equations (5.1.70), which form the foundation of the intermediate-coupling theory.

The Strong-coupling Case The approximate Eliashberg equations (5.1.70), which provide the basis for the theory of superconductors with intermediate coupling, were derived from the more general equations (5.1.54) and (5.1.55): the thermal phonons were neglected and $\tanh\frac{\beta_K z'}{2}$ substituted by unity in several expressions. This means that temperature effects are neglected when they are of the order of $(k_B T/(\leftarrow \omega_D))^2$ (for example in $\mathrm{Im}Z'(\omega)$) and retained when they are of the order of $(k_B T/(\leftarrow \omega_D))^2 \ln(\leftarrow \omega/(k_B T))$ (for instance in $\mathrm{Re}Z'(\omega)$). The occurrence of this dependence of $\mathrm{Re}Z'(0)$ on the temperature, incidentally, could result in a magnitude of $\frac{2\Delta}{k_B T}$ different from the universal value of 3.52 which follow from the BCS theory.

3) Note that the result of Eq. (5.1.76) for $\omega_{\log}$ does not depend on the frequency unit ω_e, because of Eq. (5.1.59).

The complete Eliashberg equations (5.1.54) and (5.1.55), which involve terms describing thermal phonons and include the imaginary parts of $\Delta(\omega)$ and $Z'(\omega)$, provide an accuracy allowing for retention not only of the terms $\propto \left(\dfrac{k_B T_C}{\hbar \omega_D}\right)^2 \ln\left(\dfrac{\hbar \omega_D}{k_B T_C}\right)$, but of the terms proportional to $\left(\dfrac{k_B T}{\hbar \omega_D}\right)^2$ as well.

In the event of strong coupling (in fact, for $k_B T/(\leftarrow \omega_D)/0.1$), when this exactness is required, it is an arduous task to obtain an analytical expression for T_C like that given in Eq. (5.1.73). The Eliashberg equations in this situation need to be solved numerically. Such calculations were performed in 1968 by McMillan on the basis of some special assumption on the character of the spectral function $a^2(\omega) \cdot F(\omega)$ (McMillan, 1967). As $F(\omega)$ he used the phonon spectrum of niobium measured by inelastic neutron scattering. This spectrum is fairly typical of a number of bcc metals, thus the formulae obtained for T_C are likely to be applicable at least for certain of the bcc metals. Additionally, in his calculations the effective matrix element $a^2(\omega)$ of the electron-phonon interaction is assumed to be frequency-independent. This fact is also an approximate consequence of the comparison between $a^2(\omega) \cdot F(\omega)$ obtained in tunneling experiments, and $F(\omega)$ derived from the neutron scattering data.

For every given set of a^2 and μ^* the Eliashberg equations yield a value of T_C. The resultant numerical data for the values of T_C as a function of a^2 and μ^* can be juxtaposed with the data of T_C as a function of the two other variables λ and μ^*, because λ and a^2 are inter-related: $\lambda = 2a^2 \displaystyle\int_0^\infty \dfrac{d\omega}{\omega} F(\omega)$. McMillan (1967) suggested that T_C should depend on $a^2(\omega) \cdot F(\omega)$ only through λ. The analytical function $T_C = T_C(\lambda, \mu^*)$ showing the best correlation with the calculated T_C values for given λ and μ^* turns out to be as follows:

$$T_C = \frac{\theta_D}{1.45} \exp\left[-\frac{1.04(1+\lambda)}{\lambda - \mu^*(1+0.62\lambda)}\right] \tag{5.1.77}$$

(θ_D is the Debye temperature).

This T_C expression, known as the *McMillan formula*, is widely used in the analytical treatment of strong-coupled superconductors. We should dwell once again on the particular assumptions that underlie its derivation and prohibit its use as a universal formula for T_C. In the first place, a niobium phonon spectrum is used; second, it is presupposed that the effective matrix element of the electron-phonon interaction $a^2(\omega)$ does not depend on the frequency ω; and finally it is assumed that the functional dependence of T_C on $a^2(\omega) \cdot F(\omega)$ is determined only by λ. Nonetheless, the McMillan formula gives good agreement with experimental T_C data, if λ and μ^* are evaluated from independent experiments.

McMillan also suggested that, instead of the Debye temperature θ_D, a frequency average over the spectrum $F(\omega)$ can be employed as a pre-exponent in Eq. (5.1.77):

$$\langle\omega\rangle = \frac{\int_0^\infty \omega \cdot F(\omega) d\omega}{\int_0^\infty F(\omega) d\omega}$$

For niobium the following relation holds:

$$\frac{k_B \theta_D}{1.45} = \frac{\hbar\langle\omega\rangle}{1.2}$$

Assuming that for other superconductors the pre-exponent is also determined by $\langle\omega\rangle$, instead of Eq. (5.1.190), we have the second McMillan formula:

$$k_B T_C = \frac{\hbar\langle\omega\rangle}{1.2} \exp\left[-\frac{1.04(1+\lambda)}{\lambda - \mu^*(1+0.62\lambda)}\right] \tag{5.1.78}$$

As already noted above, the McMillan formulae were derived for the niobium phonon spectrum with $\lambda \lesssim 1.5$. Allen and Dynes (1975) carried out similar calculations for large λ and discovered a discrepancy between the exact results and those derived from the McMillan formula when $\lambda > 2$. Furthermore, they studied the T_C dependence on the shape of the phonon spectrum by solving the Eliashberg equations with the phonon spectrum of Pb and Hg. In fact, the functional dependence of T_C on λ turned out to be the same for all three materials: Pb, Hg and Nb. Thus the question of the applicability of McMillan's formulae to metals of various crystal structure (or even to the amorphous case) is no longer valid. Consequently, by means of McMillan's formulae we can calculate T_C for strong-coupled materials ($\lambda \lesssim 2$) with good accuracy.

5.2
The Electron-phonon Coupling Constant and the Superconducting Transition Temperature for Simple Amorphous Metals

As discussed above, the electron-phonon coupling constant is defined by Eq. (5.1.59) as

$$\lambda = 2 \int_0^\infty dz \frac{\alpha^2(z) F(z)}{z} \tag{5.2.1}$$

where according to Eq. (5.1.53)

$$\alpha^2(z)F(z) = \left(\int\limits_{S_F} \frac{d^2k}{\hbar v_{\vec{k}}}\right)^{-1} \int\limits_{S_F} \frac{d^2k}{\hbar v_{\vec{k}}} \int \frac{d^2k'}{\hbar v_{k'}} \frac{|\tilde{w}(\vec{k}-\vec{k}')|^2}{(2\pi)^3} \sum_\lambda d^3k_1 \frac{\hbar[(\vec{k}-\vec{k}')\cdot\vec{\varepsilon}(\vec{k}_1,\lambda)]^2}{2M\omega_\lambda(\vec{k}_1)}$$
$$\times\, S_0(\vec{k}_1+\vec{k}-\vec{k}')\cdot\delta[z-\hbar\omega_\lambda(\vec{k}_1)]$$

is the Eliashberg function when $V, N \to \infty$, $\left(\frac{V}{N} = v_0 = \text{const.}\right)$ (see Eq. (D.15) in Appendix D).

Now the relation

$$S_0(\vec{q}) = \frac{(2\pi)^3}{v_0}\delta(\vec{q}) + a(\vec{q})\,,$$

where $a(\vec{q})$ is the so-called *structure factor*, allows the decomposition of Eq. (5.2.1) as follows:

$$\lambda = \lambda_0 + \lambda_1 \tag{5.2.2}$$

where

$$\lambda_0 = \left(\int\limits_{S_F} \frac{d^2k}{\hbar v_{\vec{k}}}\right)^{-1} \cdot \int\limits_{S_F} \frac{d^2k}{\hbar v_{\vec{k}}} \int\limits_{S_F} \frac{d^2k'}{\hbar v_{k'}} \frac{|\tilde{w}(\vec{k}-\vec{k}')|^2}{M} \sum_\lambda \frac{[(\vec{k}-\vec{k}')\cdot\vec{\varepsilon}(\vec{k}-\vec{k}',\lambda)]^2}{v_0\cdot\omega_\lambda^2(\vec{k}-\vec{k}')} \tag{5.2.3}$$

$$\lambda_1 = \left(\int\limits_{S_F} \frac{d^2k}{\hbar v_{\vec{k}}}\right)^{-1} \cdot \int\limits_{S_F} \frac{d^2k}{\hbar v_{\vec{k}}} \int\limits_{S_F} \frac{d^2k'}{\hbar v_{k'}} \frac{|\tilde{w}(\vec{k}-\vec{k}')|^2}{M}$$
$$\times \sum_\lambda \int d^3k_1 \frac{[(\vec{k}-\vec{k}')\cdot\vec{\varepsilon}(\vec{k}_1,\lambda)]^2}{(2\pi)^3\omega_\lambda^2(\vec{k}_1)} a(\vec{k}_1+\vec{k}-\vec{k}') \tag{5.2.4}$$

In the latter equation the integration with respect to $\vec{k}_1$ for crystals is carried out over the volume of the first (basic) Brillouin zone of the reciprocal lattice, and for amorphous solids over a sphere with radius k_D, where k_D is the Debye wavenumber.

In crystalline metals, owing to the lattice periodicity, the structure factor $a(\vec{q})$ is

$$a(\vec{q}) = \frac{(2\pi)^3}{v_0} \sum_{\vec{k}_n \neq 0} \delta(\vec{q}-\vec{k}_n) \tag{5.2.5}$$

where the $\vec{k}_n$ are the reciprocal space vectors.

Because for $\vec{k}_n \neq 0$ the vector $\vec{k}_1 = \vec{k}_n + \vec{k}' - \vec{k}$ lies outside of the first Brillouin zone of the reciprocal lattice, the contribution λ_1 in Eq. (5.2.2) vanishes (i.e. the so-called *Umklapp processes* play just a minor role in superconductivity), and we arrive at the expression obtained by Scalapino (1969) for crystalline metals within the framework of perturbation theory with respect to the weak electron-ion coupling; so in crystalline material we have $\lambda = \lambda_0$.

However, because of violation of long-range order, in *amorphous* metals the structure factor $a(\vec{k})$ differs from that defined by Eq. (5.2.5), and λ_1 is non-zero.

Thus the electron-phonon coupling constant λ increases upon the transition from the crystalline state to the amorphous one. There are two reasons for this increase. First, the phonon spectrum in amorphous metals softens and a significant dip appears in the region of large wavenumbers of dispersion curve (see Chapter 2). This tends to increase the first contribution λ_0 in Eq. (5.2.1). Secondly, due to the violation of long-range order in amorphous metals, the quasi-momentum $\vec{k}$ of the electron becomes a 'poor quantum number', resulting in diffuse electron scattering and the appearance of a non-vanishing second contribution λ_1 in Eq. (5.2.1).

For a numerical estimation of the expressions (5.2.3) and (5.2.4), we rewrite them in a form convenient for integration, taking into account the *isotropy* of averages in bulk amorphous metals.

Let us begin with the integral over k_1 in the expression for λ_1:

$$I(\vec{q}) = \sum_{\lambda} \int \mathrm{d}^3 k_1 \frac{[(\vec{q}) \cdot \vec{\varepsilon}(\vec{k}_1, \lambda)]}{(2\pi)^3 \omega_{\lambda}^2(k_1)} a(\vec{k}_1 + \vec{q})$$

$$= \frac{1}{(2\pi)^3} \int_0^{k_D} k_1^2 \mathrm{d}k_1 \int \mathrm{d}\Omega_1 \cdot a\left(\sqrt{q^2 + k_1^2 + 2qk_1 \cos\theta_1}\right) q^2 \left[\frac{\cos^2\theta_1}{\omega_l^2} + \frac{\sin^2\theta_1}{\omega_t^2}\right]$$

It has been taken into account here that for the polarization of longitudinal (l) and transversal (t) phonons the following relationships apply:

$$\begin{cases} \vec{\varepsilon}(\vec{k}, 1) = \vec{\varepsilon}(\vec{k}, l) || \vec{k}; \ \omega_1(k) = \omega_l(k) \\ \vec{\varepsilon}(\vec{k}, 2) \perp \vec{\varepsilon}(\vec{k}, 3) \perp \vec{k}; \ \omega_2(k) = \omega_3(k) = \omega_t(k) \end{cases}$$

the z-axis being aligned with vector $\vec{q}$. Thus we have

$$\vec{q} \cdot \vec{\varepsilon}(\vec{k}_1, 1) = q \cdot \cos\theta_1; \ \vec{q} \cdot \vec{\varepsilon}(\vec{k}_1, 2) = q \cdot \sin\theta_1 \cdot \cos\varphi_1; \ \vec{q} \cdot \vec{\varepsilon}(\vec{k}_1, 3) = q \cdot \sin\theta_1 \cdot \sin\varphi_1$$

Then the integration over the angle φ_1 with the substitution $x = \cos\theta$ gives

$$I(q) = \frac{q^2}{4\pi^2} \int_0^{k_D} k_1^2 \mathrm{d}k_1 \int_{-1}^{+1} \mathrm{d}x \cdot a\left(\sqrt{q^2 + k_1^2 + 2qk_1 x}\right) \left[\frac{x^2}{\omega_l^2(k_1)} + \frac{1 - x^2}{\omega_t^2(k_1)}\right]$$

We now change over to integrals with respect to $\vec{k}$ and $\vec{k}'$. We have to make allowance for the fact that these integrals are taken over the solid angles, where the magnitude of the vectors is equal to k_F:

$$|\vec{k}| = |\vec{k}'| = k_F$$

Hence

$$\begin{aligned}
\int \frac{d^2k'}{\hbar v_{\vec{k}'}} f(\vec{k} - \vec{k}') &= \frac{1}{(2\pi)^3} \frac{mk_F}{\hbar^2} \int d\Omega' f(|\vec{k} - \vec{k}'|) \\
&= \frac{1}{(2\pi)^3} \frac{mk_F}{\hbar^2} \int\limits_0^{2\pi} d\varphi' \int\limits_0^{\pi} \sin\theta' d\theta' f\left(\sqrt{2k_F^2(1 - \cos\theta')}\right) \\
&= \frac{4}{4\pi^2} \frac{mk_F}{\hbar^2} \int\limits_0^{\pi} \sin\frac{\theta}{2} \cdot \cos\frac{\theta}{2} \cdot d\frac{\theta}{2} \cdot f\left(2k_F \cdot \sin\frac{\theta}{2}\right) \\
&= \frac{1}{4\pi^2} \frac{m}{k_F \hbar^2} \int\limits_0^{2k_F} q dq \cdot f(q) = \frac{1}{\pi^2} \frac{mk_F}{\hbar^2} \int\limits_0^1 z \cdot dz \cdot f(2k_F \cdot z)
\end{aligned}$$

Then we have (Krasny *et al.*, 1990b, 1993, 1995, 1996; Krasny and Kovalenko, 1990) the following Eqs (5.2.6) and (5.2.7) for λ_0 rsp. λ_1:

$$\lambda_0 = \frac{m}{4\pi^2 k_F \cdot \hbar^2} \int\limits_0^{2k_F} \frac{q^3 |\tilde{w}(q)|^2 dq}{v_0 \cdot M \cdot \omega_l^2(q)} = 12 \frac{m}{M} z_0 \int\limits_0^1 \frac{z^3 dz |\tilde{w}(2k_F z)|^2}{v_0^2 \cdot \hbar^2 \cdot \omega_l^2(2k_F z)} \tag{5.2.6}$$

$$\begin{aligned}
\lambda_1 &= \frac{m}{4\pi^2 k_F \cdot \hbar^2} \int\limits_0^{2k_F} \frac{q^3 |\tilde{w}(q)|^2 dq}{M} \cdot \frac{1}{4\pi^2} \int\limits_0^{k_D} k_1^2 dk_1 \int\limits_{-1}^{+1} dx \cdot a\left(\sqrt{q^2 + k_1^2 + 2qk_1 x}\right) \\
&\times \left[\frac{x^2}{\omega_l^2(k_1)} + \frac{1 - x^2}{\omega_t^2(k_1)}\right] = 12 \frac{m}{M} z_0^2 \int\limits_0^1 \frac{z^3 dz |\tilde{w}(2k_F z)|^2}{v_0^2 \hbar^2} \cdot 6 \int\limits_0^{1/\sqrt{2}\sqrt[3]{z_0}} y^2 dy \\
&\times \int\limits_{-1}^{+1} dx \cdot a\left(2k_F \sqrt{z^2 + y^2 + 2zyx}\right) \cdot \left[\frac{x^2}{\omega_l^2(2k_F y)} + \frac{1 - x^2}{\omega_t^2(2k_F y)}\right]
\end{aligned} \tag{5.2.7}$$

In Eqs (5.2.6) and (5.2.7) it has been taken into account that

$$\begin{cases}
k_D^3 \cdot v_0 = 6\pi^2 \\
k_F^3 \cdot v_0 = 3\pi^2 \cdot z_0 \\
\dfrac{k_D}{2k_F} = \dfrac{1}{\sqrt{2}\sqrt[3]{z_0}}
\end{cases}$$

In the numerical estimation of the integrals in Eqs (5.2.6) and (5.2.7), considering the average isotropy of amorphous metals, a hard-sphere structure factor $a(\vec{k})$ in the Percus-Yevick approximation is taken as structure factor of the system, and as its pseudopotential the *Ashcroft one-parameter pseudopotential* (see Chapter 2) is

Tab. 5.2.1 Parameters for the Crystalline and Amorphous states of simple metals

Metal	***Crystalline state***			***Amorphous state***					
	θ^{cr} (K)	λ^{cr}	T_C^{cr} (K)	R_C (a.u.)	θ^{cr} (K)	λ^{am}	T_C^{am} (K)	T_C^{exp} (K)	R_C^* (a.u.)
I	II	III	IV	V	VI	VII	VIII	IX	X
Li		0.41		1.06	210.1	0.8089	8.3326		
				1.1	225.0	0.6153	4.5930		
				1.32	291.6	0.1532	2.7×10^{-10}		
Na	157	0.16	0.08	1.22	55.2	1.8771	6.2847		
				1.26	70.1	1.2514	5.4659		
				1.59	113.9	0.1432	1.1×10^{-12}		
K		0.13	0.09	1.78	50.9	0.3733	0.1164		
Rb			0.011	1.8	16.5	0.8986	0.7981		
				1.9	29.5	0.3886	0.0866		
Cs				2.0	12.4	0.6703	0.3195		
				2.12	22.2	0.2662	0.0023		
				2.16	23.7	0.2087	0.00004		
Cu		0.14	0.02	0.81	102.1	0.8319	4.2793		
				0.91	118.2	0.4370	0.6559		
				1.76	158.2	0.2718	0.0214		
Ag		0.1	0.002	1.04	72.7	0.4453	0.4421		
				1.87	105.7	0.2014	0.0001		
Au		0.14	0.002	0.81	36.0	1.9072	4.1376		
				2.04	74.3	0.3143	0.0518		
Be	1390	0.24±0.01	0.026	0.64	513.3	1.4529	46.9560	9.9	0.752
				0.76	628.0	0.5019	6.4109		
Mg	390	0.24±0.08	10^{-4}	0.99	141	1.8698	15.9955		
				1.31	280.4	0.1867	0.000014	~0.7	1.16
				1.39	316.6	0.1877	0.000019		
Ba				2.03	108.6	0.2256	0.0008		
Cd	209	0.24±0.01	0.517–0.555	1.95	249.0	0.4826	2.1688		
Zn	339	0.43±0.05	0.875	0.79	115.2	2.1403	14.3388		
				1.07	196.1	0.2214	0.001		
				1.27	262.6	0.2009	0.0001	1.45	0.965
				1.29	269.1	0.2109	0.0005		
				1.35	288.2	0.2449	0.0092		
Hg	72	1.0	4.153–4.16	0.915	36.9	4.3068	6.272		
				1.62	143.0	0.2872	0.0364		
Al	426	0.39±0.02	1.19	0.82	227.1	2.0446	27.4342	5.84	0.975
				0.90	258.7	1.0538	16.1218		
				1.09	375.2	0.3568	0.6332		
				1.11	389.8	0.3501	0.5763		
				1.12	397.2	0.3488	0.5704		
				1.16	427.6	0.3535	0.6763		

Tab. 5.2.1 (continued)

Metal	*Crystalline state*			*Amorphous state*					
	θ^{cr} (K)	λ^{cr}	T_C^{cr} (K)	R_C (a.u.)	θ^{cr} (K)	λ^{am}	T_C^{am} (K)	T_C^{exp} (K)	R_C^* (a.u.)
I	II	III	IV	V	VI	VII	VIII	IX	X
Ga	325	0.42±0.03	1.078–	0.9	99.0	3.4029	15.5253		
			1.196	1.1	170.5	0.5204	2.0006	8.4	1.013
				1.2	215.0	1.4129	0.8893		
In	109	0.69	3.4	1.05	59.8	3.5012	9.4856		
				1.12	78.9	1.2867	6.3508		
				1.31	134.8	0.4667	1.0168	5.6	1.141
				1.32	138.0	0.4635	1.0085		
				1.37	153.3	0.4615	1.0988		
Sn	194	0.7	3.7–	1.297	127.2	1.0969	8.3833	6.3–6.8	
			3.732	1.3	128.3	1.0915	8.3983		
Pb	130	1.12	7.175–	1.38	101.3	1.0695	6.4432	6.5	1.38
			7.236	1.47	124.7	0.9869	7.0396		
Bi				1.49	92.7	2.7292	13.2166	5.0	

taken, with the parameter R_C determined by different experiments. The R_C values (in atomic units) for nearly all simple metals are listed in the fifth column of Table 5.2.1 (Krasny *et al.*, 1995, 1996). Also given, in column VII, is the result of a numerical calculation of the electron-phonon coupling constant, λ, is also presented. The calculations give for amorphous metals sufficiently large λ, i.e. strong coupling, hence the characteristic transition temperature T_C can be evaluated by the McMillan formula (5.1.77). The Debye temperature θ_D involved in the expression (5.1.77) for T_C is estimated from the sound velocities S_l and S_t as usual:

$$\theta_D = \frac{h}{k_B}\left[\frac{8\pi^2}{v_0}\left(\frac{2}{S_t^3}+\frac{1}{S_l^3}\right)^{-1}\right]^{1/3}$$

and tabulated for amorphous metals in column VI of Table 5.2.1. The transition temperatures T_C calculated with McMillan's formula (5.1.77) are presented in column VIII. Columns II, III and IV present the Debye temperature θ_D^{cr}, electron-phonon coupling constant λ^{cr}, and transition temperature T_C^{cr}, respectively, as determined *experimentally* for the simple *crystalline* metals. Finally, in columns IX and X, for the corresponding *amorphous* metals the experimental values of the critical temperature T_C^{exp} and the corresponding parameters R_C^* calculated by Eq. (5.1.77) are tabulated.

Table 5.2.1 demonstrates that the constant λ, and thus the temperature T_C, depend critically on the pseudopotential parameter R_C. This is also evident from

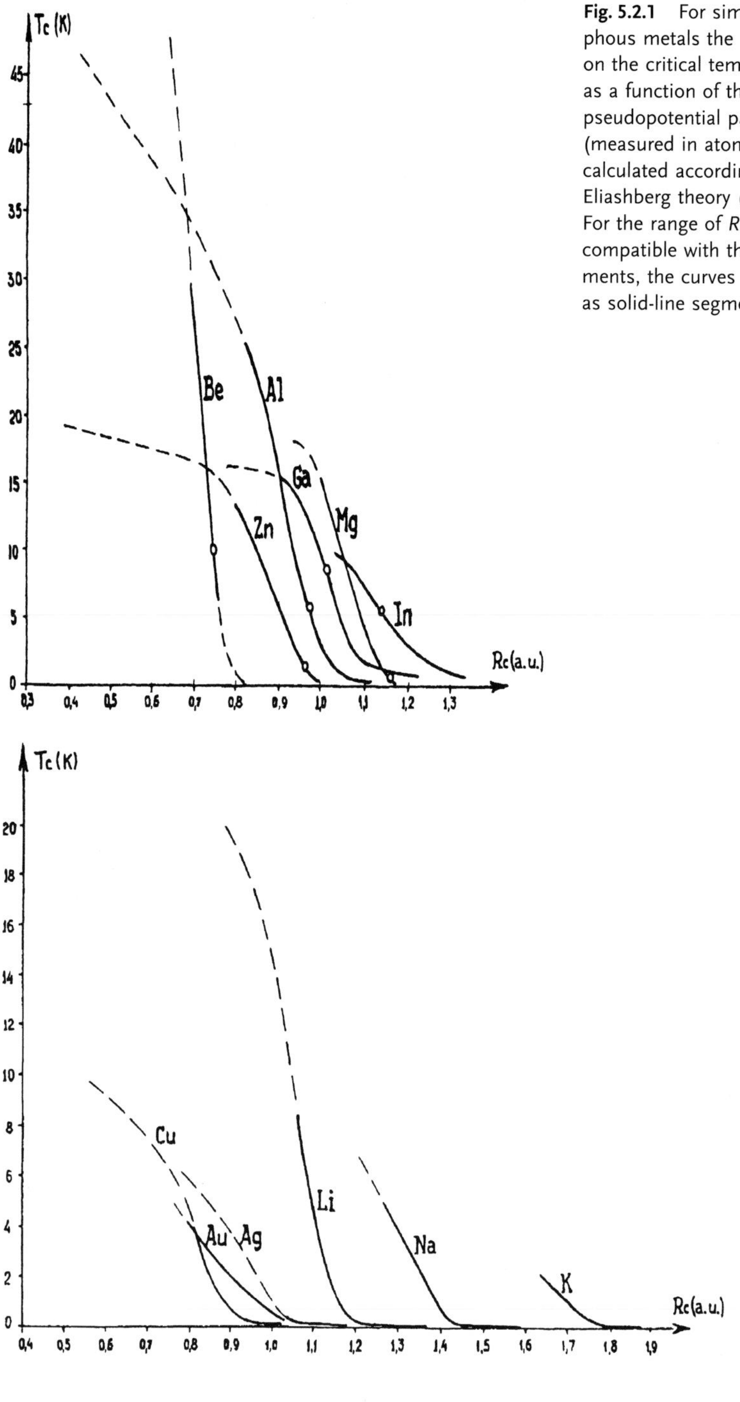

Fig. 5.2.1 For simple amorphous metals the dependence on the critical temperature T_C as a function of the Ashcroft pseudopotential parameter R_C (measured in atomic units) is calculated according to the Eliashberg theory (see text). For the range of R_C, which is compatible with the experiments, the curves are plotted as solid-line segments

Fig. 5.2.1, which plot T_C versus R_C for amorphous metals. The corresponding curves in the figure are shown as dotted lines with solid portions corresponding to experimental values of R_C. It should be noted that for nearly all amorphous metals T_C decreases sharply with increasing R_C, and that the circles, standing for the experimental values, lie along the solid lines.

The Ashcroft pseudopotential parameter R_C^*, determined from Eq. (5.1.77) on the basis of experimental transition temperatures, will later be used to also determine the concentration dependence of T_C for *binary alloys* of simple amorphous metals.

Figure 5.2.1, as already mentioned, presents the main results of the present section; it shows the dependence of calculated values of T_C on the pseudopotential parameter R_C for a range of elementary amorphous metals. At the low temperatures considered, these systems exist in the amorphous state.

5.3 Superconducting Properties of Binary Alloys of Simple Amorphous Metals

In terms of the model suggested in Chapter 2, the Hamiltonian of a two-component alloy of simple metals is

$$\hat{\mathrm{H}} = \sum_{a=1}^{2}\sum_{\nu_a=1}^{N_a}\frac{\vec{p}_{\nu_a}^{\,2}}{2M_a} + \frac{1}{2}\sum_{a,\beta}\sum_{\substack{\nu_a,\nu_\beta \\ (\nu_a\neq\nu_\beta)}}\frac{Z_a Z_\beta e^2}{|\vec{R}_{\nu_a}-\vec{R}_{\nu_\beta}|} + \sum_{j=1}^{N_e}\frac{\vec{p}_j^{\,2}}{2m} + \frac{1}{2}\sum_{\substack{j,j' \\ (j\neq j')}}\frac{e^2}{|\vec{r}_j-\vec{r}_{j'}|}$$

$$+\sum_{a=1}^{2}\sum_{\nu_a=1}^{N_a}\sum_{j=1}^{N_e} w_{0a}(\vec{r}_j-\vec{R}_{\nu_a}) \tag{5.3.1}$$

where N_a, $C_a = N_a/N$, μ_a and $Z_a e$ are respectively the number, concentration, chemical potential and charge of an a-type ion ($a = 1, 2$); $N = N_1 + N_2$ is the total number of ions; $w_{0e}(r)$ is the pseudopotential of the interaction between an electron and the ion of type a; and $N_e = Z_1N_1 + Z_2N_2$ is the total electron number. Finally we have

$$\bar{M} = M_1C_1 + M_2C_2;\ \bar{Z} = Z_1C_1 + Z_2C_2$$

We assume again that the randomly spaced ions of the amorphous metal vibrate about their quasi-equilibrium positions $R_{0\nu_a}^{(m)}$, i.e.

$$\vec{R}_{\nu_a}^{(m)} = \vec{R}_{0\nu_a}^{(m)} + \vec{\xi}_{\nu_a}^{(m)}$$

where $\vec{\xi}_{\nu_a}^{(m)} \equiv \vec{\xi}_a(\vec{R}_{0\nu_a}^{(m)})$ is the ion's displacement from the quasi-equilibrium position.

Take a particular accidentally realized mth configuration of the ions. Then, introducing again the *Nambu* operators according to Eqs (5.1.9) and (5.1.11), we employ the second-quantized representation and rewrite Eq. (5.3.1) for $\hat{\mathrm{H}}$ in the form:

$$\hat{H}^{(m)} - \sum_{a=1}^{2} \mu_a \hat{N}_a - \mu_e \hat{N}_e = \hat{H}_0^{(m)} + \hat{H}_{int}^{(m)} \tag{5.3.2}$$

where

$$\hat{H}_0^{(m)} = \hat{H}_i^{(m)} + \hat{H}_{0e} \tag{5.3.3}$$

$$\hat{H}_{int}^{(m)} = \hat{H}_{ee} + \hat{H}_{ie}^{(m)} \tag{5.3.4}$$

$$\hat{H}_i^{(m)} = \sum_{a=1}^{2} \sum_{\nu_a=1}^{N_a} \left(\frac{\vec{p}_{\nu_a}^{\,2}}{2M_a} + \frac{Z_a b_a}{v_0} - \mu_a \right) + \frac{1}{2} \sum_{a,\beta} \sum_{\substack{\nu_a,\nu_\beta \\ (\nu_a \neq \nu_\beta)}} Z_a Z_\beta V_e \left(\left| \vec{R}_{\nu_a}^{(m)} - \vec{R}_{\nu_\beta}^{(m)} \right| \right) \tag{5.3.5}$$

$$\hat{H}_{0e} = \int d^3 r \hat{\Psi}^+(\vec{r}) \tau_3 \left(-\frac{\hbar^2}{2m} \Delta - \mu_e \right) \hat{\Psi}(\vec{r}) \tag{5.3.6}$$

$$\hat{H}_{ee} = \frac{1}{2} \int d^3 r \int d^3 r' \hat{\Psi}^+(\vec{r}\,') \tau_3 \hat{\Psi}(\vec{r}\,') V_e(\vec{r}\,' - \vec{r}) \hat{\Psi}^+(\vec{r}) \tau_3 \hat{\Psi}(\vec{r}) \tag{5.3.7}$$

$$\hat{H}_{ie}^{(m)} = \int d^3 r \hat{\Psi}^+(\vec{r}) \tau_3 \hat{\Psi}(\vec{r}) \sum_{a=1}^{2} \sum_{\nu_a=1}^{N_a} w_{0a} \left(\vec{r} - \vec{R}_{\nu_a}^{(m)} \right) \tag{5.3.8}$$

The functions $V_e(\vec{r})$ and $w_{0a}(\vec{r})$ are determined by Eqs (5.1.6)–(5.1.8), and again the Nambu formalism has been used (e.g. Eq. (5.1.11)).

Let us consider, in this formalism, the one-particle electron Green's function

$$G(x - x') = \overline{G^{(m)}(x, x')} = -\overline{\langle T_\tau \{ \hat{\Psi}^{(m)}(x) \hat{\Psi}^{+(m)}(x') \} \rangle^{(m)}} \tag{5.3.9}$$

where, as for one-component systems (see Section 5.1), a statistical averaging

$$\langle \ldots \rangle^{(m)} = \frac{\mathrm{Tr}\left\{ \ldots \exp\left[-\beta \left(\hat{H}_0^{(m)} + \hat{H}_{int}^{(m)} \right) \right] \right\}}{\mathrm{Tr}\left\{ \exp\left[-\beta \left(\hat{H}_0^{(m)} + \hat{H}_{int}^{(m)} \right) \right] \right\}}$$

is carried out over the fast-relaxing electron and phonon variables; the bar in Eq. (5.3.9) denotes the averaging over the all possible quasi-equilibrium spatial arrangements of the ions.

Assuming $\hat{H}_{int}^{(m)}$ to be a small perturbation to $\hat{H}_0^{(m)}$ we can use the thermodynamic perturbation theory. Passing on to the Fourier transform $\tilde{G}(\vec{k}, \omega_n)$ of the Green's function,

$$G(x - x') = \frac{1}{\beta} \sum_n \int \frac{d^3 k}{(2\pi)^3} e^{i\vec{k} \cdot (\vec{r} - \vec{r}\,') - i\hbar\omega_n (\tau - \tau')} \tilde{G}(\vec{k}, \omega_n) \tag{5.3.10}$$

with the *Matsubara frequencies* $\omega_n = (2n+1)\cdot\pi\beta^{-1}$ with $\beta = 1/(k_B T)$, and with $x = (\vec{r}, \tau)$, where again τ is the 'thermodynamic time', with $0 \le \tau \le (1/\beta)$, after rearrangements similar to those for a one-component metal in Section 5.1, we again obtain the following expression for $\tilde{G}(\vec{k}, \omega_n)$:

$$\tilde{G}^{-1}(\vec{k}, \omega_n) = i\hbar\omega_n \cdot \tau_0 - \left(\frac{\hbar^2 k^2}{2m} - \mu_e\right)\tau_3 - \sum(\vec{k}, \omega_n) \tag{5.3.11}$$

The self-energy function $\sum(\vec{k}, \omega_n)$ is in turn expressed through $\tilde{G}(\vec{k}, \omega_n)$:

$$\sum(\vec{k}, \omega_n) = \sum_{C}(\vec{k}, \omega_n) + \sum_{Ph}(\vec{k}, \omega_n) + \sum_{Scat}(\vec{k}, \omega_n) \tag{5.3.12}$$

$$\begin{cases} \sum_{C}(\vec{k}, \omega_n) = \dfrac{1}{\beta}\sum_{n_1}\displaystyle\int \dfrac{d^3 k_1}{(2\pi)^3}\tau_3 \cdot \tilde{G}(\vec{k}_1, \omega_{n_1})\tau_3 \cdot V_e(\vec{k} - \vec{k}_1) \\ \sum_{Ph}(\vec{k}, \omega_n) = \dfrac{1}{\beta}\sum_{n_1}\displaystyle\int \dfrac{d^3 k_1}{(2\pi)^3}\tau_3 \cdot \tilde{G}(\vec{k}_1, \omega_{n_1})\tau_3 \int_{-\infty}^{+\infty} \dfrac{dz}{2\pi}\dfrac{b(\vec{k} - \vec{k}_1, z)}{z - i\hbar(\omega_n - \omega_{n_1})} \\ \sum_{scat}(\vec{k}, \omega_n) = \displaystyle\int \dfrac{d^3 k_1}{(2\pi)^3}\tau_3 \cdot \tilde{G}(\vec{k}_1, \omega_{n_1})\tau_3 \cdot \sum_{a,\beta}\dfrac{C_a\tilde{w}_a(\vec{k} - \vec{k}_1)C_\beta\tilde{w}_\beta(\vec{k} - \vec{k}_1)}{v_0} S_{a\beta}(\vec{k} - \vec{k}_1) \end{cases} \tag{5.3.13}$$

Here

$$b(\vec{q}, z) = 2\pi \sum_{a,\beta=1,2} C_a\tilde{w}_a(q)C_\beta\tilde{w}_\beta(q)\sum_{\lambda=1}^{6}\int \frac{d^3 k_1}{(2\pi)^3}\frac{\hbar[\vec{q}\cdot\vec{\varepsilon}_a(\vec{k}, \lambda)]\cdot[\vec{q}\cdot\vec{\varepsilon}_\beta(\vec{k}, \lambda)]}{2M\cdot\omega_\lambda(\vec{k})}$$
$$\times S_{a\beta}(\vec{k} + \vec{q})\cdot[\delta(z - \hbar\omega_\lambda(\vec{k})) - \delta(z + \hbar\omega_\lambda(\vec{k}))] \tag{5.3.14}$$

the function $\tilde{w}_a(\vec{q})$ is defined in Eq. (C.6) Appendix C; and the frequencies $\omega_\lambda(\vec{q})$ are determined by diagonalization of the effective quasi-phonon Hamiltonian $\overline{H_2^{(m)}}$, i.e. by recasting it into the form (see Chapter 3):

$$\overline{H_2^{(m)}} = \sum_{\lambda=1}^{6}\sum_{\vec{k}} \hbar\omega_\lambda(\vec{k})\left\{\hat{C}^+_{\vec{k},\lambda}\hat{C}_{\vec{k},\lambda} + \frac{1}{2}\right\}$$

where $\hat{C}^+_{\vec{k},\lambda}$ and $\hat{C}_{\vec{k},\lambda}$ are the quasi-phonon creation and annihilation operators with energy $\hbar\omega_\lambda(\vec{k})$ and momentum $\hbar\vec{k}$. The effective dispersion branches with $\lambda = 1, 2, 3$ are matched by acoustic quasi-phonons, and those with $\lambda = 4, 5, 6$ by the optical ones. Again we have used the same approximations as in Chapter 3, neglecting the 'broadening' of quasi-phonons, as in Chapter 3.

The diagonalization of the operator $\overline{H_2}$ is performed with the relations

$$\sum_{a'=1}^{2}\sum_{l'} \Gamma^{ll'}_{aa'}\cdot\varepsilon^{l'}_{a'}(\vec{k}, \lambda) = \omega^2_\lambda(\vec{k})\cdot C_a M_a \cdot \varepsilon^{l}_{a}(\vec{k}, \lambda) \tag{5.3.15}$$

provided that

$$\sum_{\lambda=1}^{6} \varepsilon_{a}^{l}(\vec{k},\lambda)\varepsilon_{a'}^{l'}(\vec{k},\lambda) = \delta_{ll'}\delta_{aa'} \tag{5.3.16}$$

Here $a, \beta = 1, 2;\ l, l' = x, y, z$; and

$$\begin{aligned} \Gamma_{aa'}^{ll'}(\vec{k}) &= \sqrt{C_a M_a C_{a'} M_{a'}} \cdot \tilde{\Gamma}_{aa'}^{ll'}(\vec{k}) \\ &= \frac{v_0}{(2\pi)^3} \int d^3k_1 k_1^l k_1^{l'} \Bigg\{ \tilde{\Phi}_{aa'}(\vec{k}_1) C_a C_{a'} S_{aa'}(\vec{k}_1 + \vec{k}) \\ &\quad - \delta_{aa'} \sum_{\beta=1,2} \tilde{\Phi}_{a\beta}(\vec{k}_1) C_a C_\beta S_{a\beta}(\vec{k}_1) \Bigg\} \end{aligned} \tag{5.3.17}$$

$\tilde{\Phi}_{a\beta}(\vec{k})$ is the Fourier transform of the effective potential of the interionic interaction, and the *partial structure factors* $a_{a\beta}(\vec{k})$ are defined by

$$S_{a\beta}(\vec{k}) = \frac{(2\pi)^3}{v_0}\delta(\vec{k}) + a_{a\beta}(\vec{k}) \tag{5.3.18}$$

It follows from Eqs (5.3.14) and (5.3.15) (see Chapter 2), that the frequencies of *longitudinal* waves ($\lambda = 3, 6$) are:

- for the *acoustic* wave:

$$\omega_{la}^2(\vec{k}) = \omega_3^2(\vec{k}) = \frac{1}{2}(\tilde{\Gamma}_{11}^{zz} + \tilde{\Gamma}_{22}^{zz}) - \sqrt{\frac{1}{4}(\tilde{\Gamma}_{11}^{zz} - \tilde{\Gamma}_{22}^{zz})^2 + \tilde{\Gamma}_{12}^{zz^2}}$$

- for the *optical* one:

$$\omega_{l0}^2(\vec{k}) = \omega_6^2(\vec{k}) = \frac{1}{2}(\tilde{\Gamma}_{11}^{zz} + \tilde{\Gamma}_{22}^{zz}) + \sqrt{\frac{1}{4}(\tilde{\Gamma}_{11}^{zz} - \tilde{\Gamma}_{22}^{zz})^2 + \tilde{\Gamma}_{12}^{zz^2}}$$

Whereas the frequencies of *transversal* waves ($\lambda = 1, 2, 4, 5$) are equal to

$$\omega_{ta}^2(\vec{k}) = \omega_1^2(\vec{k}) = \omega_2^2(\vec{k}) = \frac{1}{2}(\tilde{\Gamma}_{11}^{xx} + \tilde{\Gamma}_{22}^{xx}) - \sqrt{\frac{1}{4}(\tilde{\Gamma}_{11}^{xx} - \tilde{\Gamma}_{22}^{xx})^2 + \tilde{\Gamma}_{12}^{xx^2}}$$

for the *acoustic* wave, and to

$$\omega_{t0}^2(\vec{k}) = \omega_4^2(\vec{k}) = \omega_5^2(\vec{k}) = \frac{1}{2}(\tilde{\Gamma}_{11}^{xx} + \tilde{\Gamma}_{22}^{xx}) + \sqrt{\frac{1}{4}(\tilde{\Gamma}_{11}^{xx} - \tilde{\Gamma}_{22}^{xx})^2 + \tilde{\Gamma}_{12}^{xx^2}}$$

for the *optical* one.

Expressing the self-energy function $\sum(\vec{k}, \omega_n)$ as

$$\sum(\vec{k},\omega_n) = [1 - Z(\vec{k},\omega_n)]i\hbar\omega_n\tau_0 + \chi(\vec{k},\omega_n)\tau_3 + \varphi(\vec{k},\omega_n)\tau_1$$

and introducing again a *gap function* $\Delta(\omega) = \varphi(\omega)/Z'(\omega)$, with the constraint that $\Delta(\omega)$ tends to zero in the vicinity of the critical temperature T_C, we can linearize Eqs (5.3.11) and (5.3.12), and bring them into a form identical to the Eliashberg equations for one-component metals (see Section 5.1):

$$\begin{cases} [1 - Z'(\omega)]\omega = -\displaystyle\int_{-\infty}^{+\infty} dz' \cdot K_{Ph}(z',\omega) \\ Z'(\omega)\cdot\Delta(\omega) = \displaystyle\int_{-\infty}^{+\infty} dz' \cdot K_{Ph}(z',\omega)\cdot \mathrm{Re}\frac{\Delta(z')}{z'} - U_C\cdot N(0)\int_0^{\hbar\omega_C} dz' \\ \qquad \times \tanh\left(\dfrac{z'\beta_C}{2}\right)\cdot \mathrm{Re}\dfrac{\Delta(z')}{z'} \end{cases} \tag{5.3.19}$$

where

$$K_{Ph}(Z',\omega) = \int_0^{\infty} dz \cdot a^2(z)\cdot F(z)\cdot\frac{1}{2}\left\{\frac{\tanh\frac{\beta z'}{2} + \coth\frac{\beta z}{2}}{z' + z - \hbar\omega - i\delta} - \frac{\tanh\frac{\beta z'}{2} - \coth\frac{\beta z}{2}}{z' - z - \hbar\omega - i\delta}\right\}$$

The Eliashberg function (the spectral density of the electron-phonon coupling) is therewith generalized for a two-component alloy, and with the concentrations C_a it takes the form

$$\begin{aligned} a^2(z)F(z) = {} & \left(\int_{S_F}\frac{d^2k}{\hbar v_{\vec{k}}}\right)^{-1}\int_{S_F}\frac{d^2k}{\hbar v_{\vec{k}}}\int\frac{d^2k'}{\hbar v_{k'}}\cdot\sum_{a,\beta=1,2}\frac{C_a\tilde{w}_a(\vec{k}-\vec{k}')C_\beta\tilde{w}_\beta(\vec{k}-\vec{k}')}{M} \\ & \times\sum_{\lambda=1}^{6}\int\frac{d^3\vec{k}_1}{(2\pi)^3}\frac{\hbar((\vec{k}-\vec{k}')\cdot\vec{\varepsilon}_a(\vec{k}_1,\lambda))\cdot((\vec{k}-\vec{k}')\cdot\vec{\varepsilon}_\beta(\vec{k}_1,\lambda))}{2\omega_\lambda(\vec{k}_1)} \\ & \times S_{a\beta}(\vec{k}_1+\vec{k}-\vec{k}')\delta(z-\hbar\omega_\lambda(\vec{k}_1)) \end{aligned} \tag{5.3.20}$$

The solvability condition for Eqs (5.3.19) determines the characteristic transition temperature T_C, which, as with a one-component metal, is a functional of $a^2(z)F(z)$. In a strong-coupling approximation, for instance, this temperature is determined by the already known McMillan formula

$$T_C = \frac{\theta_D}{1.45}\exp\left[\frac{-1.04(1+\lambda)}{\lambda - \mu^*(1+0.62\lambda)}\right] \tag{5.3.21}$$

where again $\theta_D = (\hbar/k_B)[(18\pi^2/v_0)(2/S_l^3 + 1/S_t^3)]^{1/3}$ is the Debye temperature. Here

$$S_{\mathrm{t}} := \lim_{k\to 0}\frac{\omega_{\mathrm{ta}}(k)}{k};\quad S_{\mathrm{l}} := \lim_{k\to 0}\frac{\omega_{\mathrm{la}}(k)}{k}$$

are again the speeds of the transverse rsp. longitudinal acoustic sound waves in the amorphous alloy.

For the electron-phonon coupling constant, λ, we get with the concentrations C_α:

$$\lambda = 2\int_0^\infty \frac{\mathrm{d}z}{z}\alpha^2(z)F(z) = \left(\int_{S_F}\frac{\mathrm{d}^2k}{\hbar v_{\vec{k}}}\right)^{-1}\cdot\int_{S_F}\frac{\mathrm{d}^2k}{\hbar v_{\vec{k}}}\int_{S_F}\frac{\mathrm{d}^2k'}{\hbar v_{k'}}\sum_{\alpha,\beta=1,2}\frac{C_\alpha\tilde{w}_\alpha(\vec{k}-\vec{k}')C_\beta\tilde{w}_\beta(\vec{k}-\vec{k}')}{\bar{M}}$$
$$\times\sum_{\lambda=1}^{6}\int \mathrm{d}^3k_1\frac{((\vec{k}-\vec{k}')\cdot\vec{\varepsilon}_\alpha(\vec{k}_1,\lambda))\cdot((\vec{k}-\vec{k}')\cdot\vec{\varepsilon}_\beta(\vec{k}_1,\lambda))}{(2\pi)^3\omega_\lambda^2(\vec{k}_1)}S_{\alpha\beta}(\vec{k}_1+\vec{k}-\vec{k}') \tag{5.3.22}$$

Let us rewrite this expression taking into account Eqs (5.3.15) and (5.3.16). To do this we multiply the left- and right-hand sides of Eq. (5.3.15) by $\dfrac{\varepsilon_\beta^m(\vec{k},\lambda)}{\omega_\lambda^2(\vec{k})}$ and, after summation over λ, obtain

$$\sum_{\alpha'=1}^{2}\sum_{l'}\Gamma_{\alpha\alpha'}^{ll'}(\vec{k})\sum_{\lambda=1}^{6}\frac{\varepsilon_{\alpha'}^{l'}(\vec{k},\lambda)\varepsilon_\beta^m(\vec{k},\lambda)}{\omega_\lambda^2(\vec{k})} = \sqrt{C_\alpha M_\alpha\cdot C_\beta M_\beta}\,\delta_{\alpha\beta}\delta_{lm}$$

Hence, we have

$$D_{\alpha\beta}^{lm}(\vec{k}) = \sum_{\lambda=1}^{6}\frac{\varepsilon_\alpha^l(\vec{k},\lambda)\cdot\varepsilon_\beta^m(\vec{k},\lambda)}{\omega_\lambda^2(\vec{k})} = \sqrt{C_\alpha M_\alpha\cdot C_\beta M_\beta}(\hat{\Gamma}^{-1})_{\alpha\beta}^{lm} \tag{5.3.23}$$

The matrix $\hat{\Gamma}^{-1}$ is the matrix-inverse of $\hat{\Gamma}$. Applying standard rules for the derivation of an inverse matrix, we therefore obtain the following relationships (Smirnov, 1956):

$$\begin{cases} D_{\alpha\beta}^{lm} = D_{\alpha\beta}^{ll}\cdot\delta_{lm} \\ D_{11}^{ll} = \dfrac{C_1M_1\cdot\Gamma_{22}^{ll}}{\Gamma_{11}^{ll}\Gamma_{22}^{ll}-(\Gamma_{12}^{ll})^2} \\ D_{22}^{ll} = \dfrac{C_2M_2\cdot\Gamma_{11}^{ll}}{\Gamma_{11}^{ll}\Gamma_{22}^{ll}-(\Gamma_{12}^{ll})^2} \\ D_{12}^{ll} = D_{21}^{ll} = -\dfrac{\sqrt{C_1M_1\cdot C_2M_2}\cdot\Gamma_{12}^{ll}}{\Gamma_{11}^{ll}\Gamma_{22}^{ll}-(\Gamma_{12}^{ll})^2} \end{cases} \tag{5.3.24}$$

where $\begin{cases} l = x, y, z \\ D_{\alpha\beta}^{xx} = D_{\alpha\beta}^{yy} \end{cases}$. It follows from this that

$$\sum_{\lambda=1}^{6}\frac{(\vec{q}\cdot\vec{\varepsilon}_a(\vec{k},\lambda))(\vec{q}\cdot\vec{\varepsilon}_\beta(\vec{k},\lambda))}{\omega_\lambda^2(\vec{k})}=q^2\sin^2\theta D^{xx}_{a\beta}(\vec{k})+q^2\cos^2\theta D^{zz}_{a\beta}(\vec{k}) \qquad (5.3.25)$$

where θ is the angle between the vectors $\vec{q}$ and $\vec{k}$.

Substituting Eq. (5.3.25) into Eq. (5.3.22) and taking into account the expression (5.3.18) for $S_{a\beta}(\vec{q})$, we arrive at the expression that defines the electron-phonon coupling constant, λ, for amorphous binary compounds of simple metals as follows (Krasny *et al.*, 1996):

$$\lambda=\lambda_0+\lambda_1 \qquad (5.3.26)$$

where

$$\begin{aligned}\lambda_0&=\frac{m}{4\pi^2k_F\hbar^2}\int_0^{2k_F}\frac{q^3\mathrm{d}q}{v_0\bar{M}}\sum_{a,\beta}C_a\tilde{w}_a(q)\cdot C_\beta\tilde{w}_\beta(q)\cdot D^{zz}_{a\beta}(q)\\&=12\frac{m}{\bar{M}}\bar{Z}\int_0^1\frac{z^3\mathrm{d}z}{\hbar^2v_0^2}\sum_{a,\beta}C_a\tilde{w}_a(2k_Fz)\cdot C_\beta\tilde{w}_\beta(2k_Fz)\cdot D^{zz}_{a\beta}(2k_Fz)\end{aligned} \qquad (5.3.27)$$

$$\begin{aligned}\lambda_1&=\frac{1}{4\pi^2}\frac{m}{k_F\hbar^2}\int_0^{2k_F}\frac{q^3\mathrm{d}q}{\bar{M}\cdot4\pi^2}\int_0^{k_D}k_1^2\mathrm{d}k_1\int_{-1}^{+1}\mathrm{d}x\sum_{a,\beta}C_a\tilde{w}_a(q)\cdot C_\beta\tilde{w}_\beta(q)\\&\quad\times a_{a\beta}\left(\sqrt{q^2+k_1^2+2k_1qx}\right)[x^2\cdot D^{zz}_{a\beta}(k_1)+(1-x^2)\cdot D^{xx}_{a\beta}(k_1)]\\&=72\frac{m}{\bar{M}}\bar{Z}^2\int_0^1\frac{z^3\mathrm{d}z}{v_0^2\hbar^2}\int_0^{1/\sqrt{2}\cdot\sqrt[3]{Z}}y^2\mathrm{d}y\int_{-1}^{+1}\mathrm{d}x\sum_{a,\beta}C_a\tilde{w}_a(2k_Fz)\cdot C_\beta\tilde{w}_\beta(2k_Fz)\\&\quad\times a_{a\beta}(2k_F\sqrt{z^2+y^2+2zyx})\cdot[x^2D^{zz}_{a\beta}(2k_Fy)+(1-x^2)D^{xx}_{a\beta}(2k_Fy)]\end{aligned} \qquad (5.3.28)$$

The concentration dependence of the critical temperature T_C is obtained from Eqs (5.3.21) and (5.3.26)–(5.3.28). If hard-sphere form factors $a_{a\beta}(q)$ in the Percus-Yevick approximation are taken, μ^* is set to 0.13, and for the pseudopotential the one-parameter Ashcroft potential is selected (with a parameter R_C^* determined by experimental critical temperatures of pure metals listed in Table 5.2.1), then a fairly good agreement between theory and experiment is observed (Krasny *et al.*, 1996).

This is apparent from Figs 5.3.1 and 5.3.2 which show T_C versus the concentration C of Cu in the amorphous metallic alloys $Be_{1-x}Cu_x$ and $Al_{1-x}Cu_x$.

In Figs 5.3.1 and 5.3.2 the open circles stand for experimental values of T_C. The good agreement observed can be used for theoretical predictions of the concentration dependence of T_C for unexplored compounds. Figures 5.3.3–5.3.6, in particu-

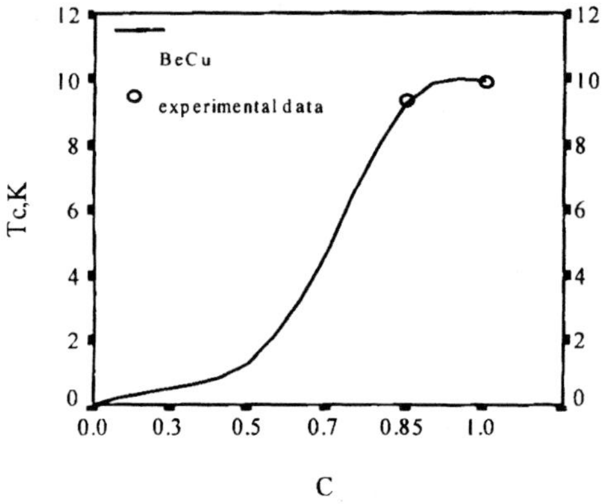

Fig. 5.3.1 The critical temperature T_C is plotted against the concentration C of the amorphous alloys $Be_{1-C}Cu_C$. The open circles denote experimental points

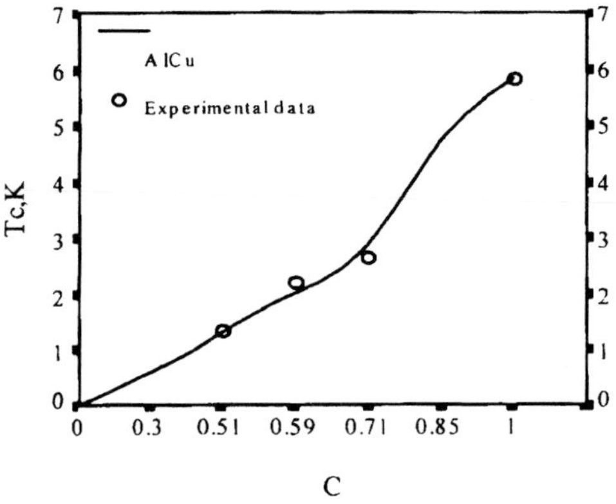

Fig. 5.3.2 The critical temperature T_C is plotted against the concentration C of the amorphous alloys $Al_{1-C}Cu_C$. The open circles denote experimental points

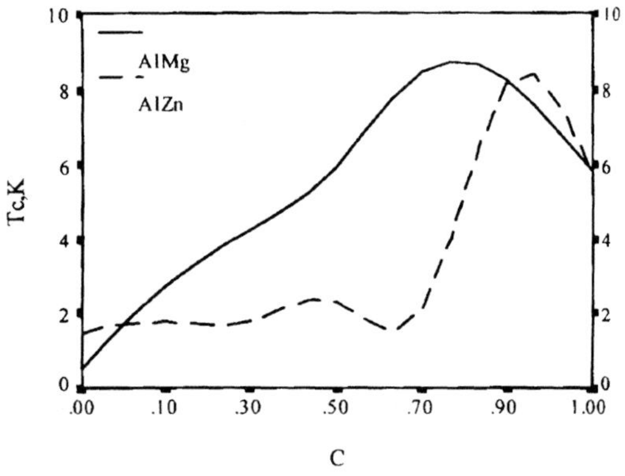

Fig. 5.3.3 The critical temperature T_C is plotted against the concentration C of amorphous AlMg and AlZn alloys

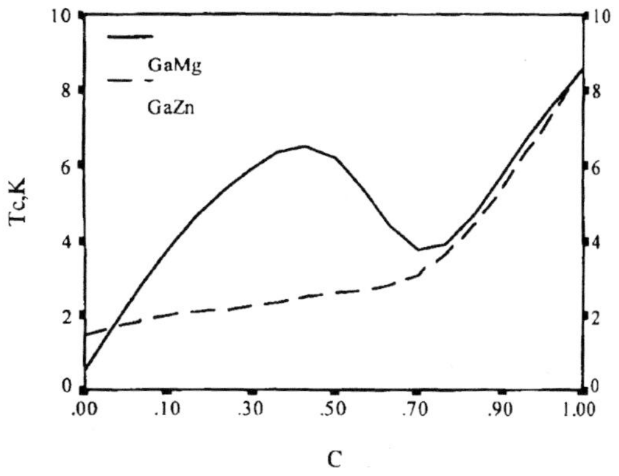

Fig. 5.3.4 The critical temperature T_C is plotted against the concentration C of amorphous GaMg and GaZn compounds

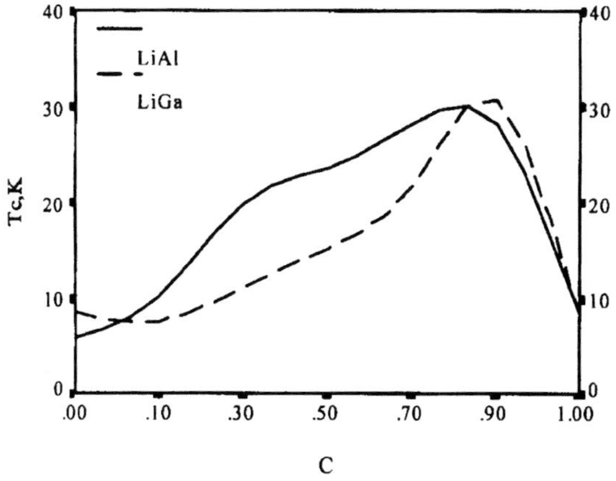

Fig. 5.3.5 The critical temperature T_C is plotted against the concentration C of amorphous alloys LiAl and LiGa

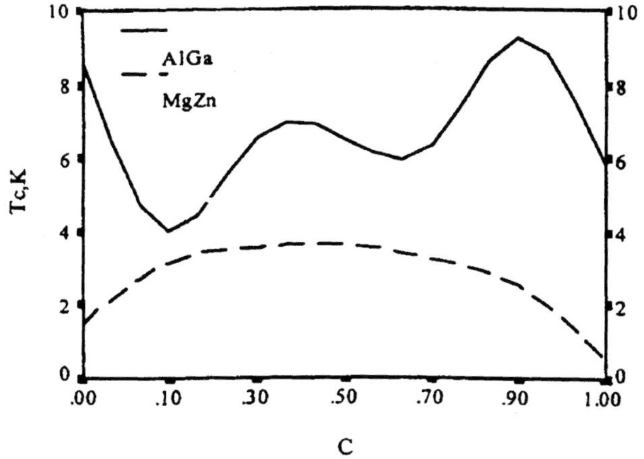

Fig. 5.3.6 The critical temperature T_C is plotted against the concentration C of amorphous alloys AlGa and MgZn

lar, show the concentration dependencies of T_C for amorphous Ga and Al alloys with Zn, Mg and Li.

The $T_C(C)$ curves for amorphous Ga_CZn_{1-C} and Ga_CMg_{1-C} compounds (Fig. 5.3.4) reach their maximal values at a Ga concentration of $C=1$ (i.e. pure gallium). For Al_CZn_{1-C} and Al_CMg_{1-C} compounds this curve peaks at an Al concentration of $C\sim 0.7$–0.9 (Fig. 5.3.3).

The $T_C(C)$ functions for Li_CGa_{1-C} and Li_CAl_{1-C} compounds have their maxima at Li concentration $C\sim 0.8$–0.9 (Fig. 5.3.5). Here the maximal transition temperatures of these amorphous compounds are 3 to 4 times the critical temperatures of pure metals. Finally, Fig. 5.3.6 demonstrates the $T_C(C)$ curves for the amorphous Al_CGa_{1-C} and Mg_CZn_{1-C} compounds, which are compounds of equal valency metals. Figure 5.3.6 shows that small Al additions to Ga drastically reduce the transition temperature (the function peaks at an Al concentration of $C\sim 0.15$), and small Ga admixtures to Al steeply increase T_C (the function $T_C(C)$ has a maximum at an Al concentration of $C\sim 0.85$). The Mg_CZn_{1-C} compound has a $T_C(C)$ curve with a flat peak within the interval $0.15 \lesssim C \lesssim 0.85$. Here T_C is almost twice as large as for pure Zn.

The results in this section have been published in Krasny *et al.* (1996), from which the figures are also taken.

6
Conclusions

We have seen that *amorphous* metals possess essentially all the properties that make *crystalline* metals of interest for many applications, in particular their *magnetic* and *superconducting* properties. Sometimes these properties are even better in the amorphous state, e.g. the critical temperature T_C for the onset of superconductivity can be much higher, as seen in Chapter 5. Also other material properties, e.g. those characterizing the *soft-magnetic behavior* of the material, may be much better in the glassy state, because in the amorphous metal there are essentially no 'dislocations' that hinder the movement of magnetic domain walls in magnetic crystals. This last aspect, i.e. internal stresses in the amorphous metal, was not treated in this text, although it is an important aspect for industrial applications. Amorphous metals, in contrast to crystalline metals, should in general also have better *corrosion properties*, again because of the absence of dislocations. It may also be an advantage that the *electrical resistivity of the amorphous metals is rather high*, typically as high as that of liquid metals, e.g. with $\rho \approx 250\ \mu\Omega\text{cm}$.

This means for example that large-scale electrical transducers made with amorphous metals of similar magnetic properties to conventional crystalline Si-Fe, which has been used in such transducers to date, would be good candidates for large-scale applications of amorphous metals, because with the amorphous material the *energy losses* would be smaller. However, here the technological details, material properties and prices become important.

There are two general advantages of the amorphous metals (i) the existence of cheap and uncomplicated preparation techniques; (ii) the possibility of preparing alloys in a *composition range*, for which crystalline counterparts do not exist. On the other hand there are two disadvantages, namely (iii) that it is difficult to prepare amorphous metals except as rather thin metallic ribbons, and (iv) the phenomenon of 'aging', i.e. the material loses its high-quality properties after about a year. The aging aspects are not treated in this text, although aging of glassy systems is studied extensively elsewhere by computer simulations (e.g. Cugliandolo, *et al.* 1994; Rieger, 1996; Kob and Barrat, 1997; Cugliandolo and Lozano, 1998; Franz *et al.*, 1998; Mussel *et al.*, 1998).

The advantages of the amorphous metals, together with their sometimes exceptional physical properties, have not yet led to a major technological breakthrough, although there *are* some applications.

In this book, we have extensively studied the physical properties of the metastable 'glassy metallic states', mainly by analytical methods. But we have also discussed ways of preparing such states numerically and studying their properties on a computer. Concerning the analytical tools, our main message is that, as we have seen, the '*many-body techniques*', i.e. temperature-dependent Green's functions, diagrams, etc. can be extended successfully from crystalline metals to amorphous ones. In fact, these diagrammatic and many-body techniques, and the results obtained with them, have led to a large number of results, which represent a decisive part of this work.

Technical aspects of the 'many-body formalism', with emphasis on amorphous metals, are presented in the following Appendices.

Finally we would like to mention that the *dynamics* of the glass forming process itself are *not* covered in this book. This is an interesting matter in itself, which is best described by the *mode coupling theory* of Götze and Sjögren (1992).

Acknowledgement

The authors would like to thank Prof. Dr. J. Zweck, University of Regensburg, for the electron-scattering micrograph on the cover of the book. They also would like to thank the DAAD, the 'Deutsche Forschungsgemeinschaft', and the Universities of Regensburg and of Odessa for hospitality and support.

Appendices

Appendix A: Calculation of the Free Energy of Amorphous Metals

Let us calculate the expression

$$F = -k_B T \cdot \overline{\ln\left\{ \mathrm{Tr}\left[\exp\left(-\frac{\hat{H}_0 + \hat{H}_{int}}{k_B T} \right) \right] \right\}} \tag{A.1}$$

where the bar denotes the configurational averaging over the ensemble of all quasi-equilibrium positions of the atoms, and the operation $\mathrm{Tr}\{\ldots\}$ applies to the collective variables of the fast-relaxing subsystems (electrons, phonons, magnons, etc.)

It is presumed therewith that only the operator $\hat{H}_{int}$ depends on the atomic coordinates. Considering $\hat{H}_{int}$ to be small, we can take advantage of the thermodynamic perturbation theory and introduce a *temperature scattering matrix* $\hat{\sigma}(\beta)$:

$$\exp\left(-\frac{\hat{H}_0 + \hat{H}_{int}}{k_B T} \right) = e^{-\beta \hat{H}_0} \cdot \hat{\sigma}(\beta) \tag{A.2}$$

where $\beta = \dfrac{1}{k_B T}$

It can readily be shown that (see Abrikosov *et al.*, 1965)

$$\hat{\sigma}(\beta) = \hat{T}_\tau \cdot \exp\left[-\int_0^\beta \hat{H}_{int}(\tau) d\tau \right] \tag{A.3}$$

where

$$\hat{H}_{int}(\tau) = \exp(\hat{H}_0 \tau) \hat{H}_{int} \exp(-\hat{H}_0 \tau) \tag{A.4}$$

and $\hat{T}_\tau$ is the 'time-ordering' operator, which permutes the arguments $\tau_1, \ldots, \tau_n$ in the operator product $\hat{A}(\tau_1) \cdot \ldots \cdot \hat{A}(\tau_n)$ in such a way that in the resulting expres-

sion, $\hat{A}(\tau_{i_1}) \cdot \ldots \cdot \hat{A}(\tau_{i_n})$, the 'thermodynamic time', τ, *decreases* as we pass from left to right from one operator to another, i.e.$\tau_{i_1} > \tau_{i_2} > \ldots > \tau_{i_n}$:
Equation (A.3) implies therefore the series

$$\hat{\sigma}(\beta) = \sum_{n=0}^{\infty} \frac{(-1)^n}{n!} \int_0^{\beta} d\tau_1 \ldots \int_0^{\beta} d\tau_n \cdot T_\tau\{\hat{\mathrm{H}}_{\mathrm{int}}(\tau_1) \ldots \hat{\mathrm{H}}_{\mathrm{int}}(\tau_n)\}$$
$$\equiv \sum_{n=0}^{\infty} (-1)^n \int_0^{\beta} d\tau_1 \int_0^{\tau_1} d\tau_2 \ldots \int_0^{\tau_{n-1}} d\tau_n \cdot \{\hat{\mathrm{H}}_{\mathrm{int}}(\tau_1) \ldots \hat{\mathrm{H}}_{\mathrm{int}}(\tau_n)\} \tag{A.5}$$

Taking into account Eq. (A.2), Eq. (A.1) can thus be rewritten as

$$\mathrm{F} = \mathrm{F}_0 - k_{\mathrm{B}} T \cdot \overline{\ln \langle \hat{\sigma}(\beta) \rangle_0} \tag{A.6}$$

Here

$$\mathrm{F}_0 = -k_{\mathrm{B}} T \cdot \overline{\ln\left\{\mathrm{Tr}\left[\exp\left(-\frac{\hat{\mathrm{H}}_0}{k_{\mathrm{B}} T}\right)\right]\right\}}$$

$$\langle \hat{\sigma}(\beta) \rangle_0 = \mathrm{Tr}\left\{\hat{\sigma}(\beta) \cdot \exp\left(\frac{\mathrm{F}_0 - \hat{\mathrm{H}}_0}{k_{\mathrm{B}} T}\right)\right\}$$

Upon partial summation of diagrams of a certain type in terms of Feynman diagrammatics, Eq. (A.6) can be represented as

$$\mathrm{F} = \mathrm{F}_0 - k_{\mathrm{B}} T \left\{\overline{\langle \hat{\sigma}(\beta) \rangle_{0\,\mathrm{con}}} - 1\right\} \tag{A.7}$$

where

$$\langle \hat{\sigma}(\beta) \rangle_{0\,\mathrm{con}} = \sum_{n=0}^{\infty} \frac{(-1)^n}{n!} \int_0^{\beta} d\tau_1 \ldots \int_0^{\beta} d\tau_n \, \langle \hat{T}_\tau\{\hat{\mathrm{H}}_{\mathrm{int}}(\tau_1) \ldots \hat{\mathrm{H}}_{\mathrm{int}}(\tau_n)\} \rangle_{0\,\mathrm{con}}$$

The operation $\langle \ldots \rangle_{0\,\mathrm{con}}$ is defined as the sum of all *connected* diagrams of the temperature scattering matrix.

Now the eigenvalues and wave functions of the operator $\hat{\mathrm{H}}_0$ are independent of the coordinates of atomic sites in the material. Therefore the averaging in Eq. (A.7) can be *reversed* in order that first the *configuration* average is carried out, and next the Gibbs *thermal averaging* over the variables of fast-relaxing subsystems (electrons, phonons, magnons...):

$$\mathrm{F} = \mathrm{F}_0 - k_{\mathrm{B}} T \left\{\langle \overline{\hat{\sigma}(\beta)} \rangle_{0\,\mathrm{con}} - 1\right\} \tag{A.8}$$

In this expression the substitution

$$\hat{\mathrm{H}}_{\mathrm{int}}(\tau) = \hat{\hat{\mathrm{H}}}_{\mathrm{int}}(\tau) + \Delta\hat{\mathrm{H}}_{\mathrm{int}}(\tau)$$

is subsequently performed, where

$$\Delta\hat{\mathrm{H}}_{\mathrm{int}}(\tau) = \hat{\mathrm{H}}_{\mathrm{int}}(\tau) - \hat{\hat{\mathrm{H}}}_{\mathrm{int}}(\tau)$$

Then, on a certain rearrangement of summands, the operator $\overline{\hat{\sigma}(\beta)}$ in Eq. (A.8) can be rewritten in the form:

$$\begin{aligned}
\overline{\hat{\sigma}(\beta)} &= \sum_{n=0}^{\infty} \frac{(-1)^n}{n!} \int_0^\beta d\tau_1 \ldots \int_0^\beta d\tau_n \, \overline{\hat{T}_\tau\{\hat{\mathrm{H}}_{\mathrm{int}}(\tau_1)\ldots\hat{\mathrm{H}}_{\mathrm{int}}(\tau_n)\}} \\
&= \sum_{n=0}^{\infty} \frac{(-1)^n}{n!} \int_0^\beta d\tau_1 \ldots \int_0^\beta d\tau_n \, \hat{T}_\tau\,\{\tilde{\mathrm{H}}_{\mathrm{int}}(\tau_1)\ldots\tilde{\mathrm{H}}_{\mathrm{int}}(\tau_n)\} \\
&= \hat{T}_\tau \cdot \left\{ \exp\left[-\int_0^\beta \tilde{\mathrm{H}}_{\mathrm{int}}(\tau) d\tau \right] \right\}
\end{aligned}$$

where

$$\begin{aligned}
\tilde{\mathrm{H}}_{\mathrm{int}}(\tau) = \hat{\hat{\mathrm{H}}}_{\mathrm{int}}(\tau) &- \frac{(-1)^2}{2^2} \int_0^\beta d\tau_1 \cdot \overline{\hat{T}_\tau\{\Delta\hat{\mathrm{H}}_{\mathrm{int}}(\tau)\cdot\Delta\hat{\mathrm{H}}_{\mathrm{int}}(\tau_1)\}} \\
&- \frac{(-1)^3}{3!} \int_0^\beta d\tau_1 \int_0^\beta d\tau_2 \, \overline{\hat{T}_\tau\{\Delta\hat{\mathrm{H}}_{\mathrm{int}}(\tau)\cdot\Delta\hat{\mathrm{H}}_{\mathrm{int}}(\tau_1)\cdot\Delta\hat{\mathrm{H}}_{\mathrm{int}}(\tau_2)\}} \\
&- \frac{(-1)^4}{4!} \int_0^\beta d\tau_1 \int_0^\beta d\tau_2 \int_0^\beta d\tau_3 \cdot \hat{T}_\tau \left\{ \overline{\Delta\hat{\mathrm{H}}_{\mathrm{int}}(\tau)\cdot\Delta\hat{\mathrm{H}}_{\mathrm{int}}(\tau_1)\cdot\Delta\hat{\mathrm{H}}_{\mathrm{int}}(\tau_2)\cdot\Delta\hat{\mathrm{H}}_{\mathrm{int}}(\tau_3)} \right. \\
&\left. - 3\cdot\overline{\Delta\hat{\mathrm{H}}_{\mathrm{int}}(\tau)\cdot\Delta\hat{\mathrm{H}}_{\mathrm{int}}(\tau_1)}\cdot\overline{\Delta\hat{\mathrm{H}}_{\mathrm{int}}(\tau_2)\cdot\Delta\hat{\mathrm{H}}_{\mathrm{int}}(\tau_3)} \right\} + \ldots
\end{aligned}$$

Thus the calculation of the system's free energy, F, leads to the following expression (Güntherodt and Beck, 1981):

$$\begin{aligned}
\mathrm{F} &= \mathrm{F}_0 - k_{\mathrm{B}} T \left\{ \left\langle \hat{T}_\tau \cdot \exp\left[-\int_0^\beta \tilde{\mathrm{H}}_{\mathrm{int}}(\tau) d\tau \right] \right\rangle_{0\,\mathrm{con}} - 1 \right\} \\
&= \mathrm{F}_0 - k_{\mathrm{B}} T \cdot \ln \left\{ \left\langle \hat{T}_\tau \cdot \exp\left[-\int_0^\beta \tilde{\mathrm{H}}_{\mathrm{int}}(\tau) d\tau \right] \right\rangle_0 \right\}
\end{aligned}$$

$$= -k_B T \cdot \ln \left\{ \mathrm{Tr} \left[\exp \left(-\frac{\hat{\mathrm{H}}_0 + \tilde{\hat{\mathrm{H}}}_{\mathrm{int}}}{k_B T} \right) \right] \right\} \tag{A.9}$$

where

$$\tilde{\hat{\mathrm{H}}}_{\mathrm{int}} = \hat{\mathrm{H}}_{\mathrm{int}} - \mathrm{e}^{-\hat{\mathrm{H}}_0 \beta} \left\{ \frac{(-1)^2}{2^2} \int_0^\beta \mathrm{d}\tau_1 \cdot \overline{\hat{T}_\tau [\Delta\hat{\mathrm{H}}_{\mathrm{int}}(\beta) \cdot \Delta\hat{\mathrm{H}}_{\mathrm{int}}(\tau_1)]} \right.$$
$$+ \frac{(-1)^3}{3!} \int_0^\beta \mathrm{d}\tau_1 \int_0^\beta \mathrm{d}\tau_2 \, \overline{\hat{T}_\tau [\Delta\hat{\mathrm{H}}_{\mathrm{int}}(\beta) \cdot \Delta\hat{\mathrm{H}}_{\mathrm{int}}(\tau_1) \cdot \Delta\hat{\mathrm{H}}_{\mathrm{int}}(\tau_2)]}$$
$$+ \frac{(-1)^4}{4!} \int_0^\beta \mathrm{d}\tau_1 \int_0^\beta \mathrm{d}\tau_2 \int_0^\beta \mathrm{d}\tau_3 \cdot \hat{T}_\tau \left[\overline{\Delta\hat{\mathrm{H}}_{\mathrm{int}}(\beta) \cdot \Delta\hat{\mathrm{H}}_{\mathrm{int}}(\tau_1) \cdot \Delta\hat{\mathrm{H}}_{\mathrm{int}}(\tau_2) \cdot \Delta\hat{\mathrm{H}}_{\mathrm{int}}(\tau_3)} \right.$$
$$\left. \left. - 3 \cdot \overline{\Delta\hat{\mathrm{H}}_{\mathrm{int}}(\beta) \cdot \Delta\hat{\mathrm{H}}_{\mathrm{int}}(\tau_1)} \cdot \overline{\Delta\hat{\mathrm{H}}_{\mathrm{int}}(\tau_2) \cdot \Delta\hat{\mathrm{H}}_{\mathrm{int}}(\tau_3)} \right] + \ldots \right\} \mathrm{e}^{\hat{\mathrm{H}}_0 \beta} \tag{A.10}$$

Appendix B: Calculation of the Free Energy of Amorphous Ferromagnets

We consider a system of N magnetic atoms fixed at the points $\vec{R}_1, \ldots, \vec{R}_N$, with their spins coupled by an exchange interaction (the Heisenberg model). The Hamiltonian of the system is

$$\hat{\mathrm{H}} = \hat{\mathrm{H}}_0 + \hat{\mathrm{H}}_{\mathrm{i}} \tag{B.1}$$

$$\hat{\mathrm{H}}_0 = -g\mu_0 H \sum_{l=1}^{N} \hat{S}_l^z = -\gamma_0 \sum_l \hat{S}_l^z \tag{B.2}$$

$$\hat{\mathrm{H}}_{\mathrm{i}} = -\frac{1}{2} \sum_{l \neq l'} J_{ll'} l \cdot l' = -\frac{1}{2} \sum_{l \neq l'} J_{ll'} [\hat{S}_l^+ \hat{S}_{l'}^- + \hat{S}_l^z \hat{S}_{l'}^z] \tag{B.3}$$

Here $J_{ll'} = J(\vec{R}_l - \vec{R}_{l'})$ is a rapidly decreasing function of the interatomic distance; $\hat{\vec{S}} = \hat{\vec{S}}(\vec{R}_l) = (\hat{S}_l^x; \hat{S}_l^y; \hat{S}_l^z)$ is the spin operator of the atom l; $\hat{S}_l^\pm = \hat{S}_l^x \pm i\hat{S}_l^y$; $\mathrm{i} = \sqrt{-1}$ is the imaginary unit; $\gamma_0 = g\mu_0 H$; $\mu_0 = \dfrac{e\hbar}{2mc}$ is the Bohr magneton; g is the Landé factor; $\vec{H} = (0, 0, H)$; is the internal magnetic field, aligned along the z-axis.

The fixed atomic sites are assumed to make up no regular crystal lattice. The spin operators obey the following commutator rules:

$$[\hat{S}_l^+, \hat{S}_{l'}^-] = 2\hat{S}_l^z \cdot \delta_{ll'}, \; [\hat{S}_l^z, \hat{S}_{l'}^\pm] = \pm\hat{S}_l^\pm \delta_{ll'} \,. \tag{B.4}$$

Moreover,

$$\hat{\vec{S}}_l^2 = (\hat{S}_l^x)^2 + (\hat{S}_l^y)^2 + (\hat{S}_l^z)^2 = S \cdot (S+1) \tag{B.5}$$

where S is the lth atom spin (l=1, 2, 3, … N).

The *free energy* of such a system is determined by the expression:

$$\mathrm{F} = -k_\mathrm{B} T \cdot \ln\left\{\mathrm{Tr}\left[\exp\left(-\frac{\hat{\mathrm{H}}}{k_\mathrm{B} T}\right)\right]\right\} = -\frac{1}{\beta}\ln\{\mathrm{Tr}[\exp(-\beta\hat{\mathrm{H}})]\} \tag{B.6}$$

where $\beta = \frac{1}{k_\mathrm{B} T}$. The energy will be calculated in terms of the thermodynamic perturbation theory. To this end we introduce the *temperature scattering matrix* $\hat{\sigma}(\beta)$ (see Barjakhtar *et al.*, 1984; Abrikosov *et al.*, 1962)

$$\exp(-\beta\hat{\mathrm{H}}) = \mathrm{e}^{-\beta\hat{\mathrm{H}}_0} \cdot \hat{\sigma}(\beta) \tag{B.7}$$

where

$$\begin{aligned}\hat{\sigma}(\beta) &= \hat{T}_\tau \cdot \exp\left\{-\int_0^\beta \hat{\mathrm{H}}_\mathrm{i}(\tau)\mathrm{d}\tau\right\} \\ &= \sum_{n=0}^{\infty}\frac{(-1)^n}{n!}\int_0^\beta \mathrm{d}\tau_1 \ldots \int_0^\beta \mathrm{d}\tau_n \cdot \hat{T}_\tau\{\hat{\mathrm{H}}_\mathrm{i}(\tau_1)\ldots\hat{\mathrm{H}}_\mathrm{i}(\tau_n)\}\end{aligned} \tag{B.8}$$

$$\hat{\mathrm{H}}_\mathrm{i}(\tau) = \mathrm{e}^{\hat{\mathrm{H}}_0\tau} \cdot \hat{\mathrm{H}}_\mathrm{i} \cdot \mathrm{e}^{-\hat{\mathrm{H}}_0\tau} \tag{B.9}$$

$\hat{T}_\tau$ is the *chronological ordering operator* rearranging the operator product $\hat{A}(\tau_1)\ldots\hat{A}(\tau_n)$ so that the 'thermodynamic time' τ in the arguments decreases from left to right: $\hat{T}_\tau\{\hat{\mathrm{H}}_\mathrm{i}(\tau_1)\ldots\hat{\mathrm{H}}_\mathrm{i}(\tau_n)\} \equiv \hat{\mathrm{H}}_\mathrm{i}(\tau_{i_1})\hat{\mathrm{H}}_\mathrm{i}(\tau_{i_2})\ldots\hat{\mathrm{H}}_\mathrm{i}(\tau_{i_n})$, with $\tau_{i_1} > \tau_{i_2} > \ldots > \tau_{i_n}$.

Then the free energy can be rewritten as

$$\mathrm{F} = -\frac{1}{\beta}\ln\{\mathrm{e}^{-\beta f_0} \cdot \mathrm{Tr}[\mathrm{e}^{\beta(f_0-\hat{\mathrm{H}}_0)} \cdot \hat{\sigma}(\beta)]\} = f_0 - \frac{1}{\beta}\ln\langle\hat{\sigma}(\beta)\rangle_0 \tag{B.10}$$

Here

$$f_0 = -\frac{1}{\beta}\ln z_0 \tag{B.11}$$

with

$$z_0 = \mathrm{Tr}[\exp(-\beta\,\hat{\mathrm{H}}_0)] \tag{B.12}$$

$$\langle\hat{\sigma}(\beta)\rangle_0 = \mathrm{Tr}[\mathrm{e}^{+\beta(F_0-\hat{\mathrm{H}}_0)}\hat{\sigma}(\beta)] = \frac{1}{z_0}\mathrm{Tr}[\mathrm{e}^{-\beta\hat{\mathrm{H}}_0}\hat{\sigma}(\beta)] \tag{B.13}$$

It was demonstrated in Appendix A (see also Abrikosov *et al.*, 1962) that $\langle\hat{\sigma}(\beta)\rangle_0$ can be represented as follows:

$$\langle\hat{\sigma}(\beta)\rangle_0 = \exp\{\langle\hat{\sigma}(\beta)\rangle_{0c} - 1\} \tag{B.14}$$

where the notation $\langle\hat{\sigma}(\beta)\rangle_{0c}$ stands for the sum of *connected* diagrams involved in the expression $\langle\hat{\sigma}(\beta)\rangle_0$. Substituting Eq. (B.14) into Eq. (B.10) we get the free energy in the form:

$$\mathrm{F} = f_0 - \beta^{-1}\{\langle\hat{\sigma}(\beta)\rangle_{0c} - 1\} \tag{B.15}$$

Averages of other physical quantities can also be represented as certain sums of connected diagrams, because they are determined through the derivatives of the free energy with respect to thermodynamic parameters.

Considering the explicit forms of the operators $\hat{\mathrm{H}}_0$ and $\hat{\mathrm{H}}_{\mathrm{i}}$ we can readily see from Eqs (B.8), (B.9) and (B.15) that the calculation of the free energy reduces to the calculation of averages of products of the spin operators, ordered with respect to τ, in the interaction representation

$$\langle T_\tau\{\hat{S}_{l_1}^{a_1}(\tau_1)\cdot\hat{S}_{l_2}^{a_2}(\tau_2)\ldots\hat{S}_{l_n}^{a_n}(\tau_n)\}\rangle_0 \tag{B.16}$$

Now, because

$$\begin{cases} [\hat{S}_l^+;\hat{S}_l^-] = 2\hat{S}_l^z \\ [[\hat{S}_l^+;\hat{S}_l^z];\hat{S}_l^-] = -2\hat{S}_l^z \\ [[[\hat{S}_l^+;\hat{S}_l^-];\hat{S}_l^+];\hat{S}_l^-] = 4\hat{S}_l^z \\ \qquad\ldots \\ \qquad\ldots \end{cases} \tag{B.17}$$

the operator $\hat{S}_z$ evidently plays a key role. Hence the averaging of the products of spin operators is a two-step procedure. First, using relations (B.17), we have to eliminate all operators $\hat{S}_l^+$ and $\hat{S}_l^-$ from the initial average, and then independently calculate averages of the operators $\hat{S}_l^z$.

Preparatory to developing the calculation algorithm for the averages of the type (B.16), we consider several examples. Let us calculate the average

$$\langle T_\tau\hat{S}_l^+(\tau)\hat{S}_{l'}^-(\tau')\rangle_0 = \begin{cases} \langle\hat{S}_l^+(\tau)\hat{S}_{l'}^-(\tau')\rangle_0 & \tau > \tau' \\ \langle\hat{S}_{l'}^-(\tau')\hat{S}_l^+(\tau)\rangle_0 & \tau < \tau' \end{cases} \tag{B.18}$$

Within the average $\langle\hat{S}_l^+(\tau)\hat{S}_{l'}^-(\tau')\rangle_0$ we carry the operator $\hat{S}_l^+(\tau)$ over the operator $\hat{S}_{l'}^-(\tau')$, employing the identity

$$\hat{S}_l^+(\tau)\hat{S}_{l'}^-(\tau') = \hat{S}_{l'}^-(\tau')\hat{S}_l^+(\tau) + [\hat{S}_l^+(\tau),\hat{S}_{l'}^-(\tau')] \tag{B.19}$$

With Eq. (B.2)

$$\hat{S}_l^+(\tau) = \mathrm{e}^{-\gamma_0\tau}\hat{S}_l^+;\ \hat{S}_l^-(\tau) = \mathrm{e}^{\gamma_0\tau}\hat{S}_l^-$$

and so we have

$$\langle \hat{S}_l^+(\tau)\hat{S}_{l'}^-(\tau')\rangle_0 = \langle \hat{S}_{l'}^-(\tau')\hat{S}_l^+(\tau)\rangle_0 + 2\delta_{ll'}\mathrm{e}^{-\gamma_0(\tau-\tau')}\langle \hat{S}_{l'}^z\rangle_0 \tag{B.20}$$

Next, taking advantage of the invariance of the trace with respect to cyclic permutation of operators, we rewrite the first term on the r.h.s. of Eq. (B.20) in the form:

$$\langle \hat{S}_{l'}^-(\tau')\hat{S}_l^+(\tau)\rangle_0 = \mathrm{Tr}\{\hat{S}_l^+(\tau)\rho_0\hat{S}_{l'}^-(\tau')\} = \mathrm{e}^{-\beta\gamma_0}\langle \hat{S}_l^+(\tau)\hat{S}_{l'}^-(\tau')\rangle_0$$

Here the following relationship was used:

$$\hat{S}_l^+(\tau)\rho_0 = \exp(-\beta\gamma_0)\rho_0\hat{S}_l^+(\tau) \tag{B.21}$$

The transformations performed express the initial average through itself with a factor $\exp(-\beta\gamma_0)$, and the commutator $[\hat{S}_l^+, \hat{S}_l^-] = 2\hat{S}_{l'}^z$:

$$\langle \hat{S}_l^+(\tau)\hat{S}_{l'}^-(\tau')\rangle_0 = \mathrm{e}^{-\beta\gamma_0}\langle \hat{S}_l^+(\tau)\hat{S}_{l'}^-(\tau')\rangle_0 + 2\delta_{ll'}\mathrm{e}^{-\gamma_0(\tau-\tau')}\langle \hat{S}_{l'}^z\rangle_0$$

or

$$\langle \hat{S}_l^+(\tau)\hat{S}_{l'}^-(\tau')\rangle_0 = 2\delta_{ll'}\langle \hat{S}_{l'}^z\rangle_0\mathrm{e}^{-\gamma_0(\tau-\tau')}(1+n_{\gamma_0}) \tag{B.22}$$

where $n_\gamma = [\exp(\beta\gamma_0) - 1]^{-1}$ is the Bose distribution function.

To calculate Eq. (B.18) at $\tau < \tau'$ we perform a circular permutation under the Tr symbol and include Eq. (B.190). So we have

$$\langle \hat{S}_{l'}^-(\tau')\hat{S}_l^+(\tau)\rangle_0 = 2\delta_{ll'}\langle \hat{S}_{l'}^z\rangle\mathrm{e}^{-\gamma_0(\tau-\tau')}\cdot n_{\gamma_0} \tag{B.23}$$

Equations (B.22) and (B.23) can be expressed in a unified manner with the function

$$K_{ll'}^0(\tau-\tau') = \delta_{ll'}K_0(\tau-\tau') \tag{B.24}$$

determined as follows:

$$K_0(\tau-\tau') = -\mathrm{e}^{-\gamma_0(\tau-\tau')}\begin{cases} n_{\gamma_0} & ;\ \tau' > \tau \\ 1+n_{\gamma_0} & ;\ \tau' < \tau \end{cases} \tag{B.25}$$

Then Eq. (B.18) can be rewritten as

$$\langle T_\tau\hat{S}_l^+(\tau)\hat{S}_{l'}^-(\tau')\rangle_0 = -2\langle \hat{S}_{l'}^z\rangle_0 K_{ll'}^0(\tau-\tau') \tag{B.26}$$

Consider yet another example: let us calculate the correlator

$$\langle T_\tau\hat{S}_l^+(\tau)\hat{S}_{l'}^z(\tau')\hat{S}_{l''}^-(\tau'')\rangle_0 \tag{B.27}$$

It should be noted that here the second operator, $\hat{S}^z_{l'}(\tau')$, is *constant* with respect to the time argument, but for the sake of convenience it is 'time-tagged', too, similar to the first and third operator, where the time dependence is non-trivial.

When $\tau > \tau' > \tau''$, the correlation function (B.27) takes the form

$$F_3 = \langle \hat{S}^+_l(\tau)\hat{S}^z_{l'}(\tau')\hat{S}^-_{l''}(\tau'')\rangle_0 \tag{B.28}$$

The operators $\hat{S}^+_l(\tau)$ and $\hat{S}^z_{l'}(\tau')$ in Eq. (B.27) can be rearranged using the equation

$$\begin{aligned}\hat{S}^+_l(\tau)\hat{S}^z_{l'}(\tau') &= \hat{S}^z_{l'}(\tau')\hat{S}^+_l(\tau) + [\hat{S}^+_l(\tau), \hat{S}^z_{l'}(\tau')] \\ &= \hat{S}^z_{l'}(\tau')\hat{S}^+_l(\tau) - \delta_{ll'}\hat{S}^+_{l'}(\tau)\cdot \mathrm{e}^{-\gamma_0(\tau-\tau')}\end{aligned}$$

The second term in this equation has already been calculated, and the first is transformed using Eqs (B.19) and (B.21):

$$\begin{aligned}\langle \hat{S}^z_{l'}(\tau')\hat{S}^+_l(\tau)\hat{S}^-_{l''}(\tau'')\rangle_0 = {} & \mathrm{e}^{-\beta\gamma_0}\langle \hat{S}^+_l(\tau)\hat{S}^z_{l'}(\tau')\hat{S}^-_{l''}(\tau'')\rangle_0 \\ & + 2\delta_{ll'}\mathrm{e}^{-\gamma_0(\tau-\tau')}\langle \hat{S}^z_{l'}(\tau')\hat{S}^z_{l''}(\tau'')\rangle_0\end{aligned} \tag{B.29}$$

The average entering into the first term on the r.h.s. of Eq. (B.29) is the initial average of Eq. (B.28). Taking into account Eq. (B.22) we finally get

$$\begin{aligned}F_3 = {} & (1 + n_{\gamma_0})\{2\delta_{ll'}\mathrm{e}^{-\gamma_0(\tau-\tau'')}\langle \hat{S}^z_{l'}\hat{S}^z_{l''}\rangle_0 \\ & - \delta_{ll'}\mathrm{e}^{-\gamma_0(\tau-\tau')}2\delta_{l'l''}\mathrm{e}^{-\gamma_0(\tau'-\tau'')}\langle \hat{S}^z_{l''}\rangle_0(1 + n_{\gamma_0})\}\end{aligned} \tag{B.30}$$

For other relations between τ, τ' and τ'' the correlation function (B.29) is calculated in much the same way. The resulting expression for the average (B.27) is

$$\begin{aligned}\langle T_\tau \hat{S}^+_l(\tau)\hat{S}^z_{l'}(\tau')\hat{S}^-_{l''}(\tau'')\rangle_0 = {} & -2K^0_{ll'}(\tau-\tau')K^0_{l'l''}(\tau' - \tau^{e'})\langle \hat{S}^z_{l''}\rangle_0 \\ & - 2K^0_{ll''}(\tau-\tau'')\langle \hat{S}^z_{l'}\hat{S}^z_{l''}\rangle_0\end{aligned} \tag{B.31}$$

Equation (B.31) can be obtained using the following formal procedure: let us define the concept of *contractions* of the operator $\hat{S}^+_l$ with operators $\hat{S}^-_{l'}$ and $\hat{S}^z_{l''}$ by the expressions

$$\langle T_\tau \ldots \underbrace{\hat{S}^+_l(\tau)\hat{S}^-_{l'}}(\tau')\ldots\rangle_0 = -2K^0_{ll'}(\tau-\tau')\langle T_\tau \ldots \hat{S}^z_{l'}(\tau')\ldots\rangle_0 \tag{B.32}$$

$$\langle T_\tau \ldots \underbrace{\hat{S}^+_l(\tau)\hat{S}^z_{l'}}(\tau')\ldots\rangle_0 = K^0_{ll'}(\tau-\tau')\langle T_\tau \ldots \hat{S}^+_{l'}(\tau')\ldots\rangle_0 \tag{B.33}$$

It is easy to verify that the relationship (B.30) results from all possible contractions of the operator $\hat{S}^+_l(\tau)$ with $\hat{S}^z_{l'}(\tau')$ and $\hat{S}^-_{l''}(\tau'')$.

This deduction is a general one, based on the fact that upon commutation of *two* spin operators only *one* operator is left over. Hence, singling out a certain operator $\hat{S}_l^+$ in the initial average, after commutating it successively with all other operators within the average, we will reduce the initial average to a sum of averages containing *one operator less* than the initial average involves. This procedure, being repeated with every summand obtained in the first step, eliminates from the averages all operators $\hat{S}_l^+$ and $\hat{S}_l^-$, expressing the initial average exclusively in terms of the operators $\hat{S}_l^z$. The relationships (B.32) and (B.33) reduce the calculation of the average of m spin operators to the average of m–1 operators, and performing all possible contractions of operators $\hat{S}_l^+$ with operators $\hat{S}_{l'}^-$ and $\hat{S}_{l'}^z$ corresponds to implementation of all successive commutations of the operators $\hat{S}_l^+$ with operators $\hat{S}_{l'}^-$ and $\hat{S}_{l'}^z$.

It is significant that – owing to the choice of Hamiltonian $\hat{H}_0$ in the form (B.2) – the only non-zero thermodynamic averages of spin variables will be those with *equal numbers of operators* $\hat{S}_l^+$ and $\hat{S}_l^-$. And although an operator $\hat{S}_l^z$ resulting from a contraction (B.32) is not necessarily involved in subsequent contractions, the operator $\hat{S}_l^+$ arising from a contraction (B.33) will certainly be involved in further ones.

As the initial one, any one of the operators $\hat{S}_l^+$ can be chosen: it is easy to show that the choice does not affect the result of the calculation. By way of example let us consider the following set of contractions:

$$\langle T_\tau \hat{S}_l^+(\tau)\hat{S}_{l'}^-(\tau')\hat{S}_{l''}^+(\tau'')\hat{S}_{l'''}^-(\tau''')\rangle_0 \tag{B.34}$$

Here the operator $\hat{S}_{l''}^+(\tau'')$ is contracted with an operator $\hat{S}_{l'}^z(\tau')$, which arises from a contraction of the operator $\hat{S}_l^+(\tau)$ with the operator $\hat{S}_{l'}^-(\tau')$. The operator $\hat{S}_{l'}^+(\tau')$, appearing as a result at this stage, is thereafter contracted with the operator $\hat{S}_{l'''}^-(\tau''')$. So the contraction (B.34) is described by the expression

$$(-2)^2 K_{ll'}^0(\tau-\tau')K_{l''l'}^0(\tau''-\tau)K_{l'l'''}^0(\tau'-\tau''')\langle \hat{S}_{l'''}^z\rangle_0 \tag{B.35}$$

This expression (B.35) is also obtained when we begin the procedure with the operator $\hat{S}_{l''}^+(\tau'')$ instead of $\hat{S}_l^+(\tau)$, i.e. if we perform a set of contractions of the following type:

$$\langle T_\tau \hat{S}_l^+(\tau)\hat{S}_{l'}^-(\tau')\hat{S}_{l''}^+(\tau'')\hat{S}_{l'''}^-(\tau''')\rangle_0 \tag{B.36}$$

The fact that the set of contractions (B.34) and (B.36) lead to the same result reflects the *arbitrariness in the choice of the initial operator* $\hat{S}_l^+$. That is why we will write

$$\langle T_\tau \hat{S}_l^+(\tau)\hat{S}_{l'}^-(\tau')\hat{S}_{l''}^+(\tau'')\hat{S}_{l'''}^-(\tau''')\rangle_0 \tag{B.37}$$

instead of (B.34) and (B.36), thus making the equivalence of these contractions obvious also in the graphical representation.

Using (B.32) and (B.33) in succession, we can get all possible types of contractions for spin operators. Application of these contractions, for example, gives

$$
\begin{aligned}
\langle T_\tau \hat{S}_l^+(\tau)\hat{S}_{l'}^-(\tau')\hat{S}_{l_1}^+(\tau_1)\hat{S}_{l_2}^-(\tau_2)\rangle_0 &= \langle T_\tau \hat{S}_l^+(\tau)\hat{S}_{l'}^-(\tau')\hat{S}_{l_1}^+(\tau_1)\hat{S}_{l_2}^-(\tau_2)\rangle_0 \\
&+ \langle T_\tau \hat{S}_l^+(\tau)\hat{S}_{l'}^-(\tau')\hat{S}_{l_1}^+(\tau_1)\hat{S}_{l_2}^-(\tau_2)\rangle_0 \\
&+ \langle T_\tau \hat{S}_l^+(\tau)\hat{S}_{l'}^-(\tau')\hat{S}_{l_1}^+(\tau_1)\hat{S}_{l_2}^-(\tau_2)\rangle_0 \\
&+ \langle T_\tau \hat{S}_l^+(\tau)\hat{S}_{l'}^-(\tau')\hat{S}_{l_1}^+(\tau_1)\hat{S}_{l_2}^-(\tau_2)\rangle_0 \\
&= (-2)^2 K_{ll'}^0(\tau-\tau')K_{l_1l_2}^0(\tau_1-\tau_2)\langle \hat{S}_{l'}^z\hat{S}_{l_2}^z\rangle_0 \\
&+ (-2)^2 K_{l_1l'}^0(\tau_1-\tau')K_{ll_2}^0(\tau-\tau_2)\langle \hat{S}_{l_2}^z\hat{S}_{l'}^z\rangle_0 \\
&+ (-2)^2 K_{ll'}^0(\tau-\tau')K_{l_1l'}^0(\tau_1-\tau')K_{l'l_2}^0(\tau'-\tau_2)\langle \hat{S}_{l_2}^z\rangle_0 \\
&+ (-2)^2 K_{ll_2}^0(\tau-\tau_2)K_{l_1l_2}^0(\tau_1-\tau_2)K_{l_2l'}^0(\tau_2-\tau')\langle \hat{S}_{l'}^z\rangle_0
\end{aligned}
\tag{B.38}
$$

These examples show that averaging of the T_τ-products reduces to the task of implementing all conceivable contractions of the operators $\hat{S}_l^+$ with the operators $\hat{S}_{l'}^-$ and $\hat{S}_{l'}^z$, with subsequent calculation of averages only of the operators $\hat{S}_l^z$.

Let us then deduce the rule of calculation of the averages $\langle \hat{S}_1^z\hat{S}_2^z\ldots\hat{S}_p^z\rangle_0$, where $1\equiv l_1, 2\equiv l_2,\ldots,p\equiv l_p$.

Because $\exp\left(\beta\gamma_0\sum_l \hat{S}_l^z\right)=\prod_l \exp(\beta\gamma_0\hat{S}_l^z)$, the calculation of an average can be *decomposed* into the calculation of the product of the averages relating to a single atom. For a single operator $\hat{S}_l^z$ we get

$$
\langle \hat{S}_l^z\rangle_0 = \sum_{m=-s}^{s}(m\,\mathrm{e}^{\beta\gamma_0 m})\Big/\sum_{m=-s}^{s}\mathrm{e}^{\beta\gamma_0 m} = sB_s(\beta\gamma_0)=b(\beta\gamma_0)
\tag{B.39}
$$

where $B_s(x)$ is the *Brillouin function* for the spin S:

$$
B_s(x)=\left(1+\frac{1}{2s}\right)\coth\left(s+\frac{1}{2}x\right)-\frac{1}{2s}\coth\frac{x}{2}
\tag{B.40}
$$

The average of any power of the operator $\hat{S}_l^z$ can be expressed through the function $b(\beta\gamma_0)$ and its derivatives. To prove this it is sufficient to note that

$$
\langle \hat{S}_1^z\hat{S}_2^z\ldots\hat{S}_n^z\rangle_0=\frac{1}{\beta^n}\left(\frac{\partial^n}{\partial\gamma_{01}\,\partial\gamma_{02}\ldots\partial\gamma_{0n}}z\right)\Big/z
$$

where $z=\prod_l z_l=\mathrm{Tr}\exp\left(\sum_l\beta\gamma_0\hat{S}_l^z\right)$. As $b(\beta\gamma_{01})\equiv b_1=\frac{z_1'}{z}$, we can write

$$
\langle \hat{S}_1^z\hat{S}_2^z\ldots\hat{S}_n^z\rangle_0=\frac{1}{\beta^{n-1}}\left(\frac{\partial^{n-1}}{\partial\gamma_{02}\ldots\partial\gamma_{0n-1}}b_1z\right)\Big/z
$$

After expressing the derivative of the product of two functions through the Leibniz formula and taking into account that $\dfrac{\partial b_1}{\beta \partial y_{0l}} = b_1' \delta_{1l}$, where δ_{lm} is the Kronecker symbol of the atomic indices, we arrive at the relationships

$$\begin{cases} \langle \hat{S}_l^z \rangle_0 = b(\beta y_0) \equiv b \\ \langle \hat{S}_1^z \hat{S}_2^z \rangle_0 = b^2(\beta y_0) + b' \delta_{12} \\ \langle \hat{S}_1^z \hat{S}_2^z \hat{S}_3^z \rangle_0 = b^3 + b'b(\delta_{12} + \delta_{13} + \delta_{23}) + b'' \delta_{12} \delta_{23} \end{cases} \tag{B.41}$$

Here we should note that the averages $\langle \hat{S}_{l_1}^z \hat{S}_{l_2}^z \ldots \hat{S}_{l_0}^z \rangle_0$ do not depend on the position of the atoms in an amorphous substance.

Taking advantage of the formulae (B.41) we work out the final expressions for the averages (B.31) and (B.38):

$$\langle T_\tau \hat{S}_l^+(\tau) \hat{S}_{l'}^z(\tau') \hat{S}_{l''}^-(\tau'') \rangle_0 = -2K_{ll'}^0(\tau - \tau') K_{l'l''}^0(\tau' - \tau'') b - 2K_{ll''}^0(\tau - \tau'') \cdot (b^2 + b' \delta_{l'l''})$$

$$\begin{aligned} \langle T_\tau \hat{S}_l^+(\tau) \hat{S}_{l'}^-(\tau') \hat{S}_{l_1}^+(\tau_1) \hat{S}_{l_2}^-(\tau_2) \rangle_0 &= +(-2)^2 K_{ll'}^0(\tau - \tau') K_{l_1 l_2}^0(\tau_1 - \tau_2)(b^2 + b' \delta_{l_1 l_2}) \\ &+ (-2)^2 K_{ll_2}^0(\tau - \tau_2) K_{l_1 l'}^0(\tau_1 - \tau')(b^2 + b' \delta_{l_2 l'}) \\ &+ (-2)^2 K_{l_1 l'}^0(\tau_1 - \tau') K_{l' l_2}^0(\tau' - \tau_2) K_{l' l_2}^0(\tau' - \tau_2) b \\ &+ (-2)^2 K_{ll_2}^0(\tau - \tau_2) K_{l_1 l_2}^0(\tau_1 - \tau_2) K_{l_2 l'}^0(\tau_2 - \tau') b \end{aligned}$$

To find the averages with respect to the spin variables upon the random distribution of atoms in the system we follow the diagrammatic technique by Vaks, Larkin and Pikin (1967), because their concepts provide a means for development of the perturbation theory with a pictorial physical interpretation.

As indicated earlier, calculation of the averages $\langle T_\tau \hat{S}_{l_1}^{a_1}(\tau_1) \ldots \hat{S}_{l_n}^{a_n}(\tau_n) \rangle_0$ reduces to the implementation of all possible contractions of the operator $\hat{S}_l^+$ with the operators $\hat{S}_{l'}^-$ and $\hat{S}_{l'}^z$ followed by thermal averaging of the operators $\hat{S}_l^z$. The contractions imply replacing a *pair* of operators in the initial average by only *one* operator, according to the equations:

$$\underbrace{\hat{S}_1^+(\tau_1) \hat{S}_2^z}_{\longrightarrow}(\tau_2) = K_{12}^0(\tau_1 - \tau_2) \hat{S}_2^+(\tau_2) \tag{B.42}$$

$$\underbrace{\hat{S}_1^+(\tau_1) \hat{S}_2^-}_{\longrightarrow}(\tau_2) = (-2) K_{12}^0(\tau_1 - \tau_2) \hat{S}_2^z(\tau_2) \tag{B.43}$$

which follow immediately from the relations (B.32) and (B.33).

Now, the index '1' and the argument τ_1 associated with the selected operator $\hat{S}_1^+(\tau_1)$ evidently cannot be involved in further pairings. Let the average contain an operator with index '2' and argument τ_2 (the operator $\hat{S}_2^+(\tau_2)$ or $\hat{S}_2^z(\tau_2)$). An intermediate operator $\hat{S}_2^+(\tau_2)$ has to be involved in further contractions, and the index '2' and τ_2 will enter only one more contraction. As a result the following sets of contractions are possible:

$$\hat{S}_1^+(\tau_1)\hat{S}_2^z(\tau_2)\hat{S}_3^-(\tau_3) = -2K_{12}^0(\tau_1-\tau_2)K_{23}^0(\tau_2-\tau_3)\hat{S}_3^z(\tau_3) \tag{B.44}$$

$$\hat{S}_1^+(\tau_1)\hat{S}_2^z(\tau_2)\hat{S}_3^z(\tau_3)\ldots = K_{12}^0(\tau_1-\tau_2)K_{23}^0(\tau_2-\tau_3)\hat{S}_3^+(\tau_3) \tag{B.45}$$

Note that in (B.45) the sequence of contractions is unfinished and the operator $\hat{S}^+$, with index '3' and argument τ_3, by now has to be involved in subsequent contractions, while the operator $\hat{S}_3^z(\tau_3)$ in (B.44) can either be left intact within a complete set of contractions, or can be drawn into further contractions with a certain operator $\hat{S}_4^+(\tau_4)$. The latter case again gives the operator $\hat{S}^+$ with the index '3' and argument τ_3. This operator must enter yet another contraction, i.e. here the index '3' and argument τ_3 are implicated in three (and only in three) contractions. In this way the sets of contractions given below arise:

$$\hat{S}_1^+(\tau_1)\hat{S}_2^-(\tau_2)\hat{S}_3^+(\tau_3)\hat{S}_4^-(\tau_4)$$
$$= (-2)^2K_{12}^0(\tau_1-\tau_2)K_{32}^0(\tau_3-\tau_2)K_{24}^0(\tau_2-\tau_4)\hat{S}_4^z(\tau_4) \tag{B.46}$$

$$\hat{S}_1^+(\tau_1)\hat{S}_2^-(\tau_2)\hat{S}_3^+(\tau_3)\hat{S}_4^z(\tau_4)\ldots$$
$$= (-2)K_{12}^0(\tau_1-\tau_2)K_{32}^0(\tau_3-\tau_2)K_{24}^0(\tau_2-\tau_4)\hat{S}_4^+(\tau_4)\ldots \tag{B.47}$$

Thus the index associated with the operator $\hat{S}^+$ enters only one contraction; the index corresponding to the operator $\hat{S}^-$ is involved either in one contraction, remaining with the operator $\hat{S}^z$, or in three of them; and the index at the operator $\hat{S}^z$ either is not implicated in contractions at all, or enters two of them.

In view of the foregoing considerations let us introduce a system of plane-wave space and (imaginary) 'thermodynamic time' coordinates, of which the operators $\hat{S}^a$ involved in the expression $\langle T_\tau \hat{S}_1^{a_1}(\tau_1)\ldots\hat{S}_n^{a_n}(\tau_n)\rangle$ are dependent. The operators $\hat{S}^+$ are therewith matched by 'points' •; the operators $\hat{S}^-$ by 'points with three rays' —•<; and the operators $\hat{S}^z$ and $\hat{S}^-$ (see below) are represented by 'circles ∘. The elementary Green's function $K_{ll'}^0(\tau-\tau')$ is depicted by a 'solid-line arrow' with initial and final coordinates (the so-called *Green's arrow*) '$l\tau \to l'\tau'$'. The circles corresponding to the operators $\hat{S}^z$ are either detached, or the lines enter *into* and go *out* of them. When a line only *enters* a circle, the circle stands for $\hat{S}^-$ (see above).

These definitions allow us to display the contractions (B.43)–(B.47) graphically. If the Green's function $K_{12}(\tau_1-\tau_2)$ results from the contraction (B.43) and the operator $\hat{S}_2^z(\tau_2)$, arisen therewith, does not enter further contractions, the corresponding graph is as follows:

(In the diagrams it is handy to combine space and time variables by using 1 or 2 as one common index.) Here the circle with coordinates '2' and τ_2 comes from the *interlocking* of two rays of the point corresponding to the operator $\hat{S}^-$. This kind of representation of the operator $\hat{S}^-$ is simultaneously indicative of the resulting operator $\hat{S}^z$ (evolved from the contraction between $\hat{S}^+$ and $\hat{S}^-$) appearing in the average.

Equations (B.44)–(B.47) are 'portrayed' by the following four diagrams:

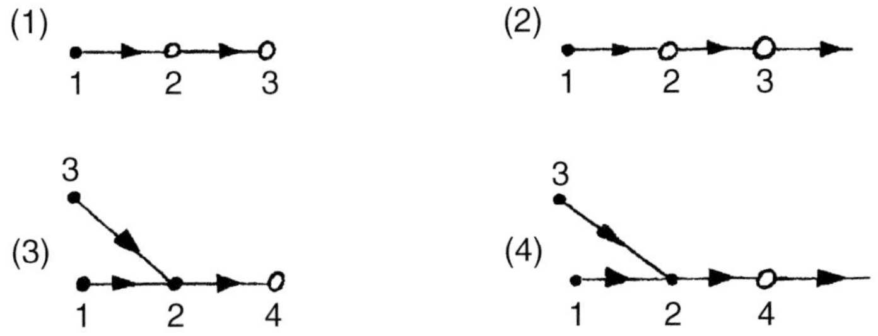

The first diagram shows that index '2' at the operator $\hat{S}^z$ and the variable τ_2, being arguments of two Green's functions, are involved in two contractions, and the resulting operator $\hat{S}_3^z$ is not involved in subsequent contractions. The third diagram points out that the arguments of three Green's functions contain the index '2' and the variable τ_2, while the resulting operator $\hat{S}_4^z$ remains 'uncontracted'. The second and fourth diagrams, corresponding to the contractions (B.45) and (B.47), show such unfinished contraction processes, where index '2' and variable τ_2 will not participate in further contractions. Thus the diagrams can incorporate only the five following types of vertices: (a), (b), (c), (d), (e) (see Fig. B.1).

Here the (a)-type vertex corresponds to the operator $\hat{S}^+$; the (b) and (c) vertices to the operator $\hat{S}^-$; and the vertices of kind (d) and (e) represent the operator $\hat{S}^z$. The graphs that can be constructed from the elements of Fig. B.1 will be called *skeleton graphs.*

Now we have sufficient diagrammatic elements to depict the first step of calculations of thermal averages of spin operators, which leaves behind the averages of the operators $\hat{S}^z$. In Fig. B.1 these operators are matched by vertices (b) and (d). Let us additionally present diagrams for the averages of products of operators $\hat{S}$: $\langle \hat{S}_1^z \hat{S}_2^z \ldots \hat{S}_n^z \rangle_0$. Such an average is broken into a sum of terms incorporating δ-symbols of spatial coordinates of operators $\hat{S}^z$.

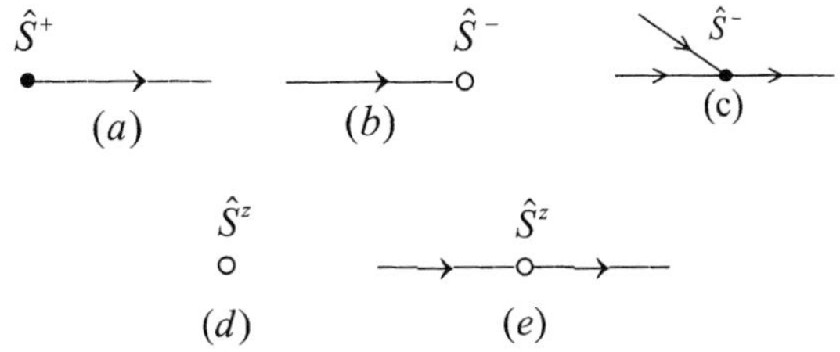

Fig. B.1 a) A vertex with one outgoing Green's arrow; b) a vertex with one ingoing Green's arrow; c) a vertex with two in- and one outgoing Green's arrows; d) an isolated vertex without any Green's arrow; e) a vertex with one in- and one outgoing Green's arrow

The Kronecker symbols in the diagrams are represented by *dotted* lines connecting the points, which depict operators $\hat{S}^z$. The ordering of the vertices $\hat{S}^z$ along the dotted line makes no difference. A set of operators $\hat{S}_l^z$ joined by a dotted line is called a *block*.

Each block is matched by a factor $b^{[f-1]}\delta_{(\ldots)}\ldots\delta_{(\ldots)}$, where f is the number of operators $\hat{S}^z$ in the block, and the deltas account for the equality of spatial variables in it. Each of the (b)- and (d)-type vertices outside the block is matched by the function $b(\beta y_0) = \langle \hat{S}_l \rangle_0$. Let us consider a number of specific examples.

The contraction

$$\langle T_\tau \hat{S}_l^+(\tau)\hat{S}_{l'}^-(\tau')\rangle_0 = -2\langle \hat{S}_{l'}^z\rangle K_{ll'}^0(\tau - \tau')$$

is depicted by the simplest diagram

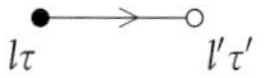

(B.48)

The correlation function $\langle T_\tau \hat{S}_l^+(\tau)\hat{S}_{l'}^z(\tau')\hat{S}_{l''}^-(\tau'')\rangle_0$ is portrayed as follows:

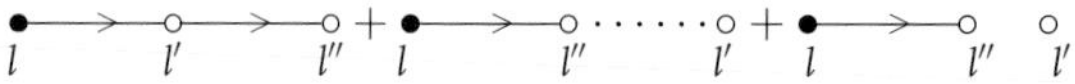

(B.49)

This graph is analytically expressed by:

$$\begin{aligned}\langle T_\tau \hat{S}_l^+(\tau)\hat{S}_{l'}^z(\tau')\hat{S}_{l''}^-(\tau'')\rangle_0 &= (-2)K_{ll'}^0(\tau - \tau')K_{l'l''}^0(\tau' - \tau'')\langle \hat{S}_{l''}^z\rangle_0 \\ &+ (-2)K_{ll''}^0(\tau - \tau'')\langle \hat{S}_{l''}^z \hat{S}_{l'}^z\rangle = (-2)K_{ll'}^0(\tau - \tau')K_{l'l''}^0(\tau' - \tau'')b \\ &+ (-2)K_{ll''}^0(\tau - \tau'')b'\delta_{l'l''} + (-2)K_{ll''}^0(\tau - \tau'')b^2\end{aligned}$$

Interpreting the average $\langle T_\tau \hat{S}_l^+(\tau)\hat{S}_{l'}^-(\tau')\hat{S}_1^z(\tau_1)\hat{S}_2^z(\tau_2)\rangle_0$ in terms of the Wick's theorem for spin operators we get

$$\begin{aligned}\langle T_\tau \hat{S}_l^+(\tau)\hat{S}_{l'}^-(\tau')\hat{S}_1^z(\tau_1)\hat{S}_2^z(\tau_2)\rangle_0 &= (-2)K_{ll'}^0(\tau - \tau')\langle \hat{S}_{l'}^z\hat{S}_1^z\hat{S}_2^z\rangle_0 \\ &+ (-2)K_{l1}^0(\tau - \tau_1)K_{1l'}^0(\tau_1 - \tau')\langle \hat{S}_{l'}^z\hat{S}_2^z\rangle_0 \\ &+ (-2)K_{l2}^0(\tau - \tau_1)K_{1l'}^0(\tau_1 - \tau')\langle \hat{S}_{l'}^z\hat{S}_1^z\rangle_0 \\ &+ (-2)K_{l1}^0(\tau - \tau_1)K_{12}^0(\tau_1 - \tau_2)K_{2l'}^0(\tau_2 - \tau')\langle \hat{S}_{l'}^z\rangle_0 \\ &+ (-2)K_{l2}^0(\tau - \tau_2)K_{21}^0(\tau_2 - \tau_1)K_{1l'}^0(\tau_1 - \tau')\langle \hat{S}_{l'}^z\rangle_0\end{aligned} \quad \text{(B.50)}$$

The first summand in (B.50) is depicted by the skeleton diagrams:

(B.51)

Calculation of the average $\langle \hat{S}_{l'}^z\hat{S}_1^z\hat{S}_2^z\rangle_0$ in view of (B.41) gives the expression

$$\langle \hat{S}^z_{l'} \hat{S}^z_1 \hat{S}^z_2 \rangle_0 = b^3 + bb'(\delta_{l'1} + \delta_{l'2} + \delta_{12}) + b''\delta_{l'1}\delta_{12}$$

which will be displayed graphically, if the vertices $\hat{S}^z$ are combined within the skeleton graph (B.51) by *blocks* containing one, two or three vertices at a time. As a consequence, the skeleton graph multiplies into five diagrams with isolated blocks:

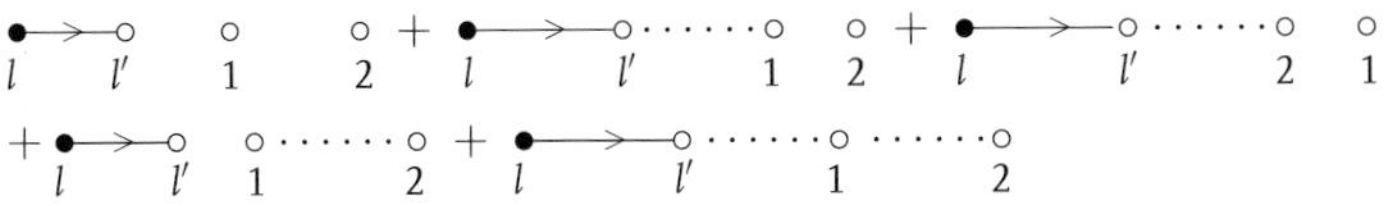

The second and third summands in (B.50) are represented by the skeleton graphs

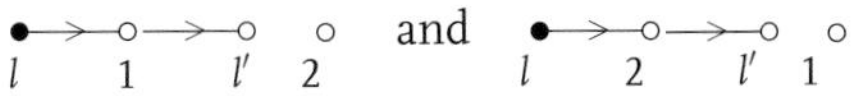

combined into the blocks:

and

Finally, the last two terms in (B.50) are graphically displayed as:

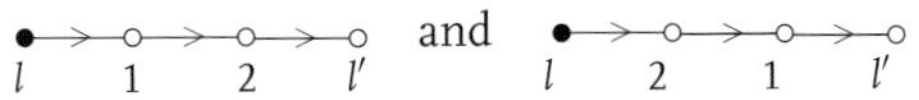

Hence the calculation of the products of operators $\hat{S}^z$, arising from contractions, multiplies the skeleton graph to a set of diagrams with various blocks.

Therein lies a distinction between the diagrammatics for spin operators and the standard technique used in the theory of interacting Bose or Fermi particles.

For a Heisenberg ferromagnet under an external field H, described by the Hamiltonian $\hat{\mathrm{H}} = \hat{\mathrm{H}}_0 + \hat{\mathrm{H}}_{\mathrm{i}}$, with

$$\hat{\mathrm{H}}_0 = -g\mu_{\mathrm{B}} H \sum_{l=1}^{N} \hat{S}^z_l$$

$$\hat{\mathrm{H}}_{\mathrm{i}} = -\frac{1}{2} \sum_{l \neq l'} J_{ll'} \vec{S}_l \cdot \vec{S}_{l'} = -\frac{1}{2} \sum_{l \neq l'} J_{ll'} (\hat{S}^+_l \hat{S}^-_{l'} + \hat{S}^z_l \hat{S}^z_{l'})$$

the spin–spin exchange interaction $J_{ll'}$ will be depicted by a *wavy line* l •〰• l' (an *interaction line*). The presence of the factor $J_{ll'}$ in the Hamiltonian $\hat{\mathrm{H}}_{\mathrm{i}}$ implies that vertices l and l' have to be connected with an interaction line. The (a)-vertices can be connected by interaction lines only with vertices (b) and (c), and the vertices (d) and (e) are connected exclusively with other vertices of the type (d) and (e).

Let us state the diagrammatical rules for the calculation of the spin-operator Green's functions and make up the conventions for matching graphs with analytic expressions.

To calculate graphically the correlation function

$$\langle T_\tau \hat{S}_l^{a}(\tau)\hat{S}_{l'}^{a'}(\tau')\hat{S}_{l_1}^{a_1}(\tau_1)\ldots\hat{S}_{l_p}^{a_p}(\tau_p)\rangle_0$$

which incorporates n_+ operators $\hat{S}^+$, n_- operators $\hat{S}^-$, and n_z operators $\hat{S}^z$, we should represent all operators $\hat{S}^+$ by *points* •, all $\hat{S}^-$ operators by the *three-ray points* —•<, and all operators $\hat{S}^z$ by *circles* ○. Then these graphical elements (•, —•<, ○) should be linked together by all conceivable solid-line *arrows*. In this case an operator $\hat{S}^-$ will be depicted by the vertex (b) if one *Green line* comes into the operator's point (such a vertex results from the *interlocking* of two rays); or otherwise would be the vertex (c) (see Fig. B.1).

Only circles may be detached. If all vertices of a certain Hamiltonian $\hat{\mathrm{H}}_\mathrm{i}$ are connected in pairs by wavy interaction lines, we will get a *skeleton graph*. Then the initial skeleton graph should be multiplied into a set of diagrams with various blocks.

A distinction should be made between diagrams *with* indices and those *without* indices: in an *indexed diagram* all vertices are labeled with the indices of the related operators. There are internal and external indices. As external ones serve the indices of operators involved in the initial average, the indices on vertices of the operators appearing in a certain Hamiltonian $\hat{\mathrm{H}}_\mathrm{i}$ become the internal ones.

Now we introduce the concept of *topologically discernible* and *indiscernible* diagrams with indices. The diagrams will be referred to as topologically indiscernible, or identical, if they are interconvertible by a continuous, planar or spatial transfer of vertices and lines (i.e. if they can be superimposed so that not only the diagrams coincide, but the indices on vertices as well). All topologically indiscernible diagrams depict the same analytical expression. Hence it is obvious that only discernible diagrams should be considered: those that cannot be identically interconverted.

Among the topologically discernible diagrams with indices there are examples differing from each other only by interchanged indices on internal vertices, i.e. by a redistribution of various $\hat{\mathrm{H}}_\mathrm{i}$. Such diagrams are said to be topologically equivalent. Because the time variables of internal vertices are integrated over, and a sum is taken with respect to spatial variables, the contributions corresponding to topologically equivalent diagrams are the same.

As the topologically equivalent diagrams give equal contributions, only one of all discernible topologically equivalent diagrams needs to be considered, and the related analytical expressions should be multiplied by the *number* of discernible topologically equivalent diagrams. Graphically in this case the whole set of discernible topologically equivalent diagrams *with* indices should be matched by one diagram *without* indices.

In the diagrammatics of spin-operators, compared to the diagram techniques for Bose and Fermi variables, there are some peculiar features concerning the counting of the discernible topologically equivalent diagrams with indices related to the particular diagram without them.

This is due to the fact that the operators $\hat{S}^z$ integrated into a *block* are time-independent and have identical spatial variables. As a consequence, any transfer of the vertices (b) and (d) included into a block produces topologically indiscernible diagrams. The totality of all transpositions of $\hat{\mathrm{H}}_{\mathrm{i}}(\tau)$ includes among them those of vertices (b) and (d) enclosed in a block. A transposition will produce a topologically indiscernible diagram if the transposition of operators $\hat{\mathrm{H}}_{\mathrm{i}}(\tau)$ can be obtained in such a way that the diagram resulting from the transfer of vertices (b) and (d) within the block coincides with the initial one not only graphically, but also in the indices of the vertices.

Let us explain the above in terms of examples. The diagrams

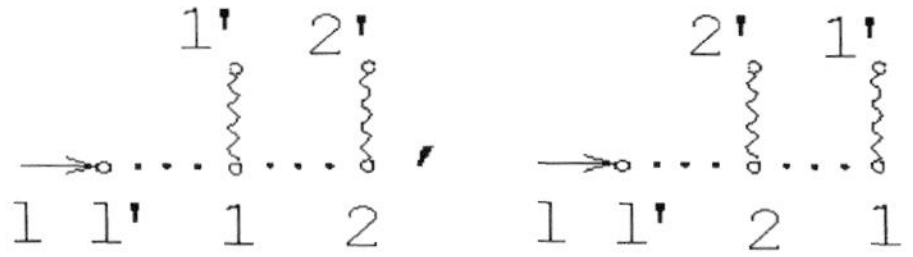

(B.52)

are topologically indiscernible, because the transposition of $\hat{\mathrm{H}}_{\mathrm{i}}$ mentioned above reduces to the transfer of vertices within the block.

As another example:

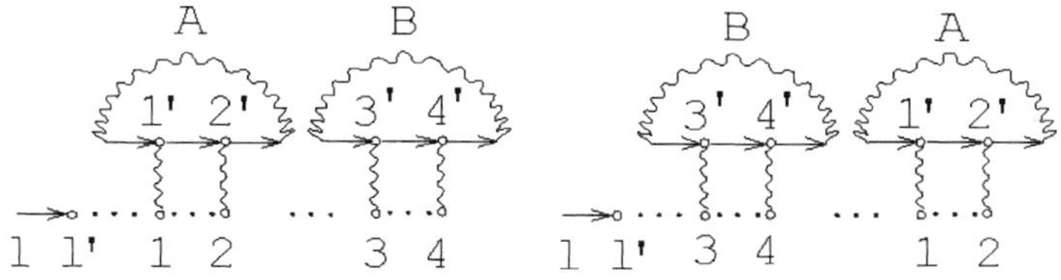

(B.53)

The transpositions of parts A and B in the diagrams are indiscernible.

One further type of indiscernible transposition is

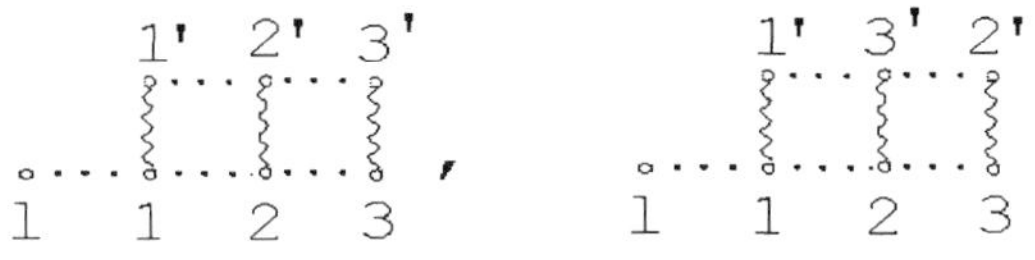

(B.54)

There are 3! such transpositions for this diagram.

Preparatory to stating the diagrammatic rules let us take into account the following considerations. The numerical factor at each diagram of the Green's function $\langle T_\tau \hat{S}_l^+(\tau)\hat{S}_{l'}^-(\tau')\rangle_0$ with external vertices $\hat{S}_l^+$ and $\hat{S}_{l'}^-$ incorporates the common multiplier (–2) resulting from the contraction (B.26). It is convenient to render such a factor independent of the external vertices. This can be arranged by introducing into the definition of the Green's function the multiplier $(-2)^{-m_+}$, where m_+ is the number of external operators $\hat{S}^+$.

Let us now state the general diagrammatic rules obtained in this way for the Heisenberg ferromagnet under an external field. To develop the nth order correction with respect to $\hat{\mathrm{H}}_{\mathrm{i}}$ to the correlation function,

$$(-2)^{-m_+}\langle T_\tau \hat{S}_l^a(\tau)\hat{S}_{l'}^{a'}(\tau')\hat{\sigma}(\beta)\rangle_0, \; (a, a' = \pm; z)$$

the following has to be done:

1. Construct all topologically inequivalent diagrams without indices incorporating $2n$ internal and two external vertices, using the elements depicted above.
2. Arrange anyhow the indices of internal vertices.
3. Match every *solid line* with the Green's function $K_{ll'}^0(\tau - \tau') = \delta_{ll'}K^0(\tau - \tau')$, where l, τ and l', τ' are the respective spatial and (imaginary) time coordinates of its initial and final points.
4. Match every *wavy line* ('interaction line') with a factor $J_{ll'}$.
5. Match every *block* with the factor $b^{(f-1)}\delta_{(\ldots)}\ldots\delta_{(\ldots)}$, where f is the number of (b)- and (d)-type vertices in the block. The equality of spatial coordinates of the vertices is taken into account by the $(f-1)$ delta-functions in the product.
6. Sum up with respect to the spatial variables and integrate over the (imaginary) times of all internal vertices, keeping in mind that internal vertices linked by 'interaction lines' share a common time.
7. Multiply the obtained analytical expression of the diagram by a factor

$$\frac{(-1)^{m'_+}}{P_n} \tag{B.55}$$

where m'_+ is the number of internal vertices $\hat{S}^+$; P_n is the number of transpositions of indiscernible elements in the diagram.

The coefficient (B.55) has the following origin: the analytical expression of each diagram is multiplied by a factor

$$(-2)^{m'_+}\frac{P_n}{2^n \cdot n!}$$

where P_n is the number of discernible topologically equivalent diagrams.

The factor $\frac{1}{n!}$ comes from the expansion of the scattering matrix, the multiplier 2^{-n} evolves from operators $\hat{\mathrm{H}}_{\mathrm{int}}$, the origin of the coefficient $(-2)^{m'_+}$ is clear from (B.27).

The coefficient P_n is determined as follows. The nth order correlation contains the product of n operators $\hat{\mathrm{H}}_{\mathrm{i}}$. The total number of permutations of integration variables is therefore $n!$. Furthermore, a 'turnover' of interaction lines connecting the two vertices corresponding to operators $\hat{S}^z$ also results in topologically equivalent diagrams, because it is assumed in $\hat{\mathrm{H}}_{\mathrm{i}}$ that $J_{ll'} = J_{l'l}$. The number of such 'turnovers' equals 2^{m_z}, where m_z is the number of z–z-interactions in the diagram. However the coefficient $2^{m_z}n!$ also includes the indiscernible transpositions. They may be excluded by the introduction of the factor P_n.

Adduce examples.

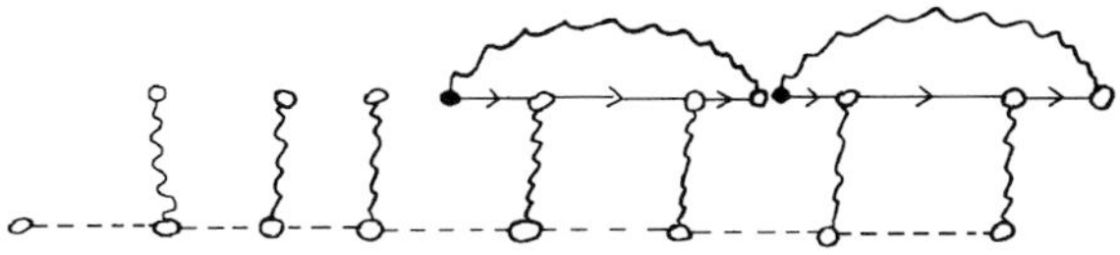

This sketch features the numerical factor $P_n = (3!(2!)$ related to two types of indiscernible parts within a block. The diagram

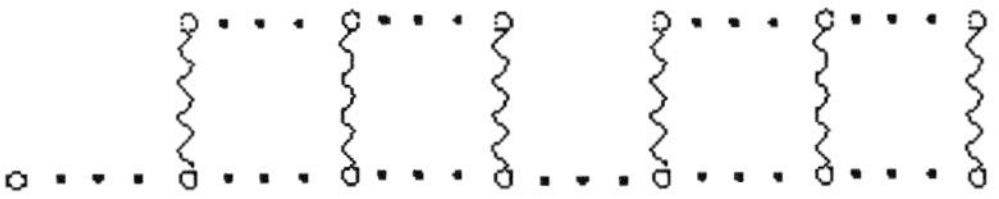

presents $P_n = (2!(3!)^2)$ evolved from two indiscernible parts of the same type within a block, which in turn contain three indiscernible parts each.

Hence the final numerical coefficient of the diagram becomes:

$$(-2)^{m'_+}\frac{1}{2^n n!}n!\frac{1}{P_n}2^{m_z} = (-1)^{m'_+}\frac{1}{P_n}$$

Here $m_z + m'_+ = n$ is taken into account.

The following simple examples (a), (b) and (c) in Fig. B.2 will make the diagrammatic rules clearer.

Figure B.2 shows a selection of diagrams, of the second (a, b) and third order (c) in the interaction, for the correlation function

$$K_{ll'}(\tau-\tau') \equiv K_{ll'}^{+-}(\tau-\tau') = -\frac{1}{2}\langle T\hat{S}_l^+(\tau)\hat{S}_{l'}^-(\tau')\sigma(\beta)\rangle_0 \tag{B.56}$$

The corresponding analytical expressions are as follows:

$$K_b^{+-}(\vec{R}_l, \vec{R}_{l'}, \tau-\tau')$$
$$= -(b')^2\int_0^\beta d\tau_1 \sum_{1\neq1'}\sum_{2\neq2'} K_{l1}^0(\tau-\tau_1)\times K_{1'l'}^0(\tau_1-\tau')\cdot K_{11'}J_{22'}\cdot\delta_{12}\delta_{l'2}$$

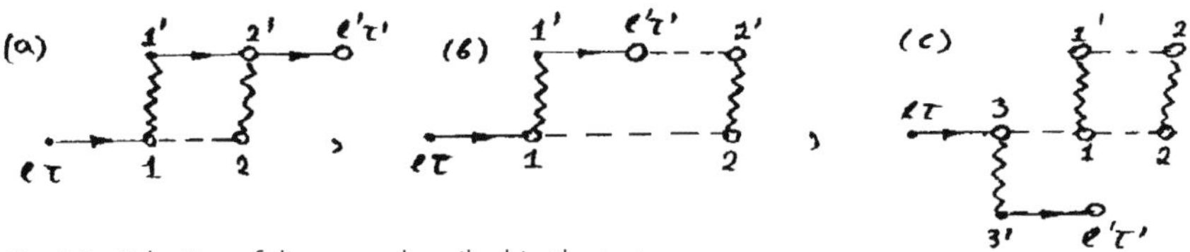

Fig. B.2 Selection of diagrams described in the text

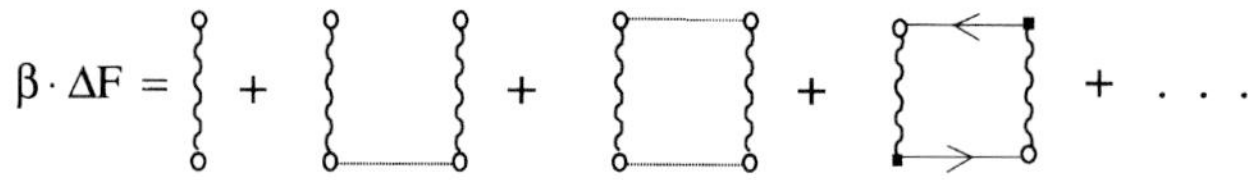

Fig. B.3 Diagrams for the Free Energy F

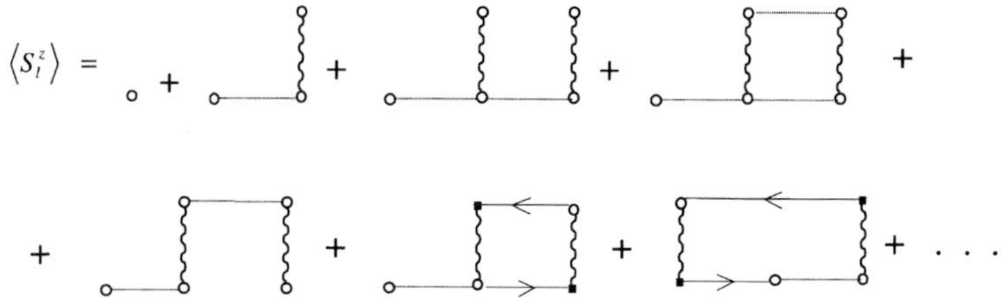

Fig. B.4 Diagrams for the Magnetic Moment $\langle S_l^z \rangle$

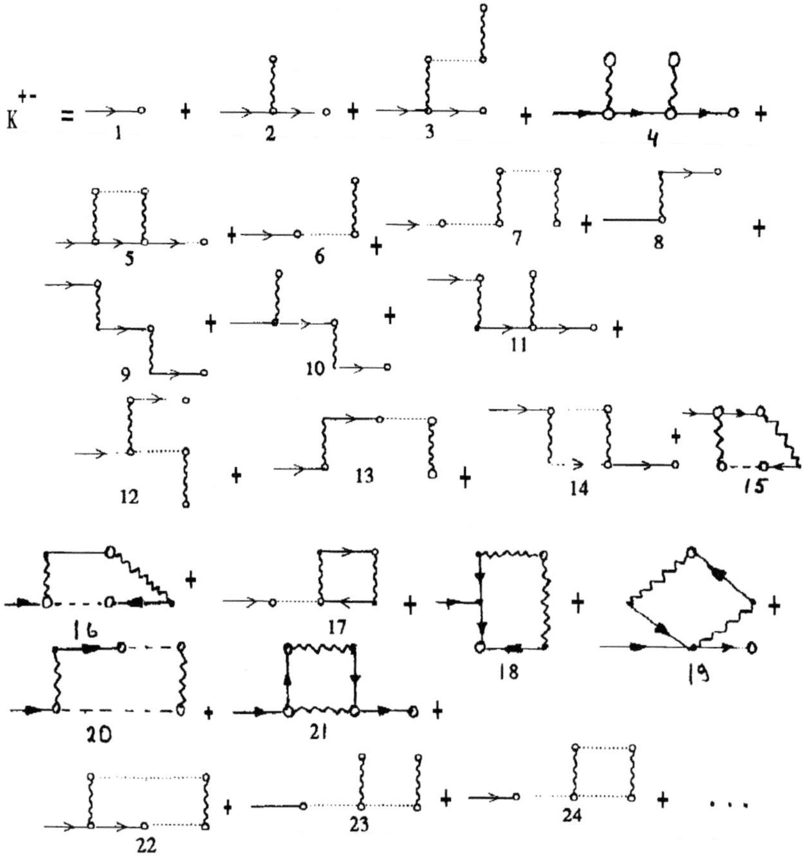

Fig. B.5 Diagrams for the Green's Function K^{+-} (see text on p. 237)

$$K_c^{+-}(\vec{R}_l, \vec{R}_{l'}, \tau - \tau') = -\frac{1}{2!} b \cdot b' \cdot b'' \int_0^\beta d\tau_1 \int_0^\beta d\tau_2 \int_0^\beta d\tau_3$$

$$\times \sum_{1 \neq 1'} \sum_{3 \neq 3'} \sum_{2 \neq 2'} K_{l3}^0(\tau - \tau_3) \cdot K_{3'l'}^0(\tau_3 - \tau') \cdot J_{33'} J_{11'} J_{22'} \cdot \delta_{1'2'} \delta_{12} \delta_{13}$$

The last equation takes into consideration the two indiscernible transpositions $1'1 \rightleftarrows 2'2$.

In conclusion let us introduce the *connectivity* concept for the diagrammatics of the spin-operators:

A diagram will be referred to as a *disconnected* one, if it incorporates elements that are *not* connected with an external vertex by Green's lines, interaction lines, or by blocks. And a diagram is termed connected when all its elements are linked together by Green's lines, interaction lines, or by blocks.

In what follows we shall not write down the disconnected diagrams, but amplify the aforementioned diagrammatic rules by the requirement to construct a complete set of all discernible topologically inequivalent connected graphs when the contributions to the Green's function or other physical quantities are to be depicted. We then present the diagram series for the free energy of a Heisenberg ferromagnet with randomly distributed atoms, correct to the second-order terms in $\hat{\mathrm{H}}_{\mathrm{i}}$ (see Fig. B.3); the diagram series for $\langle \hat{S}_l^z \rangle$, correct to the second-order terms in $\hat{\mathrm{H}}_{\mathrm{i}}$ (see Fig. B.4); and the diagram series for the Green's function

$$K^{+-}(\vec{R}_l, \vec{R}_{l'}, \tau) - \frac{1}{2} \langle T_\tau \hat{S}_l^+(\tau) \hat{S}_{l'}^-(0) \rangle = -\frac{1}{2} \langle T_\tau \hat{S}_l^+(\tau) \hat{S}_{l'}^-(0) \hat{\sigma}(\beta) \rangle_{0c}$$

correct to the second-order terms in $\hat{\mathrm{H}}_{\mathrm{i}}$ (see Fig. B.5).

The series in Fig. B.4 is associated with the following analytical expression:

$$\langle \hat{S}_l^z \rangle = b + b' \cdot \beta \cdot \sum_{l'} J_{ll'} \cdot b + \frac{1}{2!} b'' \cdot \left(\beta \sum_{l'} J_{ll'} \cdot b \right)^2$$

$$+ \frac{1}{2!} b'' \cdot b' \sum_{l'} (\beta J_{ll'})^2 + (b')^2 \cdot b \cdot \left(\beta \sum_{l'} J_{ll'} \right)^2$$

$$+ b \cdot b'' \int_0^\beta d\tau_1 \int_0^\beta d\tau_2 \sum_{\substack{l_1 \neq l'_1 \\ l_2 \neq l'_2}} K_{l_1 l_2}^0(\tau_1 - \tau_2) \cdot K_{l'_1 l'_2}^0(\tau_2 - \tau_1) \cdot J_{l_1 l'_1} \cdot J_{l_2 l'_2} \cdot \delta_{ll_1}$$

$$+ b^2 \int_0^\beta d\tau_1 \int_0^\beta d\tau_2 \int_0^\beta d\tau_3 \sum_{\substack{l_1, l_2, l_3, l'_3 \\ (l_3 \neq l'_3)}} K_{l_1 l_2}^0(\tau_1 - \tau_2) \cdot K_{l_2 l_3}^0(\tau_2 - \tau_3)$$

$$\times K_{l'_3 l}^0(\tau_3 - \tau_1) \cdot J_{ll_1} \cdot J_{l_3 l'_3} + \cdots$$

This concludes Appendix B on the diagrammatic calculation of the free energy of amorphous ferromagnets, correct to within the second order of $\hat{\mathrm{H}}_{\mathrm{i}}$.

Appendix C: Derivation of the Eliashberg Equation for Amorphous Metals

Let us perform a diagrammatic analysis of the translation-invariant full *causal Green's function* of the electron,

$$G(x-x') = G(\vec{r}-r',\tau-\tau') = \overline{\langle G_{\mathrm{e}}^{(m)}(x,x')\rangle_{\mathrm{Ph}}} \tag{C.1}$$

averaged first over the fast vibrational degrees of freedom (the 'phonons'), which is the so-called *annealed* short-time average represented by $\langle\ldots\rangle_{\mathrm{Ph}}$, and afterwards over the fixed configuration $^{(m)}$ of the set of ions, after the vibrations have been subtracted. This *quenched* configuration average is represented by the overbar, as usual (see also Appendix A). Thereby it is

$$G_{\mathrm{e}}^{(m)}(x,x') = -\langle T_\tau[\hat{\Psi}_0(x)\hat{\Psi}_0^+(x')\hat{S}^{(m)}(\beta)]\rangle_{0\,\mathrm{ec}} \tag{C.2}$$

Here $\langle\ldots\rangle_{\mathrm{oe}}$ means a thermal average with respect to $\hat{\mathrm{H}}_0^{(m)} := \hat{\mathrm{H}}^{(m)} - \hat{\mathrm{H}}_{\mathrm{int}}^{(m)}$, i.e. in the *interaction representation* with respect to the electron-electron interaction and the imaginary 'thermodynamic time' τ, as in Appendix A; also the index $_0$ at the electron creation and destruction operators is a hint to this representation; finally $\langle\ldots\rangle_{\mathrm{c}}$ means the sum of the *connected* diagrams of the 'temperature scattering matrix' $\hat{S}^{(m)}(\beta)$(see below). The index $^{(m)}$, as already mentioned, refers to a particular configuration of the amorphous sample.

Let us expand this S-matrix into the pseudopotential series:

$$\hat{S}^{(m)}(\beta) = \hat{S}_{\mathrm{e}}(\beta) + \sum_{n=1}^{\infty}\frac{(-1)^n}{n!}\int_0^\beta \mathrm{d}\tau_1\ldots\int_0^\beta \mathrm{d}\tau_n T_\tau\left\{\hat{\mathrm{H}}_{\mathrm{int}}^{(m)}(\tau_1)\ldots\hat{\mathrm{H}}_{\mathrm{int}}^{(m)}(\tau_n)\hat{S}_{\mathrm{e}}(\beta)\right\} \tag{C.3}$$

where the contribution from $n=0$, $\hat{S}_{\mathrm{e}}(\beta) = T_\tau\left\{\exp\left[-\int_0^\beta \mathrm{d}\tau\,\hat{\mathrm{H}}_{\mathrm{ee}}(\tau)\right]\right\}$, results from the 'trivial' electron interaction in a homogeneous neutralizing background.

Putting (C.3) into (C.2) we have:

$$G_{\mathrm{e}}^{(m)}(x,x') = \sum_{n=0}^{\infty} G_{n\mathrm{e}}^{(m)}(x,x') \tag{C.4}$$

where again the $n=0$ contribution

$$G_{0\mathrm{e}}^{(m)}(x,x') \equiv G_{0\mathrm{e}}(x-x') = -\langle T_\tau[\hat{\Psi}_0(x)\hat{\Psi}_0^+(x')\hat{S}_{\mathrm{e}}(\beta)]\rangle_{0\,\mathrm{ec}}$$

is the 'trivial' one-particle Green's function of interacting electrons in the homogeneous neutralizing background, whereas for $n \geq 1$

$$G_{ne}^{(m)}(x,x') = -\frac{(-1)^n}{n!}\left\langle T_\tau\left[\hat{\Psi}_0(x)\cdot\hat{\Psi}_0^+(x')\right.\right.$$
$$\left.\left.\times\int_0^\beta d\tau_1\ldots\int_0^\beta d\tau_n \hat{H}_{ie}^{(m)}(\tau_1)\ldots\hat{H}_{ie}^{(m)}(\tau_n)\,\hat{S}_e\right]\right\rangle_{0\,ec}$$

The rules governing the calculation of corrections of any order are:

1. All topologically inequivalent diagrams with $2n$ vertices and two external ends at the points x and x' are depicted.
2. Every internal vertex is matched by a matrix τ_3; two *solid lines* and a *wavy line* converge to the vertex.
3. An external vertex is designated by a *cross* $\times$. To this vertex, matched by $\hat{W}^{(m)}(x)\cdot\tau_3$, converge two solid lines (see Eq. (5.1.29)).
4. The Green's function $G_0(x,x')$ corresponds to the solid line.
5. Each wavy line is matched by the potential $V_e(x-x') = V_e(\vec{r}-\vec{r}\,')\,\delta(\bar{t}-\bar{t}')$
6. An integration over $d^4x = d\vec{r}\cdot dt$ is carried out with respect to the coordinates of all internal vertices.
7. The expression obtained is multiplied by a factor $(-1)^{n+F}$, where n is the order of the diagram and F is the number of closed fermion loops included.

Then $G_{ne}(x,x')$ can be represented by the following diagram expansion

$$G_{0e}(x,x') = -\langle T_\tau[\hat{\Psi}_0(x)\hat{\Psi}_0^+(x')\hat{S}_e]\rangle_{0\,ec}$$

which is written down to the second order, and the one-particle Green's function

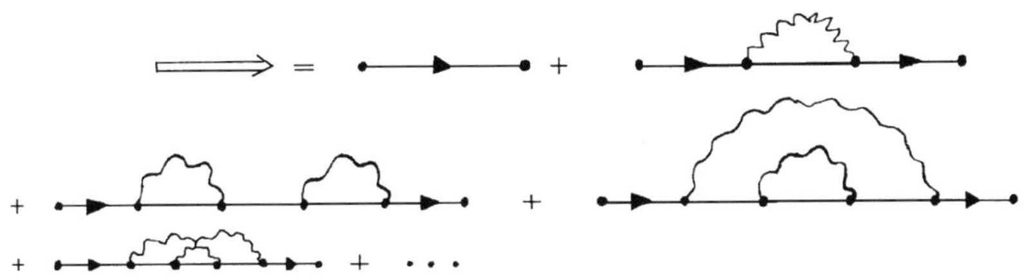

$$G_{1e}^{(m)}(x,x') = -\frac{(-1)}{1!}\int dx_1\langle T_\tau[\hat{\Psi}_0(x)\hat{\Psi}_0^+(x')\hat{\Psi}_0^+(x_1)\tau_3\hat{W}^{(m)}(x_1)\hat{\Psi}_0(x_1)S_e]\rangle_{0\,ec}$$

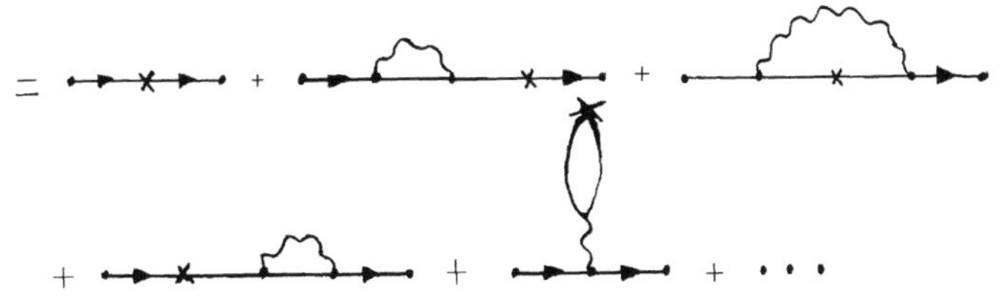

$$=$$ 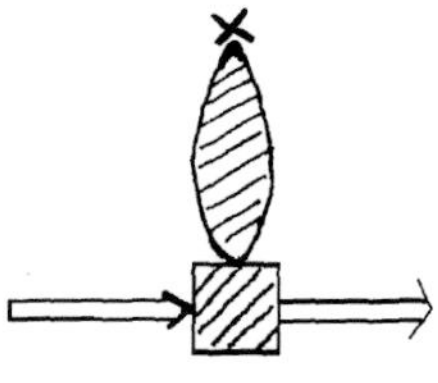

$$= G_{0e}\,\Gamma_e \cdot \varepsilon^{-1}\tau_3 \hat{W}^{(m)} G_{0e}$$

Finally the two-particle Green's function is

$$G_{2e}^{(m)}(x, x') = -\frac{(-1)^2}{2!}\langle T_\tau[\hat{\Psi}_0(x)\hat{\Psi}_0^+(x')\int dx_1 \int dx_2$$
$$\times\, \hat{\Psi}_0^+(x_1)\tau_3\hat{W}^{(m)}(x_1)\hat{\Psi}_0(x_1)\hat{\Psi}_0^+(x_2)\tau_3\hat{W}^{(m)}(x_2)\hat{\Psi}_0(x_2)S_e]\rangle_{0\,ec}$$

$$=$$ 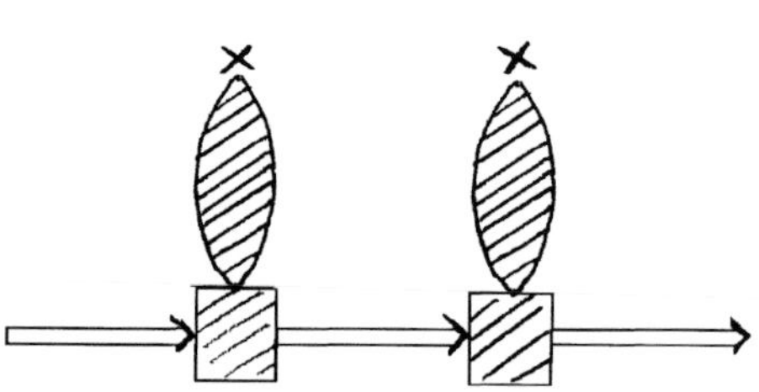

$$= T_\tau G_{0e}\Gamma_e\varepsilon^{-1}\tau_3\hat{W}^{(m)}G_{0e}\Gamma_e\varepsilon^{-1}\tau_3\hat{W}^{(m)}G_{0e}$$

and so on (the final results are written in terms of operators). Here $\Gamma_e(x - x')$ is the *vertex function* including the Coulomb interaction, $\varepsilon(x - x')$ is the electrons' *permittivity*. All these functions are determined by the properties of the uniform interacting electron gas. Introduce the abbreviation

$$\hat{\tilde{W}}^{(m)}(x - x') = \Gamma_e\varepsilon^{-1}\hat{W}^{(m)}$$
$$= e^{\hat{H}_i^{(m)}(\tau-\tau')}\int \Gamma_e(\vec{r} - \vec{r}_1)\varepsilon^{-1}(\vec{r}_1 - \vec{r}_2)\hat{W}^{(m)}(\vec{r}_2 - \vec{r}')d^3r_1 d^3r_2 e^{-\hat{H}_i^{(m)}(\tau-\tau')}$$
$$= e^{\hat{H}_i^{(m)}(\tau-\tau')}\hat{\tilde{W}}^{(m)}(\vec{r} - \vec{r}')e^{-\hat{H}_i^{(m)}(\tau-\tau')}$$

where

$$\hat{\tilde{W}}^{(m)}(\vec{r}) = \sum_\nu \hat{\tilde{w}}^{(m)}(\vec{r} - \vec{R}_\nu^{(m)}) = \frac{1}{v_0}\sum_{\vec{q}} e^{i\vec{q}\cdot\vec{r}}\rho^{(m)}(\vec{q})\tilde{w}(\vec{q}) \tag{C.6}$$

Here $\rho^{(m)}(\vec{q}) = \frac{1}{N}\sum_\nu e^{-i\vec{q}\cdot\vec{R}_\nu^{(m)}}$, and

$$\tilde{w}(\vec{q}) = \int e^{-i\vec{q}\cdot\vec{r}}\tilde{w}(\vec{r})d^3r = \frac{\Gamma_e(\vec{q})w(\vec{q})}{\varepsilon(\vec{q})} \tag{C.6}$$

In this case $\tilde{w}(0) = 0$ due to electrical neutrality (because $w(0) = 0$). However

$$\lim_{q\to 0} \tilde{w}(q) = \lim_{q\to 0} \frac{\Gamma_e(q)w(q)}{\varepsilon(q)} = \lim_{q\to 0} \frac{w(q)}{\varepsilon(q)} = \frac{z}{\pi(0)} \tag{C.7}$$

Here we made allowance (see Chapter 2) for

$$\varepsilon(q) = 1 + \frac{4\pi e^2 \pi(q)}{q^2} \tag{C.8}$$

$\pi(q)$ being the *polarization operator.*

Thus the electron Green's function $G_e^{(m)}$ takes the form

$$\begin{aligned} G_e^{(m)} &= G_{0e} + G_{0e}\tau_3 \hat{\tilde{W}}^{(m)} G_{0e} + T_\tau G_{0e}\tau_3 \hat{\tilde{W}}^{(m)}_{0e} G_{0e}\tau_3 \hat{\tilde{W}}^{(m)} G_{0e} \\ &+ T_\tau G_{0e}\tau_3 \hat{\tilde{W}}^{(m)} G_{0e}\tau_3 \hat{\tilde{W}}^{(m)} G_{0e}\tau_3 \hat{\tilde{W}}^{(m)} G_{0e} + \ldots \end{aligned} \tag{C.9}$$

where T_τ is the 'time ordering operator' for the τ-arguments in $\hat{\tilde{W}}(x)$.

As mentioned above, the full electron Green's function will be derived by averaging (C.9) over all phonon variables and all quasi-equilibrium configurations

$$\begin{aligned} G &= \overline{\langle G_e^{(m)} \rangle_{Ph}^{(m)}} = G_{0e} + G_{0e}\overline{\langle \tau_3 \hat{\tilde{W}}^{(m)} \rangle_{Ph}^{(m)}} G_{0e} \\ &+ G_{0e}\overline{\langle T_\tau \tau_3 \hat{\tilde{W}}^{(m)} G_{0e}\, \tau_3 \hat{\tilde{W}}^{(m)} \rangle_{Ph}^{(m)}} G_{0e} + \ldots \end{aligned} \tag{C.10}$$

Thus it can readily be seen that the determination of the full Green's function $G(x - x')$ amounts to finding out the following expressions

$$\overline{\langle T_\tau [\hat{\tilde{W}}^{(m)}(x_1) \ldots \hat{\tilde{W}}^{(m)}(x_n)] \rangle_{Ph}^{(m)}} \tag{C.11}$$

It can be assumed in the adiabatic approximation (see Eq. (5.1.33)) that

$$\begin{aligned} \hat{\tilde{W}}^{(m)}(x) &= e^{\hat{H}_i^{(m)}\tau} \hat{\tilde{W}}^{(m)}(\vec{r}) e^{-\hat{H}_i^{(m)}\tau} \\ &\approx e^{\hat{H}_i^{(m)}\tau} \cdot Z_e^{-1(m)}(\tau) \hat{\tilde{W}}^{(m)}(\vec{r}) Z_e^{(m)}(\tau) e^{-\hat{H}_i^{(m)}\tau} \\ &= e^{\hat{H}_i^{*(m)}\tau} \cdot \hat{\tilde{W}}^{(m)}(\vec{r}) e^{-\hat{H}_i^{*(m)}\tau} \end{aligned} \tag{C.12}$$

Make another approximation: suppose that the vibrations of the ions near their equilibrium positions are small, i.e. $|\vec{\xi}| \ll a$, where a is the distance between nearest-neighbor ions in the crystalline or amorphous system. Then we can expand the expressions $\hat{H}_i^{*(m)}$ and $\hat{\tilde{W}}^{(m)}(\vec{r})$(or $\rho(q)$ on passing to a Fourier transform) in terms of small displacements of the ions from the (quasi-)equilibrium locations and restrict ourselves to the quadratic terms (*harmonic approximation*):

$$\hat{H}_i^{*(m)} = U_0^{(m)} + \hat{H}_2^{(m)} + \ldots \tag{C.13}$$

with

$$U_0^{(m)} = N\left(-\mu_i + \frac{zb}{v_0}\right) + U(\vec{R}_{01}^{(m)}, \vec{R}_{02}^{(m)}, \ldots \vec{R}_{0N}^{(m)}) \tag{C.14}$$

$$\hat{H}_2^{(m)} = \sum_\nu \frac{1}{2} M\left(\dot{\hat{\vec{\xi}}}_\nu^{(m)}\right)^2 + \frac{1}{2}\sum_{\nu\neq\nu'} \hat{\vec{\xi}}_\nu^{(m)} \cdot \overleftrightarrow{A}_{\nu\nu'}^{(m)} \cdot \hat{\vec{\xi}}_{\nu'}^{(m)} + \ldots \tag{C.15}$$

where $\overleftrightarrow{A}_{\nu\nu'}^{(m)}$ is the so-called *dynamical matrix* describing the potential energy of the vibrational excitations of the system in this approximation, and

$$\rho^{(m)}(\vec{q}) = \rho_0^{(m)}(\vec{q}) + \rho_1^{(m)}(\vec{q}) + \rho_2^{(m)}(\vec{q}) + \ldots \tag{C.16}$$

$$\rho_0^{(m)}(\vec{q}) = \frac{1}{N}\sum_\nu e^{-i\vec{q}\cdot\vec{R}_{0\nu}^{(m)}} \tag{C.17}$$

$$\rho_1^{(m)}(\vec{q}) = \frac{1}{N}\sum_\nu (-i\vec{q}\cdot\hat{\vec{\xi}}(\vec{R}_{0\nu}^{(m)}))e^{-i\vec{q}\cdot\vec{R}_{0\nu}^{(m)}} \tag{C.18}$$

According to Eq. (C.16) the operator $\hat{\hat{W}}(x)$ can be written as

$$\hat{\hat{W}}(x) = \hat{\hat{W}}_0(x) + \hat{\hat{W}}_1(x) + \ldots \tag{C.19}$$

The Hamiltonian $\hat{H}_2^{(m)}$ can thus be written in the form (see Chapter 2):

$$\hat{H}_2^{(m)} = \overline{\hat{H}_2^{(m)}} + (\hat{H}_2^{(m)} - \overline{\hat{H}_2^{(m)}}) \tag{C.20}$$

The first term is the Hamiltonian of the fictitious ideal gas of 'quasi-phonons' that results from averaging the Hamiltonian over all ion positions (*annealed approximation*) in the ensemble of ion configurations corresponding to the amorphous system considered, this term being diagonalized with the following well-known relationship

$$\hat{\vec{\xi}}(\vec{R}) = \sum_{\vec{k}\lambda}\left(\frac{\hbar}{2NM\omega_\lambda(k)}\right)^{1/2} (\hat{C}_{-\vec{k}\lambda} + \hat{C}^+_{+\vec{k}\lambda})\,\vec{\varepsilon}\,(\vec{k},\lambda)\, e^{-i\vec{k}\cdot\vec{R}} \tag{C.21}$$

which takes the form (see Chapter 2)

$$\overline{\hat{H}_2^{(m)}} = \sum_{\vec{k}\lambda} \hbar\omega_\lambda(\vec{k})\left[\hat{C}^+_{\vec{k}\lambda}\,\hat{C}_{\vec{k}\lambda} + \frac{1}{2}\right]$$

where $\omega_\lambda(\vec{k})$ is the dispersion relation of the quasi-phonons, and λ distinguishes their polarization direction ($\lambda = 3$: longitudinal quasi-phonons: $\vec{\xi}$ parallel $\vec{k}$; $\lambda = 1,2$: transversal quasi-phonons: $\vec{\xi}$ perpendicular to $\vec{k}$).

Here $\hat{C}_{\vec{k}\lambda}$ and $\hat{C}^+_{\vec{k}\lambda}$ are the (bosonic) creation and annihilation operators of the quasi-phonons, and the corresponding excitation energy $\hbar\omega_\lambda(\vec{k})$ is determined by the relationship

$$\omega_\lambda^2(\vec{k}) = \frac{1}{NM} \sum_{\substack{\vec{k}_1 \\ (\lambda=1,2,3)}} [\vec{k}_1 \cdot \vec{\varepsilon}(\vec{k}, \lambda)]^2 \; \tilde{\Phi}(\vec{k}_1)^2 [S_0(\vec{k}_1 + \vec{k}) - S_0(\vec{k}_1)] \tag{C.22}$$

with a pairwise interaction between the ions, $\tilde{\Phi}(\vec{k})$ being the Fourier transform of the potential energy of the two-ion interaction; $S_0(\vec{q}) = \frac{(2\pi)^3}{v_0}\delta(\vec{q}) + a(\vec{q})a(\vec{q})$ is the *structure factor* of the material, which should actually depend only on the *magnitude* of $\vec{q}$, if the amorphous material is structurally isotropic on average. If this is true, then also ω_λ^2 depends only on the magnitude of $\vec{k}$.

The calculations show a minimum in the dispersion curve (see Chapter 2) around $q \approx 2\pi/r_0$, where r_0 is the mean value of the nearest-neighbor distance of the ions in the amorphous systems; at the same value of q the structure factor has a pronounced peak.

The second summand in Eq. (C.20) describes the interaction of these 'quasi-phonons' with the structural micro-irregularities in the material. Assuming that this interaction has only a slight effect on superconductivity, this second term in Eq. (C.20) is usually neglected, i.e.

$$\hat{H}_2^{(m)} \approx \overline{\hat{H}_2^{(m)}}$$

A similar approximation is possible in the expressions for $\hat{\tilde{W}}^{(m)}(x)$ or $\hat{\vec{\xi}}(\vec{R}_{0v}^{(m)}, \tau)$:

$$\begin{aligned} \hat{\vec{\xi}}(\vec{R}_{0v}^{(m)}, \tau) &= e^{\hat{H}_2^{(m)}\tau}\, \hat{\vec{\xi}}(\vec{R}_{0v}^{(m)})\, e^{-\hat{H}_2^{(m)}\tau} \approx \exp\left(\overline{\hat{H}_2^{(m)}} \cdot \tau\right) \hat{\vec{\xi}}\,(\vec{R}_{0v}^{(m)}) \exp\left(-\overline{\hat{H}_2^{(m)}} \cdot \tau\right) \\ &= \sum_{\vec{k}\lambda} \left(\frac{\hbar}{2NM\omega_\lambda(k)}\right)^{1/2} \left[\hat{C}_{\vec{k}\lambda}(\tau) + \hat{C}^+_{-\vec{k}\lambda}(\tau)\right] \vec{\varepsilon}\,(\vec{k}, \lambda)\, e^{i\vec{k}\cdot\vec{R}_{0v}^{(m)}} \end{aligned} \tag{C.23}$$

with

$$\begin{cases} \hat{C}_{\vec{k}\lambda}(\tau) = e^{\overline{\hat{H}_2^{(m)}}\tau}\, \hat{C}_{\vec{k}\lambda}\, e^{-\overline{\hat{H}_2^{(m)}}\tau} = \hat{C}_{\vec{k}\lambda}\, e^{-\hbar\omega(\vec{k},\lambda)\cdot\tau} \\ \hat{C}^+_{\vec{k}\lambda}(\tau) = e^{\overline{\hat{H}_2^{(m)}}\tau}\, \hat{C}^+_{\vec{k}\lambda}\, e^{-\overline{\hat{H}_2^{(m)}}\tau} = \hat{C}^+_{\vec{k}\lambda}\, e^{-\hbar\omega(\vec{k},\lambda)\cdot\tau} \end{cases} \tag{C.24}$$

The above approximations enable again a reversal of averaging in Eq. (C.11), i.e. we can initially perform the configurational average over the positions of the ions, and then the thermal average over the phonon variables:

$$\overline{\left\langle T_\tau \left[\hat{\tilde{W}}^{(m)}(x_1) \ldots \hat{\tilde{W}}^{(m)}(x_n) \right] \right\rangle_{\text{Ph}}^{(m)}}$$

$$= \frac{\overline{\text{Tr}_{\text{Ph}} \left\{ e^{-\hat{H}_2^{*(m)} \cdot \beta} T_\tau \left[\hat{\tilde{W}}^{(m)}(x_1) \ldots \hat{\tilde{W}}^{(m)}(x_n) \right] \right\}}}{\text{Tr}_{\text{Ph}} \left\{ \exp\left(-\hat{H}_2^{*(m)} \cdot \beta \right) \right\}}$$

$$= \frac{\overline{\text{Tr}_{\text{Ph}} \left\{ \exp\left(-\hat{H}_2^{(m)} \cdot \beta \right) T_\tau \left[\hat{\tilde{W}}^{(m)}(x_1) \ldots \hat{\tilde{W}}^{(m)}(x_n) \right] \right\}}}{\text{Tr}_{\text{Ph}} \left\{ \exp\left(-\hat{H}_2^{*(m)} \cdot \beta \right) \right\}}$$

$$\approx \frac{\text{Tr}_{\text{Ph}} \left\{ \exp\left(-\overline{\hat{H}_2^{(m)}} \cdot \beta \right) \overline{T_\tau \left[\hat{\tilde{W}}^{(m)}(x_1) \ldots \hat{\tilde{W}}^{(m)}(x_n) \right]} \right\}}{\text{Tr}_{\text{Ph}} \left\{ \exp\left(-\overline{\hat{H}_2^{(m)}} \cdot \beta \right) \right\}}$$

$$= \left\langle \overline{T_\tau \left\{ \hat{\tilde{W}}^{(m)}(x_1) \ldots \hat{\tilde{W}}^{(m)}(x_n) \right\}} \right\rangle_{\text{Ph}} \tag{C.25}$$

First we consider the averaging over configurations: any average of products of functions depending only on single-particle coordinates has the following properties:

$$\overline{f_1 f_2 \ldots f_n} \to \overline{f_1} \cdot \overline{f_2} \ldots \overline{f_n}\,, \quad \text{when} \quad |\vec{R}_i - \vec{R}_j| \to \infty$$

i.e. infinitely distant particles are uncorrelated. This is also true when a cluster of m particles of the n-particle system is moved off to infinity, e.g.

$$\overline{f_1 f_2 \ldots f_n} \to \overline{f_1 \ldots f_m} \cdot \overline{f_{m+1} \ldots f_n}\,, \text{ etc.}$$

Hence the averages of this type can always be represented as follows:

$$\begin{cases} \overline{f_1} = \overline{f_1} \\ \overline{f_1 f_2} = \overline{f_1} \cdot \overline{f_2} + \overline{f_1 f_2}^C \\ \overline{f_1 f_2 f_3} = \overline{f_1} \cdot \overline{f_2} \cdot \overline{f_3} + \overline{f_1} \cdot \overline{f_2 f_3}^C + \overline{f_2} \cdot \overline{f_1 f_3}^C + \overline{f_3} \cdot \overline{f_1 f_2}^C + \overline{f_1 f_2 f_3}^C \text{ etc.} \\ \overline{f_1 f_2 f_3 f_4} = \overline{f_1} \cdot \overline{f_2} \cdot \overline{f_3} \cdot \overline{f_4} + \overline{f_1} \cdot \overline{f_2 f_3 f_4}^C + \overline{f_2} \cdot \overline{f_1 f_3 f_4}^C + \overline{f_3} \cdot \overline{f_1 f_2 f_4}^C \\ +\overline{f_4} \cdot \overline{f_1 f_2 f_3}^C + \overline{f_1} \cdot \overline{f_2} \cdot \overline{f_3 f_4}^C + \overline{f_1 f_2}^C \cdot \overline{f_3 f_4}^C + \overline{f_1} \cdot \overline{f_3} \cdot \overline{f_2 f_4}^C \\ +\overline{f_1 f_3}^C \cdot \overline{f_2 f_4}^C + \overline{f_1} \cdot \overline{f_4} \cdot \overline{f_2 f_3}^C + \overline{f_1 f_4}^C \cdot \overline{f_2 f_3}^C + \overline{f_1 f_2 f_3 f_4}^C \end{cases} \tag{C.26}$$

Here the irreducible part of the correlation function $\overline{f_1 f_2 \ldots f_e}^C$ vanishes for $|\vec{R}_i - R_j| \to \infty$.

With Eq. (C.26) taken into account, we can express

$$\overline{T_\tau \left\{ \hat{\tilde{W}}^{(m)}(x_1) \ldots \hat{\tilde{W}}^{(m)}(x_n) \right\}}$$

as follows:

$$1)\ \overline{\hat{\hat{W}}^{(m)}(x_1)} = \exp\left(\overline{\hat{H}_2^{(m)}}\tau\right)\overline{\hat{\hat{W}}^{(m)}(x_1)}\exp\left(-\overline{\hat{H}_2^{(m)}}\tau\right)$$

$$= \exp\left(\overline{\hat{H}_2^{(m)}}\tau\right)\sum_{\vec{q}}\frac{1}{v_0}\,e^{i\vec{q}\cdot\vec{r}}\tilde{w}(\vec{q})\overline{\rho^{(m)}(\vec{q})}\exp\left(-\overline{\hat{H}_2^{(m)}}\tau\right)$$

$$= \exp\left(\overline{\hat{H}_2^{(m)}}\tau\right)\sum_{\vec{q}}\frac{1}{v_0}\,e^{i\vec{q}\cdot\vec{r}}\tilde{w}(\vec{q})\,\delta_{\vec{q}0}\exp\left(-\overline{\hat{H}_2^{(m)}}\tau\right) \equiv 0 \qquad \text{(C.27)}$$

$$2)\ \overline{T_\tau\hat{\hat{W}}^{(m)}(x_1)\,\hat{\hat{W}}^{(m)}(x_2)} = T_\tau\overline{\hat{\hat{W}}^{(m)}(x_1)}\cdot\overline{\hat{\hat{W}}^{(m)}(x_2)} + \overline{T_\tau\hat{\hat{W}}^{(m)}(x_1)\cdot\hat{\hat{W}}^{(m)}(x_2)}^{C}$$

$$= \overline{T_\tau\hat{\hat{W}}^{(m)}(x_1)\cdot\hat{\hat{W}}^{(m)}(x_2)}^{C} \qquad \text{(C.28)}$$

$$3)\ \overline{T_\tau\hat{\hat{W}}^{(m)}(x_1)\,\hat{\hat{W}}^{(m)}(x_2)\,\hat{\hat{W}}^{(m)}(x_3)} = \overline{T_\tau\hat{\hat{W}}^{(m)}(x_1)\,\hat{\hat{W}}^{(m)}(x_2)\,\hat{\hat{W}}^{(m)}(x_3)}^{C}$$

$$4)\ \overline{T_\tau\hat{\hat{W}}^{(m)}(x_1)\,\hat{\hat{W}}^{(m)}(x_2)\,\hat{\hat{W}}^{(m)}(x_3)\,\hat{\hat{W}}^{(m)}(x_4)}$$

$$= T_\tau\left\{\overline{\hat{\hat{W}}^{(m)}(x_1)\hat{\hat{W}}^{(m)}(x_2)}^{C}\cdot\overline{\hat{\hat{W}}^{(m)}(x_3)\hat{\hat{W}}^{(m)}(x_4)}^{C}\right\}$$

$$+ T_\tau\left\{\overline{\hat{\hat{W}}^{(m)}(x_1)\hat{\hat{W}}^{(m)}(x_3)}^{C}\cdot\overline{\hat{\hat{W}}^{(m)}(x_2)\hat{\hat{W}}^{(m)}(x_4)}^{C}\right\}$$

$$+ T_\tau\left\{\overline{\hat{\hat{W}}^{(m)}(x_1)\hat{\hat{W}}^{(m)}(x_4)}^{C}\cdot\overline{\hat{\hat{W}}^{(m)}(x_2)\hat{\hat{W}}^{(m)}(x_3)}^{C}\right\}$$

$$+ \overline{T_\tau\hat{\hat{W}}^{(m)}(x_1)\hat{\hat{W}}^{(m)}(x_2)\hat{\hat{W}}^{(m)}(x_3)\hat{\hat{W}}^{(m)}(x_4)}^{C} \quad \text{etc.} \qquad \text{(C.29)}$$

Every ion coordinate x_i in $\hat{\hat{W}}^{(m)}(x_i)$ is matched by a cross $\times_i$ (when the expansion (C.19) is used, $\times$ will correspond to $\hat{\hat{W}}_0(x)$, $\otimes$ to $\hat{\hat{W}}_1(x)$, etc.). Then the *irreducible* correlation functions $\overline{\hat{\hat{W}}^{(m)}(x_1)\ldots\hat{\hat{W}}^{(m)}(x_n)}^{C}$ will be depicted as

$$\overline{\hat{\hat{W}}^{(m)}(x_1)\ldots\hat{\hat{W}}^{(m)}(x_n)}^{C} = \{x_1 \times x_1 x_2 \ldots x_n\}$$

Thus, in a graphical representation Eqs (C.27)–(C29) take the form:

$$\overline{T_\tau\hat{\hat{W}}^{(m)}(x_1)\hat{\hat{W}}^{(m)}(x_2)} = T_\tau\{x_1x_2\}^C \qquad \text{(C.30)}$$

$$\overline{T_\tau\hat{\hat{W}}^{(m)}(x_1)\hat{\hat{W}}^{(m)}(x_2)\hat{\hat{W}}^{(m)}(x_3)} = T_\tau\{x_1x_2x_3\}^C \qquad \text{(C.31)}$$

$$\overline{T_\tau\hat{\hat{W}}^{(m)}(x_1)\hat{\hat{W}}^{(m)}(x_2)\hat{\hat{W}}^{(m)}(x_3)\hat{\hat{W}}^{(m)}(x_4)} = T_\tau\Big\{\{x_1x_2\}^C\{x3x_4\}^C$$

$$+\{x_1x_4\}^C\{x_2x_3\}^C + \{x_1x_3\}^C\{x_2x_4\}^C + \{x_1x_2x_3x_4\}^C\Big\} \text{ etc.} \qquad \text{(C.32)}$$

Now turn to the averages over phonon variables

$$\left\langle T\left\{ \hat{\tilde{W}}^{(m)}(x_1)\ldots\hat{\tilde{W}}^{(m)}(x_n)\right\}\right\rangle_{\text{Ph}} \tag{C.33}$$

Take into consideration that

$$\hat{\tilde{W}}^{(m)}(x) = \hat{\tilde{W}}_0^{(m)} + \hat{\tilde{W}}_1^{(m)} + \ldots \tag{C.34}$$

$$\hat{\tilde{W}}_0^{(m)}(x) = \hat{\tilde{W}}_0^{(m)}(\vec{r}) = \frac{1}{v_0}\sum_{\vec{q}} \mathrm{e}^{i\vec{q}\cdot\vec{r}}\tilde{w}(\vec{q})\frac{1}{N}\sum_{\nu} \mathrm{e}^{-i\vec{q}\cdot\vec{R}_{0\nu}^{(m)}} \tag{C.35}$$

$$\begin{aligned}\hat{\tilde{W}}_1^{(m)}(x) &= \frac{-1}{v_0}\sum_{\vec{q}} \mathrm{e}^{i\vec{q}\cdot\vec{r}}\tilde{w}(\vec{q})\frac{1}{N}\sum_{\nu}(i\vec{q}\cdot\vec{\xi}\,(\vec{R}_{0\nu}^{(m)}))\,\mathrm{e}^{-i\vec{q}\cdot\vec{R}_{0\nu}^{(m)}}\\ &= \frac{-i}{v_0}\sum_{\vec{q}}\sum_{\vec{k}\lambda}\left(\frac{\hbar}{2NM\omega_\lambda(k)}\right)^{1/2}[\vec{q}\cdot\vec{\varepsilon}\,(\vec{k},\,\lambda)]\tilde{w}(\vec{q})\mathrm{e}^{i\vec{q}\cdot\vec{r}}\\ &\quad\times\left[\hat{C}_{\vec{k}\lambda}(\tau)+\hat{C}^{+}_{-\vec{k}\lambda}(\tau)\right]\frac{1}{N}\sum_{\nu}\mathrm{e}^{+i(\vec{k}-\vec{q})\cdot\vec{R}_{0\nu}^{(m)}}\end{aligned} \tag{C.36}$$

It is evident that only the function $\hat{\tilde{W}}_1(x)$ depends on phonon operators. Hence we actually have to be able to calculate just the expressions

$$\left\langle T\left\{ \hat{\tilde{W}}_1^{(m)}(x_1)\ldots\hat{\tilde{W}}_1^{(m)}(x_n)\right\}\right\rangle_{\text{Ph}} \tag{C.37}$$

When calculating this expression, using Wick's theorem, we can represent it as a sum of the products of all possible two-operator pairings. In this case for each pairing there is an average of two operators:

$$\begin{aligned}&\left\langle T_\tau\left\{ \hat{\tilde{W}}_1^{(m)}(x_1)\,\hat{\tilde{W}}_1^{(m)}(x_2)\right\}\right\rangle_{\text{Ph}}\\ &= \frac{(-i)^2}{v_0^2}\sum_{\vec{q}_1}\sum_{\vec{k}_1\lambda_1}\sum_{\vec{q}_2}\sum_{\vec{k}_2\lambda_2}\left(\frac{\hbar}{4N^2M^2\omega_{\lambda_1}(\vec{k}_1)\omega_{\lambda_2}(\vec{k}_2)}\right)^{1/2}(\vec{q}_1\cdot\vec{\varepsilon}\,(\vec{k}_1\,\lambda_1))\cdot(\vec{q}_2\cdot\vec{\varepsilon}\,(\vec{k}_2\,\lambda_2))\\ &\quad\times\tilde{w}(\vec{q}_1)\cdot\tilde{w}(\vec{q}_2)\,\mathrm{e}^{i\vec{q}_1\cdot\vec{r}_1}\cdot\mathrm{e}^{i\vec{q}_2\cdot\vec{r}_2}\frac{1}{N^2}\sum_{\nu_1}\sum_{\nu_2}\mathrm{e}^{-i(\vec{k}_1+\vec{q}_1)\cdot\vec{R}_{0\nu_1}^{(m)}}\cdot\mathrm{e}^{-i(\vec{k}_2+\vec{q}_2)\cdot\vec{R}_{0\nu_2}^{(m)}}\\ &\quad\times\left\langle T_\tau\left\{\left[\hat{C}_{-\vec{k}_1\lambda_1}(\tau_1)+\hat{C}^{+}_{\vec{k}_1\lambda_1}(\tau_1)\right]\left[\hat{C}_{-\vec{k}_2\lambda_2}(\tau_2)+\hat{C}^{+}_{\vec{k}_2\,\lambda_2}(\tau_2)\right]\right\}\right\rangle_{\text{Ph}}\end{aligned} \tag{C.38}$$

Because

$$\begin{aligned}&\left\langle T_\tau\left\{\left[\hat{C}_{-\vec{k}_1\lambda_1}(\tau_1)+\hat{C}^{+}_{\vec{k}_1\lambda_1}(\tau_1)\right]\left[\hat{C}_{-\vec{k}_2\lambda_2}(\tau_2)+\hat{C}^{+}_{\vec{k}_2\,\lambda_2}(\tau_2)\right]\right\}\right\rangle_{\text{Ph}}\\ &\quad=\delta_{\lambda_1\lambda_2}\,\delta_{\vec{k}_1;-\vec{k}_2}\{[n(\omega_{\lambda_1}(\vec{k}_1))+1]\exp\left(-\hbar\omega_{\lambda_1}(\vec{k}_1)(\tau_1-\tau_2)\right)\\ &\qquad+n(\omega_{\lambda_1}(\vec{k}_1))\exp\left(\hbar\omega_{\lambda_1}(\vec{k}_1)(\tau_1-\tau_2)\right]\}\end{aligned} \tag{C.39}$$

so

$$\left\langle T_\tau\left\{ \hat{\tilde{W}}_1^{(m)}(x_1)\hat{\tilde{W}}_1^{(m)}(x_2)\right\}\right\rangle_{\text{Ph}}$$
$$= -\frac{1}{v_0^2}\sum_{\vec{q}_1}\sum_{\vec{q}_2}\sum_{\vec{k}\lambda}\frac{\hbar}{2NM\omega_\lambda(\vec{k})}\cdot(\vec{q}_1\vec{\varepsilon}(\vec{k}\lambda))\cdot(\vec{q}_2\vec{\varepsilon}(\vec{k}\lambda))\,\tilde{w}(\vec{q}_1)\,\tilde{w}(\vec{q}_2)\,\mathrm{e}^{i\vec{q}_1\cdot\vec{r}_1}\,\mathrm{e}^{i\vec{q}_2\cdot\vec{r}_2}$$
$$\times\{[n(\omega_\lambda(\vec{k}))+1]\exp[-\hbar\omega_\lambda(\vec{k})(\tau_1-\tau_2)]\exp[\hbar\omega_\lambda(\vec{k})(\tau_1-\tau_2 t)]\}$$
$$\times\frac{1}{N^2}\sum_{\nu_1}\sum_{\nu_2}\mathrm{e}^{-i(\vec{k}+\vec{q}_1)\cdot\vec{R}_{0\nu_1}^{(m)}}\cdot\mathrm{e}^{i(\vec{k}-\vec{q}_2)\cdot\vec{R}_{0\nu_2}^{(m)}}$$

Here it is taken into account that $\omega_\lambda(\vec{k})=\omega_\lambda(-\vec{k})$; $\vec{\varepsilon}(\vec{k}\lambda)=\vec{\varepsilon}(-\vec{k}\lambda)$; $\tilde{w}(\vec{q})=\tilde{w}^*(-\vec{q})$ and

$$n_\lambda(\omega_\lambda(\vec{k}))=\langle\hat{C}^+_{\vec{k}\lambda}\hat{C}_{\vec{k}\lambda}\rangle=\left\{\exp\left(\frac{\hbar\omega_\lambda(\vec{k})}{k_\text{B}T}\right)-1\right\}^{-1}\tag{C.41}$$

And, finally, preparatory to stating the graphical rules of calculation of the complete electron Green's function $G(x-x')$, we calculate the following expressions:

$$\overline{\left\langle T_\tau\left\{ \hat{\tilde{W}}_0^{(m)}(x_1)\hat{\tilde{W}}_0^{(m)}(x_2)\right\}\right\rangle_{\text{Ph}}}=\overline{\hat{\tilde{W}}_0^{(m)}(r_1)\,\hat{\tilde{W}}_0^{(m)}(r_2)}$$
$$=\frac{1}{v_0^2}\sum_{\vec{q}_1}\sum_{\vec{q}_2}\mathrm{e}^{i\vec{q}_1\cdot\vec{r}_1}\,\mathrm{e}^{i\vec{q}_2\cdot\vec{r}_2}\,\tilde{w}(\vec{q}_1)\tilde{w}(\vec{q}_2)\frac{1}{N^2}\overline{\sum_{\nu_1\nu_2}\mathrm{e}^{-i\vec{q}_1\cdot\vec{R}_{0\nu_1}^{(m)}-i\vec{q}_2\cdot\vec{R}_{0\nu_2}^{(m)}}}$$
$$=\frac{1}{V}\sum_{\vec{q}}\mathrm{e}^{i\vec{q}\cdot(\vec{r}_1-\vec{r}_2)}\frac{|\tilde{w}(\vec{q})|^2}{v_0}S_0(\vec{q})\tag{C.42}$$

$$\overline{\left\langle T_\tau\left\{ \hat{\tilde{W}}_1^{(m)}\cdot(x_1)\hat{\tilde{W}}_1^{(m)}(x_2)\right\}\right\rangle_{\text{Ph}}}=\frac{1}{V^2}\sum_{\vec{q}}\sum_{\vec{k}\lambda}\frac{\hbar}{2M\omega_\lambda(\vec{k})}\times[\vec{q}\cdot\vec{\varepsilon}(\vec{k}\lambda)]^2|\tilde{w}(\vec{q})|^2\,\mathrm{e}^{i\vec{q}\cdot(\vec{r}_1-\vec{r}_2)}$$
$$\times\{[n(\omega_\lambda(\vec{k}))+1]\,\mathrm{e}^{-\hbar\omega_\lambda(\vec{k})(\tau_1-\tau_2)}+n(\omega_\lambda(\vec{k}))\,\mathrm{e}^{\hbar\omega_\lambda(\vec{k})\cdot(\tau_1-\tau_2)}\}\cdot S_0(\vec{k}+\vec{q})$$
$$=\frac{1}{V^2}\sum_{\vec{q}}\sum_{\vec{k}\lambda}\frac{\hbar}{2M\omega_\lambda(\vec{k})}[\vec{q}\cdot\vec{\varepsilon}(\vec{k}\lambda)]^2|\tilde{w}(\vec{q})|^2\,\mathrm{e}^{i\vec{q}\cdot(\vec{r}_1-\vec{r}_2)}\cdot S_0(\vec{k}+\vec{q})$$
$$\times\frac{1}{\beta}\sum_n\left\{\frac{1}{\hbar\omega_\lambda(\vec{k})-i\hbar\omega_n}+\frac{1}{\hbar\omega_\lambda(\vec{k})+i\hbar\omega_n}\right\}\mathrm{e}^{-i\hbar\omega_n(\tau_1-\tau_2)}$$
$$=\frac{1}{V}\sum_{\vec{q}}\frac{1}{\beta}\sum_n\int_{-\infty}^{+\infty}\frac{\mathrm{d}z}{2\pi}\frac{b(\vec{q},z)}{z-i\hbar\omega_n}\,\mathrm{e}^{i\vec{q}\cdot(\vec{r}_1-\vec{r}_2)-i\hbar\omega_n(\tau_1-\tau_2)}\tag{C.43}$$

where

$$b(\vec{q},z)=2\pi\frac{|\tilde{w}(\vec{q})|^2}{V}\sum_{\vec{k}\lambda}\frac{\hbar[\vec{q}\cdot\vec{\varepsilon}(\vec{k}\lambda)]^2\,S_0(\vec{k}+\vec{q})}{2M\omega_\lambda(\vec{k})}[\delta(z-\hbar\omega_\lambda(\vec{k}))-\delta(z+\hbar\omega_\lambda(\vec{k}))]\tag{C.44}$$

Here allowance is made for

$$\overline{\frac{1}{N}\sum_{\nu_1\nu_2} e^{-i\vec{q}_1\cdot\vec{R}^{(m)}_{0\nu_1}}\cdot e^{-i\vec{q}_2\cdot\vec{R}^{(m)}_{0\nu_2}}} = \delta_{\vec{q}_1,-\vec{q}_2}\cdot S_0(\vec{q}) \tag{C.45}$$

where $S_0(\vec{q}) = \frac{(2\pi)^3}{v_0}\delta(\vec{q}) + a(\vec{q})$ is again the geometrical structure factor of the system, while

$$n(\omega)\,e^{\hbar\omega\tau} = \frac{1}{\beta}\sum_n \frac{e^{-i\hbar\omega_n\tau}}{\hbar\omega + i\hbar\omega_n} \quad \text{and } \omega_n = \frac{2n\pi}{\beta\hbar} \tag{C.46}$$

Now let us state the rules of calculation of *the corrections of arbitrary order* to the complete electron Green's function G, matching the Feynman diagrams to analytical expressions in the series (C.10):

1) Every analytical expression is depicted by a topologically inequivalent diagram with $2n$ *internal* and any number of *external* vertices; the diagram has two solid external ends at the points x and x'. Only *connected diagrams* are considered.
2) Any *internal vertex* is represented by $\otimes$ and matched by a matrix τ_3; two *solid* lines and a *dotted* one converge to/from the vertex:

   $\Rightarrow \otimes \Rightarrow$
   $\vdots$

3) An *external vertex* is represented by a cross $\times$. Matched by $\hat{\hat{W}}_0(x)\tau_3$, this vertex is a meeting point for two solid lines:

   $\Rightarrow \times \Rightarrow$

4) The electronic Green's function $G_{0e}\,(x - x')$ corresponds to a solid (double) line with arrow:

   $\Rightarrow$

5) Every dotted line is matched by the function $\langle T_\tau\{\hat{\hat{W}}^{(m)}_1(x_1)\hat{\hat{W}}^{(m)}_1(x_2)\}\rangle_{\text{Ph}}$:

   $---$

6) An integration is carried out with respect to coordinates of all internal vertices $\otimes$ and external ones, $\times$.
7) Upon the diagrams' development, a configurational averaging according to Eq. (C.30) is carried out.
8) The *complete electron Green's* function G is matched by a *heavy* solid line:

   ➡

   Then every summand in the expression (C.10) is matched by the following diagrams:

$$G_{0e} \Longrightarrow \tag{C.47}$$

$$G_{0e}\overline{\left\langle\left\{\tau_3\hat{\hat{W}}^{(m)}\right\}\right\rangle^{(m)}_{\text{Ph}}}G_{0e} = 0 \tag{C.48}$$

$$G_{0e}\overline{\left\langle T_\tau\left\{\tau_3\hat{\hat{W}}^{(m)}G_{0e}\tau_3\hat{\hat{W}}^{(m)}\right\}\right\rangle^{(m)}_{\text{Ph}}}G_{0e} = \tag{C.49}$$

Here the elliptical box denotes the cumulant average $\{x_1 x_2\}^c$ over the ionic configurations, to second order. To next highest order we have diagrams 3 to 6:

$$G_{0e}\overline{\left\langle T_\tau\left\{\tau_3 \hat{\hat{W}}^{(m)} G_{0e}\tau_3 \hat{\hat{W}}^{(m)} G_{0e}\tau_3 \hat{\hat{W}}^{(m)}\right\}\right\rangle_{\mathrm{Ph}}^{(m)}} G_{0e} = \qquad \text{(C.50)}$$

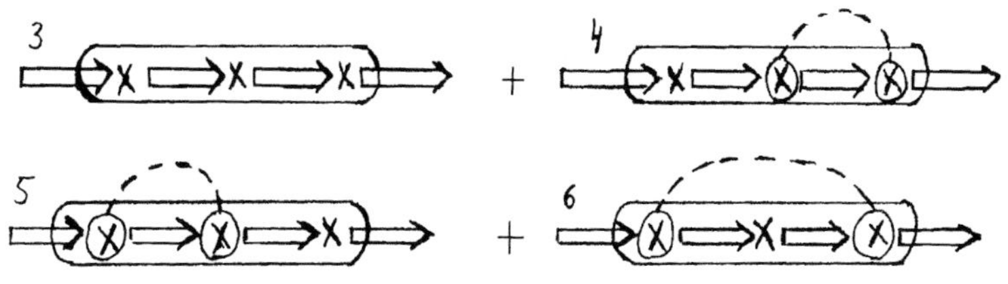

In fourth order follow 40 additional diagrams (diagrams 7 to 46):

$$G_{0e}\overline{\left\langle T_\tau\left\{\tau_3 \hat{\hat{W}}^{(m)} G_{0e}\tau_3 \hat{\hat{W}}^{(m)} G_{0e}\tau_3 \hat{\hat{W}}^{(m)} G_{0e}\tau_3 \hat{\hat{W}}^{(m)}\right\}\right\rangle_{\mathrm{Ph}}^{(m)}} G_{0e} = (\text{diagrams 7 to 46}) =$$

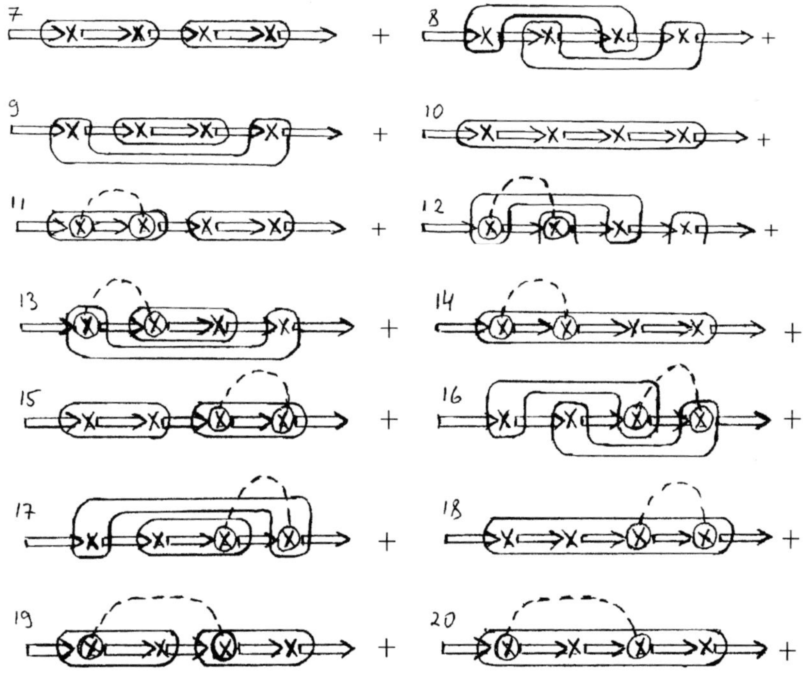

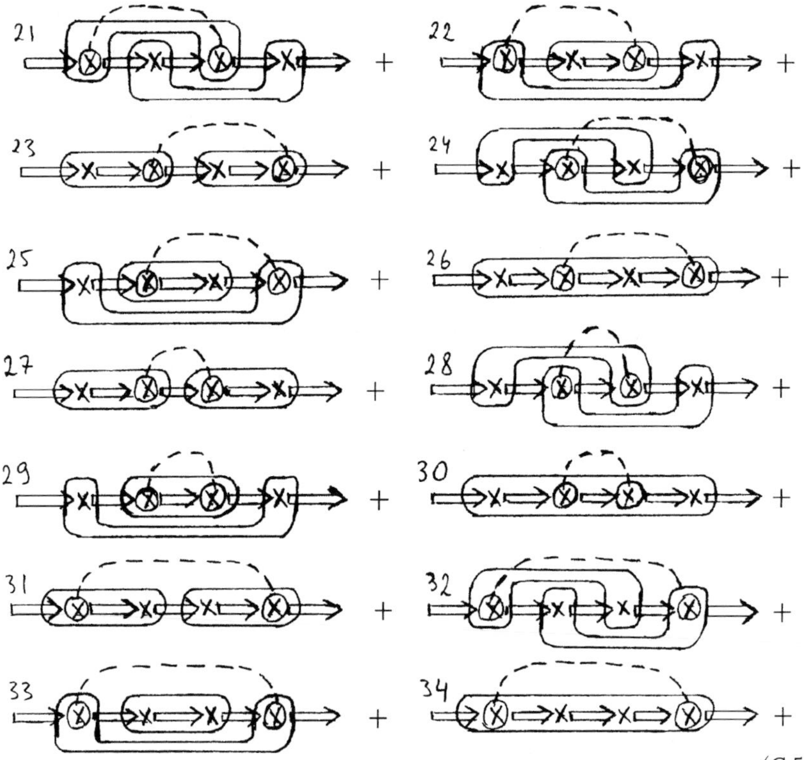
21
22
23
24
25
26
27
28
29
30
31
32
33
34

(C.51)

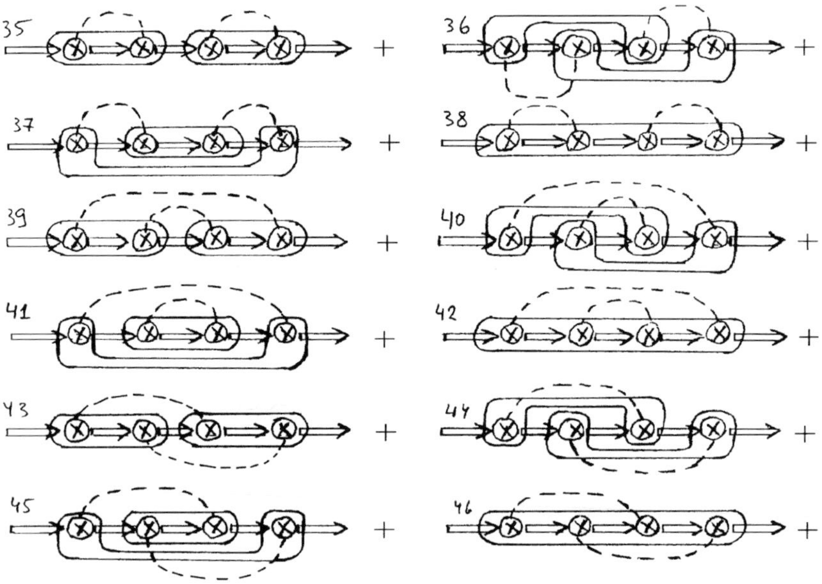
35
36
37
38
39
40
41
42
43
44
45
46

Among the diagrams of (C.51), as in all diagrammatic expansions for the electronic Green's function written down before in this appendix, there are two distinct classes: *tightly-* and *loosely-*connected ones. A diagram is said to be *loosely-connected* when it can be separated into two *disconnected* parts by a cut through *only one* electron line, leaving the vertices intact. A loosely-connected diagram may be *partitioned* into tightly-connected diagrams in this way. In the expression (C.51), for example, the diagrams 7, 11, 15 and 35 are loosely-connected. The remaining diagrams are the tightly-connected ones of the diagrammatic order considered in (C.51).

Let us denote the sum of all tightly-connected diagrams (not including external electronic lines) by $\sum'$, the *electronic self-energy*. To be precise, $\sum'$ represents the *part* of the electron's self-energy resulting from the electron's interaction with quasi-phonons and randomly positioned ions, whereas the contribution from the electron-electron interaction itself, $\hat{\Psi}(x')$, will be considered later.

The diagrammatic expression for $\sum'$ is given in Eq. (C.52).

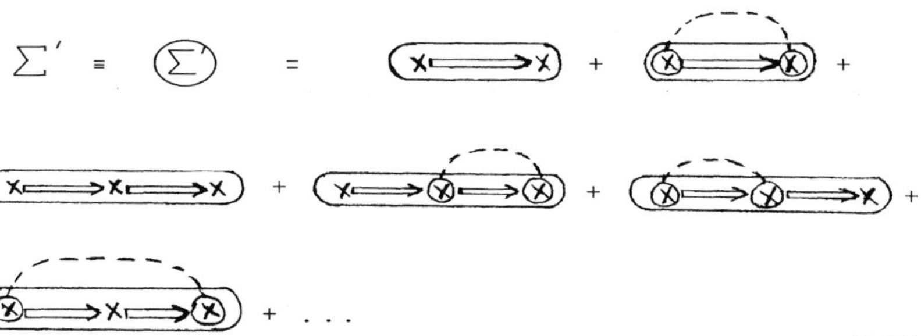

(C.52)

Grouping all tightly-connected diagrams together in the series (C.10) we can rewrite it as the 'Dyson equation' (C.53) for the electronic Green's function G:

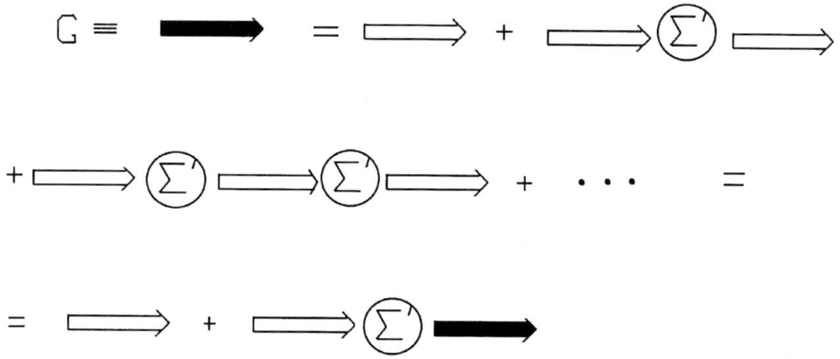

(C.53)

To this diagrammatic equation, where the 'black' electron line G is the 'unknown quantity', corresponds the (Nambu-) matrix operator equation

$$G = G_{0e} + G_{0e} \sum{}' G \tag{C.54}$$

where the electron Green's function G_{0e}, which corresponds to the 'unfilled' electron line in (C.53), satisfies the following Dyson equation:

$$G_{0e} = G_0 + G_{0e} \sum_C{}' G_{0e} \tag{C.55}$$

Here G_0 is the electron Green's function of the ideal electron gas in a homogeneous neutralizing background, and $\sum_C'$, as already mentioned, is the electronic self-energy of the interacting electron gas.

From Eqs (C.54) and (C.55) the reciprocal Green functions G^{-1} and G_0^{-1} are readily obtained:

$$\begin{cases} G_{0e}^{-1} = G_0^{-1} - \sum_C{}' \\ G^{-1} = G_{0e}^{-1} - \sum{}' = G_0^{-1} - \sum \end{cases} \tag{C.56}$$

where

$$\sum = \sum_C{}' + \sum{}' \tag{C.57}$$

is the self-energy of the full electron Green's function. Thus, in an operator (Nambu matrix) form we have

$$G = G_0 + G_0 \sum G \tag{C.58}$$

In the coordinate representation this equation becomes:

$$G(x - x') = G_0(x - x') + \int dx_1 \int dx_2 G_0(x - x_1) \sum(x_1 - x_2) G(x_2 - x') \tag{C.59}$$

Going to the Fourier transforms

$$G(x - x') = \frac{1}{\beta} \sum_n \int \frac{d^3k}{(2\pi)^3} e^{i\vec{k}\cdot(\vec{r}-\vec{r}') - i\hbar\omega_n(\tau-\tau')} G(\vec{k}, \omega_n) \tag{C.60}$$

$$G_0(x - x') = \frac{1}{\beta} \sum_n \int \frac{d^3k}{(2\pi)^3} e^{i\vec{k}\cdot(\vec{r}-\vec{r}') - i\hbar\omega_n(\tau-\tau')} G_0(\vec{k}, \omega_n) \tag{C.61}$$

$$\sum(x - x') = \frac{1}{\beta} \sum_n \int \frac{d^3k}{(2\pi)^3} e^{i\vec{k}\cdot(\vec{r}-\vec{r}') - i\hbar\omega_n(\tau-\tau')} \sum(\vec{k}, \omega_n) \tag{C.62}$$

where ω_n are the Fermion Matsubara frequencies, i.e. $\hbar\omega_n = (2n+1)\frac{\pi}{\beta}$, we have the function $G(\vec{k}, \omega_n)$ satisfying the following algebraic *Dyson equation*:

$$G(\vec{k}, \omega_n) = G_0(\vec{k}, \omega_n) + G_0(\vec{k}, \omega_n) \sum(\vec{k}, \omega_n) G(\vec{k}, \omega_n) \tag{C.63}$$

Hence

$$G(\vec{k}, \omega_n) = \frac{1}{G_0^{-1}(\vec{k}, \omega_n) - \sum(\vec{k}, \omega_n)} \tag{C.64}$$

The quantities in this expression are each two-by-two matrices and therefore can be expanded in terms of the Pauli matrices τ_i, including the unit matrix τ_0, i.e.

$$\begin{cases} G_0^{-1}(\vec{k}, \omega_n) = i\hbar\omega_n\tau_0 - \left(\dfrac{\hbar^2 k^2}{2m} - \mu_e\right)\tau_3 \\ \sum(\vec{k}, \omega_n) = [1 - Z(\vec{k}, \omega_n)]\, i\hbar\omega_n\tau_0 + \chi(\vec{k}, \omega_n)\tau_3 + \varphi(\vec{k}, \omega_n)\tau_1 \end{cases} \tag{C.65}$$

This expansion does not include the matrix τ_2, because its coefficient (in the absence of a magnetic field) can always be made zero through an appropriate choice of phase. It is now easy to obtain – substituting (C.65) into (C.64) – the expression for $G(\vec{k}, \omega_n)$ in terms of the expansion coefficients Z, χ and φ:

$$G(\vec{k}, \omega_n) = \frac{G(\vec{k}, \omega_n) i\hbar\omega_n\tau_0 + \varepsilon(\vec{k}, \omega_n) \cdot \tau_3 + \varphi(\vec{k}, \omega_n)\tau_1}{[Z(\vec{k}, \omega_n) i\hbar\omega_n]^2 - \varphi^2(\vec{k}, \omega_n) - \varepsilon^2(\vec{k}, \omega_n)} \tag{C.66}$$

where

$$\varepsilon(\vec{k}, \omega_n) = \frac{\hbar^2 k^2}{2m} - \mu_e + \chi(\vec{k}, \omega_n) \tag{C.67}$$

An explicit form of $\sum(x - x')$ is found by summing up the diagrams of (C.68).

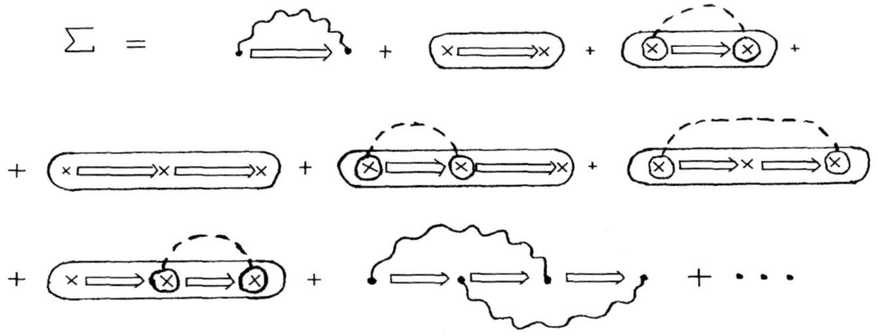

(C.68)

Here, as before, the *wavy line* represents the Coulomb interaction between different electrons, whereas the *dashed line* represents the interaction with quasi-phonons, and the *crosses* represent the positional disorder. The elliptical blocks represent the cumulant level with respect to the disorder average.

It is evident from Eq. (C.68) that the higher-order graphs (higher than second order) make either more complex Greenian lines or more complex vertices, or both. The electron-phonon interaction in a normal metal has been studied by Migdal (1958), and the electron-ion one by Abrikosov and Gorkov (1958, 1960). It was shown that the *corrections* of a vertex with phonon lines are of the order of the square root of the ratio between the Debye energy ($\hbar\omega_D$) and the Fermi energy (E_F), which is proportional to $\sqrt{\frac{m}{M}}$, where m and M are the respective masses of an electron and ion. The vertices incorporating many-particle correlations between ions gives rise to corrections of the order of $\frac{a}{l}$, where a is the average interatomic distance, and l is the electron's mean-free path determined by electron scattering from randomly positioned ions. Consequently, within such an accuracy, for $\sum$ we can take an infinite sequence of graphs derived from (C.68) by just complicating the electronic Green's lines, and leaving the vertices non-renormalized. Eliashberg (1960), see also Vonsovskij *et al.* (1977), demonstrated that this situation is also valid for superconducting metals.

Thus as $\sum(x - x')$ we can take in the series (C.68) only the diagrams 1, 2 and 3 replacing there the electronic lines $\Rightarrow$ with the corresponding black (filled) lines, see above, so

Σ = 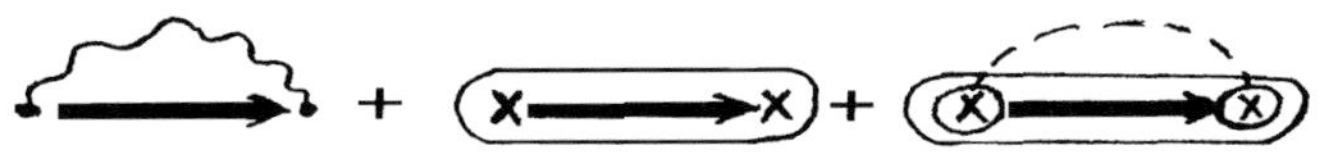 (C.69)

or

$$\sum(x - x') = \sum_{C}(x - x') + \sum_{Scat}(x - x') + \sum_{Ph}(x - x') \tag{C.70}$$

where

$$[\hat{H}_i^{(m)};\ Z_e^{(m)}] \approx 0 \tag{C.71}$$

$$\sum_{Scat}(x - x') = \tau_3 G(x - x')\tau_3 \overline{\hat{\hat{W}}_0^{(m)}(\vec{r})\ \hat{\hat{W}}_0^{(m)}(\vec{r}\,')} \tag{C.72}$$

$$\sum_{Ph}(x - x') = \tau_3 G(x - x')\tau_3 \overline{\left\langle T_\tau\left\{ \hat{\hat{W}}_1^{(m)}(x_1)\ \hat{\hat{W}}_1^{(m)}(x_2)\right\}\right\rangle_{Ph}} \tag{C.73}$$

Using (C.60)–(C.62), (C.42), (C.43) and (C.6) we find the Fourier transform for the self-energy part $\sum$ of the full electron Green's function:

$$\sum(\vec{k},\omega_n)=\sum_{\mathrm{C}}(\vec{k},\omega_n)+\sum_{\mathrm{Ph}}(\vec{k},\omega_n)+\sum_{\mathrm{Scat}}(\vec{k},\omega_n) \tag{C.74}$$

where

$$\sum_{\mathrm{C}}(\vec{k},\omega_n)=\frac{-1}{\beta}\sum_{n_1}\int\frac{d^3k_1}{(2\pi)^3}\tau_3 G(\vec{k}_1,\omega_{n_1})\tau_3 V_e(\vec{k}-\vec{k}_1) \tag{C.75}$$

is the *Coulomb part* of the electron self-energy;

$$\sum_{\mathrm{Ph}}(\vec{k},\omega_n)=\frac{1}{\beta}\sum_{n_1}\int\frac{d^3k_1}{(2\pi)^3}\tau_3 G(\vec{k}_1,\omega_{n_1})\tau_3\int_{-\infty}^{+\infty}\frac{dz}{2\pi}\frac{b(\vec{k}-\vec{k}_1,z)}{z-i\hbar(\omega_n-\omega_{n_1})} \tag{C.76}$$

is the *phonon part* of the electron self-energy;

$$\sum_{\mathrm{Scat}}(\vec{k},\omega_n)=\int\frac{d^3k_1}{(2\pi)^3}\tau_3 G(\vec{k}_1,\omega_{n_1})\tau_3\frac{|\tilde{w}(\vec{k}-\vec{k}_1)|^2}{v_0}S_0(\vec{k}-\vec{k}_1) \tag{C.77}$$

is the part of the self-energy due to *electron scattering from disordered ions.*

Here the substitution $\frac{1}{V}\sum_{\vec{k}}\rightarrow\int\frac{d^3k_1}{(2\pi)^3}$ is made. Because $\tau_3\tau_1\tau_3=-\tau_1$ and $\tau_3\tau_3=1$, we have from (C.66)

$$\tau_3 G(\vec{k},\omega_n)\tau_3=\frac{Z(\vec{k},\omega_n)i\hbar\omega_n\tau_0+\varepsilon(\vec{k},\omega_n)\tau_3-\varphi(\vec{k},\omega_n)\tau_1}{[Z(\vec{k},\omega_n)i\hbar\omega_n]^2-\varphi^2(\vec{k},\omega_n)-\varepsilon^2(\vec{k},\omega_n)} \tag{C.78}$$

It is clear from this expression that $\sum(\vec{k},\omega_n)$ actually involves a term proportional to τ_1, which is attributed to the development of a superconducting state.

To sum over frequencies in the expressions (C.75)–(C.77), the spectral representation of the electron Green's function is used for convenience:

$$G(\vec{k},\omega_n)=\frac{1}{2\pi}\int_{-\infty}^{+\infty}dz'\frac{a(\vec{k},z')}{i\hbar\omega_n-z'} \tag{C.79}$$

Upon substitution of this expression into Eqs (C.75) and (C.76), the sums over frequencies can be readily calculated:

$$\frac{1}{\beta}\sum_{n_1}\frac{1}{i\hbar\omega_n-z'}\cdot\frac{1}{z-i\hbar(\omega_n-\omega_{n_1})}=\frac{1}{2}\frac{\tanh\left(\frac{z'\beta}{2}\right)+\coth\left(\frac{z\beta}{2}\right)}{i\hbar\omega_n-z-z'} \tag{C.80}$$

$$\frac{1}{\beta}\sum_{n_1}\frac{1}{i\hbar\omega_n-z'}=-\frac{1}{2}\tanh\left(\frac{z'\beta}{2}\right) \tag{C.81}$$

Now let us take into consideration that the spectral density $a(\vec{k}, z)$ of the diagrammatic finite-temperature Green's function is related to the imaginary part of the *retarded* Green's function $G_R(\vec{k}, z)$ by

$$a(\vec{k}, z') = -2 \text{ Im } G_R(\vec{k}, z)$$

If in Eqs (C.75)–(C.77) the analytic continuation from the imaginary axis to the real axis is carried out by substitution $i\omega_n \to \omega + i\delta$, we arrive at equations for the self-energy of the *retarded* Green's function.

Finally, instead of expressions (C.75)–(C.77), we have

$$\sum_{\text{Ph}}(\vec{k}, \omega) = -\int \frac{d^3k'}{(2\pi)^3} \int_{-\infty}^{+\infty} \frac{dz}{2\pi} \int_{-\infty}^{+\infty} \frac{dz'}{2\pi} b(\vec{k} - \vec{k}', z)\tau_3 \text{ Im } G_R(\vec{k}', z')\tau_3 \frac{\tanh\left(\frac{\beta z'}{2}\right) + \coth\left(\frac{\beta z}{2}\right)}{\hbar\omega - z - z' + i\delta} \tag{C.82}$$

$$\sum_{\text{C}}(\vec{k}, \omega) = -\int \frac{d^3k'}{(2\pi)^3} V_e(\vec{k} - \vec{k}') \int_{-\infty}^{+\infty} \frac{dz'}{2\pi} \tau_3 \text{ Im } G_R(\vec{k}', z')\tau_3 \tanh\left(\frac{\beta z'}{2}\right) \tag{C.83}$$

$$\sum_{\text{Scat}}(\vec{k}, \omega) = \int \frac{d^3k'}{(2\pi)^3} \frac{|\tilde{w}(\vec{k} - \vec{k}_1)|^2}{\nu_0} S_0(\vec{k} - \vec{k}')\tau_3 G_R(\vec{k}', \omega)\tau_3 \tag{C.84}$$

To these equations the relations (C.78) and (C.74) should be added, analytically continued to the real axis,

$$\tau_3 G_R(\vec{k}, \omega)\tau_3 = \frac{Z(\vec{k}, \omega)\hbar\omega\tau_0 + \varepsilon(\vec{k}, \omega)\tau_3 - \varphi(\vec{k}, \omega)\tau_1}{|Z(\vec{k}, \omega)\hbar\omega|^2 - \varphi^2(\vec{k}, \omega) - \bar{\varepsilon}^2(\vec{k}, \omega)} \tag{C.85}$$

$$\begin{aligned} \sum(\vec{k}, \omega) &= \sum\nolimits_{\text{Ph}}(\vec{k}, \omega) + \sum\nolimits_{\text{C}}(\vec{k}, \omega) + \sum\nolimits_{\text{Scat}}(\vec{k}, \omega) \\ &= [1 - Z(\vec{k}, \omega)]\hbar\omega\tau_0 + \chi(\vec{k}, \omega)\tau_3 + \varphi(\vec{k}, \omega)\tau_1 \end{aligned} \tag{C.86}$$

Equations (C.82)–(C.86) constitute a single-matrix equation determining finally the three quantities $Z(\vec{k}, \omega), \varphi(\vec{k}, \omega)$ and $\chi(\vec{k}, \omega)$.

Appendix D: Simplification of the Eliashberg Equation

The equations determining the self-energy part of the retarded Green's function have the following form:

$$\sum(\vec{k},\omega) = \sum_{\text{Ph}}(\vec{k},\omega) + \sum_{\text{C}}(\vec{k},\omega) + \sum_{\text{Scat}}(\vec{k},\omega)$$
$$= [1 - Z(\vec{k},\omega)]\hbar\omega\tau_0 + \chi(\vec{k},\omega)\tau_3 + \varphi(\vec{k},\omega)\tau_1 \tag{D.1}$$

$$\sum_{\text{Ph}}(\vec{k},\omega) = -\int \frac{d^3k'}{(2\pi)^3} \int_{-\infty}^{+\infty} \frac{dz}{2\pi} \int_{-\infty}^{+\infty} \frac{dz'}{2\pi} b(\vec{k}-\vec{k}',z)\tau_3 \mathrm{Im}G_R(\vec{k}',z')\tau_3$$
$$\times \frac{\tanh\left(\dfrac{\beta z'}{2}\right) + \coth\left(\dfrac{\beta z}{2}\right)}{\hbar\omega - z - z' + i\delta} \tag{D.2}$$

$$\sum_{\text{C}}(\vec{k},\omega) = -\int \frac{d^3k'}{(2\pi)^3} V_e(\vec{k}-\vec{k}') \int_{-\infty}^{+\infty} \frac{dz'}{2\pi} \tau_3 \mathrm{Im}G_R(\vec{k},z')\tau_3 \tanh\left(\frac{\beta z'}{2}\right) \tag{D.3}$$

$$\sum_{\text{Scat}}(\vec{k},\omega) = \int \frac{d^3k'}{(2\pi)^3} \frac{|\tilde{w}(\vec{k}-\vec{k}_1)|^2}{v_0} S_0(\vec{k}-\vec{k}')\tau_3 G_R(\vec{k}',\omega)\tau_3 \tag{D.4}$$

$$\tau_3 G_R(\vec{k},\omega_n)\tau_3 = \frac{Z(\vec{k},\omega_n)i\hbar\omega_n\tau_0 + \varepsilon(\vec{k},\omega_n)\tau_3 - \varphi(\vec{k},\omega_n)\tau_1}{[Z(\vec{k},\omega_n)i\hbar\omega_n]^2 - \varphi^2(\vec{k},\omega_n) - \varepsilon^2(\vec{k},\omega_n)} b(\vec{q},z) \tag{D.5}$$

$$b(\vec{q},z) = \frac{|\tilde{w}(\vec{q})|^2}{V} \sum_{\vec{k}\lambda} \frac{\hbar[\vec{q}\cdot\vec{\varepsilon}(\vec{k}\lambda)]^2 S_0(\vec{k}+\vec{q})}{2M\omega_\lambda(\vec{k})} [\delta(z - \hbar\omega_\lambda(\vec{k}))$$
$$- \delta(z + \hbar\omega_\lambda(\vec{k}))] \cdot 2\pi$$

These relationships comprise the matrix equation for the unknown functions $Z(\vec{k},\omega)$ and $\varphi(\vec{k},\omega)$.

Calculation of the Phonon Part $\sum_{\text{Ph}}(\vec{k},\omega)$

Because the electron-phonon interaction significantly affects the electronic states near the Fermi surface, our concern is with $\sum_{\text{Ph}}(\vec{k},\omega)$ only for momentum $\vec{k}$ in the vicinity of k_F. The denominator in Eq. (D.5) – which is involved in Eq. (5.1.120) – varies with momentum $\vec{k}'$ at least as $\xi_{\vec{k}'}^2$. As a consequence, the integrand in (D.2) demonstrates a sharp peak at $\xi_{\vec{k}'} = 0$, i.e. for $\vec{k}'$ at the Fermi surface. What this means is that in Eq. (D.2) the dependence of $Z(\vec{k},\omega)$ and $\varphi(\vec{k},\omega)$ on the momentum magnitude can be treated as weak. Hence the integration in (D.2) with respect to the *magnitude* of $\vec{k}'$ can be carried out immediately in the following way:

$$\int \frac{d^3k'}{(2\pi)^3} f(\vec{k}') = \int_0^\infty d\varepsilon'_{\vec{k}} \int \frac{k' d\Omega' \cdot f(\vec{k}')}{(2\pi)^3 \dfrac{\hbar^2}{m}} \approx \int_{-\mu_e}^{\infty} d\xi'_{\vec{k}} \int \frac{k_F d\Omega'}{(2\pi)^3 \dfrac{\hbar^2}{m}} f(k_F,\Omega',\xi)$$

$$\approx \int_{-\infty}^{\infty} d\xi_{\vec{k}'} \int_{S_F} \frac{d^2k'}{\hbar v_{\vec{k}'}} f(k_F, \Omega', \xi) \tag{D.6}$$

$(f(k', \Omega', \xi_{\vec{k}})$ peaks at $\xi_{\vec{k}} = \varepsilon(\vec{k}) - \mu_e = 0.)$

Here $v_{\vec{k}'} = \dfrac{\hbar k_F}{m}$ is the velocity at the Fermi surface S_F, $d^2k' = k_F^2 d\Omega' \dfrac{1}{(2\pi)^3}$ being its area element. The integration over ξ can be performed by extending the integration limits to the whole interval from $-\infty$ to $+\infty$, because the integrand falls off steeply for large ξ. Thus within the above approximations we get the following integral in the expression (D.2):

$$\begin{aligned}\int_{-\infty}^{+\infty} \frac{dx}{x^2 - a^2} &= 2\int_{0}^{+\infty} \frac{dx}{x^2 - a^2} = \frac{1}{a}\ln\frac{x-a}{x+a}\bigg|_0^{\infty} = \frac{\operatorname{sign}(a)}{a}\ln\frac{x-|a|}{x+|a|}\bigg|_0^{\infty} \\ &= -\frac{\operatorname{sign}(a)}{a}\ln(-1) = -i\frac{\pi}{a}\operatorname{sign}(a)\end{aligned} \tag{D.7}$$

This integral being performed, Eq. (D.2) remains an integral equation with respect to the *direction* of the momentum at the Fermi surface S_F.

With this in mind let us introduce the *directional average*

$$\sum_{\text{Ph}}(\omega) := \int_{S_F} \frac{d^2k}{\hbar v_{\vec{k}}} \sum_{\text{Ph}}(\vec{k}, \omega) \bigg/ \int_{S_F} \frac{d^2k}{\hbar v_{\vec{k}}} \tag{D.8}$$

Then Eq. (D.2) becomes

$$\begin{aligned}\sum_{\text{Ph}}(\omega) = &\int_{-\infty}^{+\infty} \frac{dz'}{2\pi} \int_{-\infty}^{+\infty} \frac{dz}{2\pi} \left(\int_{S_F} \frac{d^2\vec{k}}{\hbar v_{\vec{k}}}\right)^{-1} \int_{S_F} \frac{d^2\vec{k}}{\hbar v_{\vec{k}}} \int_{S_F} \frac{d^2\vec{k}'}{\hbar v_{\vec{k}'}} b(\vec{k} - \vec{k}', z) \cdot \operatorname{sign} z' \cdot \pi \\ &\otimes \operatorname{Re}\frac{Z(\vec{k}', z')z'\tau_0 - \varphi(\vec{k}', z')\tau_1}{\sqrt{z'^2 Z(\vec{k}', z') - \varphi^2(\vec{k}', z')}} \cdot \frac{\tanh\left(\frac{\beta z'}{2}\right) + \coth\left(\frac{\beta z}{2}\right)}{\hbar\omega - z - z' + i\delta}\end{aligned}$$

All momenta in (D.9) are lying on the Fermi surface.

We have derived Eq. (D.9) within the following assumptions:

$$\begin{aligned}&-\int_{-\infty}^{+\infty} d\xi \operatorname{Im}\frac{Z(\vec{k}, \omega)\hbar\omega\tau_0 + \xi \cdot \tau_3 - \varphi(\vec{k}, \omega)\tau_1}{[Z(\vec{k}, \omega)\hbar\omega]^2 - \varphi^2(\vec{k}, \omega) - \xi^2} \\ &\approx -\operatorname{Im}\int_{-\infty}^{+\infty} \frac{Z(\vec{k}, \omega)\hbar\omega\tau_0 - \varphi(\vec{k}, \omega)\tau_1}{[Z(\vec{k}, \omega)\hbar\omega]^2 - \varphi^2(\vec{k}, \omega) - \xi^2} d\xi\end{aligned}$$

$$= \mathrm{Im}\left\{ [Z(\vec{k},\omega)\hbar\omega\tau_0 - \varphi(\vec{k},\omega)\tau_1] \cdot \int_{-\infty}^{+\infty} \frac{d\xi}{\xi^2 - [\hbar^2\omega^2 Z^2(\vec{k},\omega) - \varphi^2(\vec{k},\omega)]} \right\}$$

$$= \mathrm{Im}\left\{ \frac{Z(\vec{k},\omega)\hbar\omega\tau_0 - \varphi(\vec{k},\omega)\tau_1}{\sqrt{\hbar^2\omega^2 Z^2(\vec{k},\omega) - \varphi^2(\vec{k},\omega)}} (-i\pi) \cdot \mathrm{sign}\left[\hbar\omega\sqrt{Z^2(\vec{k},\omega) - \frac{\varphi^2(\vec{k},\omega)}{\hbar\omega^2}} \right] \right\}$$

$$\approx \mathrm{sign}\,\omega \cdot \pi \cdot \mathrm{Re} \frac{Z(\vec{k},\omega)\hbar\omega\tau_0 - \varphi(\vec{k},\omega)\tau_1}{\sqrt{\hbar^2\omega^2 Z^2(\vec{k},\omega) - \varphi^2(\vec{k},\omega)}}$$

because the sign of the expression $\hbar\omega\sqrt{Z^2(\vec{k},\omega) - \dfrac{\varphi^2(\vec{k},\omega)}{\hbar^2\omega^2}}$ at $|\omega| \to \infty$ is determined by $\omega[Z(\vec{k},\omega) \geq 0]$ when $|\omega| \to \infty$.

If the Fermi surface is highly anisotropic, we should take into account the dependence of $Z(\vec{k}',z')$ and $\varphi(\vec{k}',z')$ on the *direction* of the vector $\vec{k}'$. However, if the anisotropy is not too high (this is the case considered), the angular dependence of $Z(\vec{k}',z')$ and $\varphi(\vec{k}',z')$ may be thought of as less pronounced than that of $b(\vec{k}-\vec{k}',z)$, which specifies the electron-phonon interaction together with the phonon spectrum: Then the functions $Z(\vec{k}',z')$ and $\varphi(\vec{k}',z')$ can be replaced by expressions of the type (D.8) *averaged* over the Fermi surface, namely $Z(z')$ and $\varphi(z')$. Within such not severely consistent approximations, Eq. (D.8) becomes an integral equation only with respect to *frequencies*. Now we can write it in the form:

$$\sum_{\mathrm{Ph}}(\omega) = \int_{-\infty}^{+\infty} dz' \int_{-\infty}^{+\infty} \frac{dz}{4\pi} \int_{S_F} \frac{d^2k}{\hbar v_{\vec{k}}} b(\vec{k}-\vec{k}',z) \left(\int_{S_F} \frac{d^2k}{\hbar v_{\vec{k}}} \right)^{-1}$$
$$\times \frac{\tanh\left(\frac{\beta z'}{2}\right) + \coth\left(\frac{\beta z}{2}\right)}{\hbar\omega - z - z' + i\delta} \cdot \mathrm{sign}\,z' \cdot \mathrm{Re}\frac{z' Z(z')\tau_0 - \varphi(z')\tau_1}{\sqrt{z'^2 Z^2(z') - \varphi^2(z')}} \qquad \text{(D.10)}$$

According to Eq. (D.1), and taking into account all previous approximations we have:

$$\sum_{\mathrm{Ph}}(\omega) = [1 - Z_{\mathrm{Ph}}(\omega)]\hbar\omega\tau_0 + \varphi_{\mathrm{Ph}}(\omega) \cdot \tau_1 \qquad \text{(D.11)}$$

Substituting (D.10) into (D.9), and equating the coefficients at the matrices τ_0 and τ_1 in the left- and right-hand sides of the resulting equation, we arrive at the following equations for $Z(\omega)$ and $\varphi(\omega)$:

$$[1 - Z_{\mathrm{Ph}}(\omega)]\hbar\omega = -\int_{-\infty}^{+\infty} dz' K_{\mathrm{Ph}}(z',\omega)\,\mathrm{Re}\frac{z'}{\sqrt{z'^2 - \Delta^2(z')}}\,\mathrm{sign}\,z' \qquad \text{(D.12)}$$

$$Z_{Ph}(\omega)\cdot\Delta_{Ph}(\omega)=\int\limits_{-\infty}^{+\infty}dz'\,K_{Ph}(z',\omega)\,\mathrm{Re}\,\frac{\Delta(z')}{\sqrt{z'^2-\Delta^2(z')}}\,\mathrm{sign}\,z' \tag{D.13}$$

where

$$\Delta_{Ph}(\omega)=\frac{\varphi_{Ph}(\omega)}{Z_{Ph}(\omega)}\quad\text{and}\quad\Delta(\omega)=\frac{\varphi(\omega)}{Z(\omega)}$$

are introduced for $\varphi_{Ph}(\omega)$ and $\varphi(\omega)$ respectively.

The quantity $\Delta(\omega)$ is said to be the *gap function,* as it defines the gap in the superconductor's quasi-particle spectrum.

The kernel $K_{Ph}(z',\omega)$ in the integral equations (D.12) and (D.13) is as follows:

$$\begin{aligned}K_{Ph}(z',\omega)=&-\int\limits_{-\infty}^{+\infty}\frac{dz}{4\pi}\int\limits_{S_F}\frac{d^2k}{\hbar v_{\vec{k}}}\int\limits_{S_F}\frac{d^2k'}{\hbar v_{\vec{k}'}}b(\vec{k}-\vec{k}',z)\frac{\tanh\left(\frac{\beta z'}{2}\right)+\coth\left(\frac{\beta z}{2}\right)}{\hbar\omega-z-z'+i\delta}\\&\times\left(\int\limits_{S_F}\frac{d^2k}{\hbar v_{\vec{k}}}\right)^{-1}=\left(\int\limits_{S_F}\frac{d^2k}{\hbar v_{\vec{k}}}\right)^{-1}\int\limits_{S_F}\frac{d^2k}{\hbar v_{\vec{k}}}\int\frac{d^2k'}{\hbar v_{k'}}\frac{|\tilde{w}(\vec{k}-\vec{k}')|^2}{v_0}\\&\times\sum_{\vec{k}_1\lambda}\frac{\hbar[(\vec{k}-\vec{k}')\vec{\varepsilon}(\vec{k}_1\lambda)]^2}{2NM\omega_\lambda(\vec{k}_1)}S_0(\vec{k}_1+\vec{k}-\vec{k}')\\&\times 2\pi\left\{\int\limits_0^{\infty}\delta(z-\omega_\lambda(\vec{k}_1))\frac{\tanh\left(\frac{\beta z'}{2}\right)+\coth\left(\frac{\beta z}{2}\right)}{z+z'-\hbar\omega-i\delta}dz\right.\\&\left.-\int\limits_0^{\infty}\delta(z+\omega_\lambda(\vec{k}_1))\frac{\tanh\left(\frac{\beta z'}{2}\right)+\coth\left(\frac{\beta z}{2}\right)}{z+z'-\hbar\omega-i\delta}dz\right\}\end{aligned}$$

or

$$\begin{aligned}K_{Ph}(z',\omega)=&\int\limits_0^{\infty}dz\cdot\alpha^2(z)F(z)\\&\times\frac{1}{2}\left\{\frac{\tanh\left(\frac{\beta z'}{2}\right)+\coth\left(\frac{\beta z}{2}\right)}{z+z'-\hbar\omega-i\delta}-\frac{\tanh\left(\frac{\beta z'}{2}\right)-\coth\left(\frac{\beta z}{2}\right)}{z'-z-\hbar\omega-i\delta}\right\}\end{aligned} \tag{D.14}$$

Here the following abbreviation is introduced:

$$\alpha^2(z)F(z) = \left(\int\limits_{S_F}\frac{d^2k}{\hbar v_{\vec{k}}}\right)^{-1}\int\limits_{S_F}\frac{d^2k}{\hbar v_{\vec{k}}}\int\frac{d^2k'}{\hbar v_{\vec{k}'}}\frac{|\tilde{w}(\vec{k}-\vec{k}')|^2}{v_0}\sum_{\vec{k}_1\lambda}\frac{\hbar[(\vec{k}-\vec{k}')\vec{\varepsilon}(\vec{k}_1\lambda)]^2}{2NM\omega_\lambda(\vec{k}_1)}$$
$$\times S_0(\vec{k}_1+\vec{k}-\vec{k}')\cdot\delta(z-\hbar\omega_\lambda(\vec{k}_1)) \qquad (D.15)$$

This function is referred to as the 'spectral density of the electron-phonon interaction', the *Eliashberg function*; $F(\omega)$ is the phonon density of states, while $\alpha^2(\omega)$ is an averaged 'constant of interaction' between electrons and phonons of a given energy $\hbar\omega$.

We can easily see that the kernel (D.14) exhibits the symmetry

$$K_{Ph}(-z', -\omega) = -K_{Ph}(z', \omega) \qquad (D.16)$$

which in turn gives the following symmetry for $Z_{Ph}(\omega)$ and $\Delta_{Ph}(\omega)$:

$$Z_{Ph}(\omega) = Z^*_{Ph}(-\omega) \qquad (D.17)$$

$$\Delta_{Ph}(\omega) = \Delta^*_{Ph}(-\omega) \qquad (D.18)$$

Calculation of the Coulomb Part $\sum_C(\vec{k}, \omega)$

Now we undertake the rearrangement of the Coulomb term (D.3). The integral over z' from $-\infty$ to $+\infty$ can be transformed into an integral with respect to only positive z'. To this end we substitute Eq. (D.5) into (D.3) and employ the evenness of the functions $\vec{H} = (0, 0, H)$ and $\varphi(\vec{k}', z')$ relative to Z'. It is easy to verify that the factor of the matrix τ_1 in the right-hand side of Eq. (D.3) reduces to zero. Thus the Coulomb interaction does not contribute to the renormalization of $Z(\vec{k}, \omega)$. We can also ignore the contribution of the Coulomb interaction to $\chi(\vec{k}, \omega)$, because, as previously demonstrated, $\chi(\vec{k}, \omega)$ renormalizes only the chemical potential of the system. Consequently, only an equation for $\varphi_C(\vec{k}, \omega)$ is left:

$$\varphi_C(\vec{k}, \omega) = \frac{1}{\pi}\int\limits_0^\infty dz' \tanh\left(\frac{z'\beta}{2}\right)\int\frac{d^3\vec{k}'}{(2\pi)^3}V_e(\vec{k}-\vec{k}')\,\mathrm{Im}\frac{\varphi(\vec{k}', z')}{z'^2Z^2(z') - \varphi^2(\vec{k}', z') - \xi^2_{\vec{k}'}} \qquad (D.19)$$

It should be noted that $Z(z')$ is independent of $\vec{k}$, because it is determined just by the electron-phonon interaction. Besides, $\varphi_C(\vec{k}, \omega)$ turns out to be frequency-independent as a consequence of the fact that the retardation of the Coulomb interaction is neglected. So we can omit the frequency variable in this function and introduce the abbreviation $\varphi_C(\vec{k})$.

Equation (D.19) is an integral equation with respect to z' and $\vec{k}'$. Let us average this expression over the energy surface S corresponding to a certain energy ξ:

$$\int\limits_{S(\xi)} \frac{\mathrm{d}^2k}{\hbar v_{\vec{k}}}\varphi_{\mathrm{C}}(\vec{k}) \Big/ \int\limits_{S(\xi)} \frac{\mathrm{d}^2k}{\hbar v_{\vec{k}}} = \frac{1}{\pi}\int\limits_0^\infty \mathrm{d}z' \tanh\frac{\beta z'}{2}\left(\int\limits_{S(\xi)} \frac{\mathrm{d}^2k}{\hbar v_{\vec{k}}}\right)^{-1}$$
$$\times \int\limits_{S(\xi)} \frac{\mathrm{d}^2k}{\hbar v_{\vec{k}}} \int \mathrm{d}\xi_{\vec{k}'} \int\limits_{S(\xi_{k'})} \frac{\mathrm{d}^2k'}{\hbar v_{\vec{k}'}} V_{\mathrm{e}}(\vec{k}-\vec{k}')$$
$$\times\, \mathrm{Im}\frac{\varphi(\vec{k}', z')}{z'^2 Z^2(z') - \varphi^2(\vec{k}', z') - \xi^2_{\vec{k}'}} \tag{D.20}$$

As for the electron-phonon interaction, 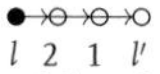 is assumed to depend on the direction of the momentum $\vec{k}'$ more weakly than the Coulomb matrix-element $V_{\mathrm{e}}(\vec{k}-\vec{k}')$. Thus in Eq. (D.20) we can put

$$\varphi(\vec{k}', z') \approx \varphi(\vec{k}', z') = \left(\int\limits_{S(\xi_{k'})} \frac{\mathrm{d}^2k'}{\hbar v_{\vec{k}'}}\right)^{-1} \int\limits_{S(\xi_{\vec{k}'})} \frac{\mathrm{d}^2k'}{\hbar v_{\vec{k}'}}\varphi(\vec{k}, z')$$

This equation becomes

$$\varphi_{\mathrm{C}}(k) = \frac{1}{\pi}\int\limits_0^\infty \mathrm{d}z' \tanh\frac{\beta z'}{2}\int \mathrm{d}\xi_{\vec{k}'} N(0) V_{\mathrm{e}}(k, k')\mathrm{Im}\frac{\varphi(k', z')}{z'^2 Z^2(z') - \varphi^2(k', z') - \xi^2_{\vec{k}'}} \tag{D.21}$$

Here the following abbreviation is introduced:

$$V_{\mathrm{e}}(k, k') = \int\limits_{S(\xi_{\vec{k}})} \frac{\mathrm{d}^2k}{\hbar v_{\vec{k}}} \int\limits_{S(\xi_{\vec{k}'})} \frac{\mathrm{d}^2k'}{\hbar v_{\vec{k}'}} V_{\mathrm{e}}(\vec{k}-\vec{k}') \Big/ \int\limits_{S(\xi_{\vec{k}})} \frac{\mathrm{d}^2k}{\hbar v_{\vec{k}}} \cdot \int\limits_{S(\xi_{\vec{k}'})} \frac{\mathrm{d}^2k'}{\hbar v_{\vec{k}'}} \tag{D.22}$$

$N(0) = \int\limits_{S(\xi_{\vec{k}'})} \frac{\mathrm{d}^2k'}{\hbar v_{\vec{k}'}} \approx \int\limits_{S_{\mathrm{F}}} \frac{\mathrm{d}^2k'}{\hbar v_{\vec{k}'}}$ is the electron density of states at the Fermi surface, considering that the integrand in (D.21) peaks at $\xi_{\vec{k}'} = 0$.

The electron-phonon interaction produces an *indirect electron-electron interaction* only within a small layer near the Fermi surface. The width of this layer is around $\hbar\omega_{\mathrm{D}}$, where ω_{D} is the Debye frequency. By contrast, the energy interval, where the Coulomb interaction is efficient, is much wider, of the order of the Fermi energy ε_{F}. As a result the contributions $\varphi_{\mathrm{C}}(k)$ and $\varphi_{\mathrm{Ph}}(\omega)$ turn out to be essential in *different* frequency ranges. To study this point it is convenient to introduce some frequency ω_{C} with the inequalities $\hbar\omega_{\mathrm{D}} \ll \hbar\omega_{\mathrm{C}} \leq \varepsilon_{\mathrm{F}}$, and split the integral over z' in (D.21) into two parts with the limits from 0 to ω_{C}, and from ω_{C} to ∞.

In the frequency range $z' > \hbar\omega_C$ the electron-phonon interaction is inefficient, so $Z(z') \to 1$ and $\varphi(p', z') \to \varphi_C(p')$. In addition $\tanh\frac{1}{2}\beta z' \to 1$, on account of the obvious inequality $k_B T_C \ll \hbar\omega_C$.

As a consequence, in this region the corresponding integral over z' can be calculated exactly:

$$\frac{1}{\pi}\int_{\hbar\omega_C}^{\infty} dz' \mathrm{Im}\frac{\varphi_C(k')}{z'^2 - \varphi_C^2(k') - \xi_{k'}^2} = \frac{\varphi_C(k')}{\pi}\int_{\hbar\omega_C}^{\infty} dz' \mathrm{Im}\frac{1}{(z' + i\delta)^2 - E_{k'}^2}$$
$$= \frac{\varphi_C(k')}{2E_{k'}}\theta(E_{k'} - \hbar\omega_C) \tag{D.23}$$

where $E_{k'} = \sqrt{\xi_{k'}^2 + \varphi_C^2(k')}$; $\theta(x)$ is the step-function.

On calculating (D.23) we take into account that

$$\mathrm{Im}\int_{\hbar\omega_C}^{\infty}\frac{1}{\xi^2 - E^2} = \frac{1}{2E}\mathrm{Im}\ln\frac{\xi - E}{\xi + E}\bigg|_{\omega_C}^{\infty} = -\frac{1}{2E}\mathrm{Im}\ln\frac{\hbar\omega_C - E}{\hbar\omega_C + E}$$
$$= -\frac{1}{2E}\begin{cases}\mathrm{Im}\ln\left|\dfrac{\hbar\omega_C - E}{\hbar\omega_C + E}\right| & \text{when } \hbar\omega_C > E \\ \mathrm{Im}\ln\left\{-\left|\dfrac{\hbar\omega_C - E}{\hbar\omega_C + E}\right|\right\} & \text{when } \hbar\omega_C < E\end{cases}$$
$$= \frac{-1}{2E}\begin{cases}0 & \text{when } \hbar\omega_C > E \\ \pi & \text{when } \hbar\omega_C < E\end{cases} = \frac{-\pi}{2E}\theta(E - \omega_C\hbar)$$

because $\ln\left|\frac{\hbar\omega_C - E}{\hbar\omega_C + E}\right|$ is real, and $\ln\left\{-\left|\frac{\hbar\omega_C - E}{\hbar\omega_C + E}\right|\right\} = \ln\left|\frac{\hbar\omega_C - E}{\hbar\omega_C + E}\right| + i\pi$.

Now Eq. (D.21) transforms into

$$\varphi_C(k) + \int\frac{d\xi_{k'}}{2E_{k'}}\theta(E_{k'} - \hbar\omega_C)N(0)V_e(k, k')\varphi_C(k') = \int d\xi_{k'}V_e(k, k')F_{k'} \tag{D.24}$$

where for the sake of brevity the notation is introduced

$$F_{k'} = N(0)\int_0^{\omega_C} dz'\frac{1}{\pi}\mathrm{Im}\frac{\varphi(k', z')}{z'^2 Z^2(z') - \varphi^2(k', z') - \xi_{k'}^2}\tanh\frac{\beta z'}{2} \tag{D.25}$$

We are looking for a solution of Eq. (D.24) in the form

$$\varphi_C(k) = \int d\xi_{k'}U_C(k, k')F_{k'} \tag{D.26}$$

This leads immediately to an integral equation for $U_C(k, k')$:

$$\mathrm{U_C}(k,k') + \int \frac{\mathrm{d}\xi_{k''}}{2 \cdot E_{k''}} \theta(E_{k'} - \hbar\omega_\mathrm{C}) N(0) \mathrm{V_e}(k,k'') \mathrm{U_C}(k'',k') = V_\mathrm{e}(k,k') \tag{D.27}$$

In its concrete meaning $\mathrm{U_C}(k,k')$ is some pseudopotential taking into account the contribution to the effective electron interaction near the Fermi surface provided by the multiple Coulomb scattering of the electrons excited into the high-energy region.

Let us now calculate the integral (D.26), as if the solution of the integral equation (D.27) were known. As follows from the definition (D.25), $F_{k'}$ goes into a sharp decrease at the energies $\xi_{\vec{k}'} \gtrsim \hbar\omega_\mathrm{C}$, i.e. the major contribution to the integral (D.26) is given by $\xi_{\vec{k}'}$ within the interval $\sim \hbar\omega_\mathrm{C}$ near the Fermi surface. The pseudopotential varies slowly in this interval, because the scale of the variation is of the order of the Fermi energy. Thus $\mathrm{U_C}$ can be factored outside the integral (D.26) at $k' \approx k_\mathrm{F}$. So, for $k \approx k_\mathrm{F}$ we have the approximation

$$\varphi_\mathrm{C}(k) \approx \mathrm{U_C} \int\limits_{-\infty}^{+\infty} \mathrm{d}\xi_{k'} F_{k'} \tag{D.28}$$

where $\mathrm{U_C} \equiv \mathrm{U_C}(k_\mathrm{F}, k_\mathrm{F})$ is the value of the pseudopotential at the Fermi surface. Now upon integration with respect to $\xi_{\vec{k}'}$ we obtain as a final expression for the Coulomb contribution:

$$\varphi_\mathrm{C}(k) = -\mathrm{U_C} N(0) \int\limits_0^{\omega_\mathrm{C}} \mathrm{d}z' \tanh\frac{\beta z'}{2} \mathrm{Re} \frac{\Delta(z')}{\sqrt{z'^2 - \Delta^2(z')}} \tag{D.29}$$

Here $\Delta(z) = \dfrac{\varphi(z)}{Z(z)}$ and $\varphi(z) = \varphi(k_\mathrm{F}, z) \approx \varphi(k,z)$ (i.e. $\varphi(k,z)$ is assumed to be independent of k and can be factored out from the integral over $\xi_{\vec{k}}$).

To find the ultimate solution of the problem we now need the pseudopotential magnitude. As an evaluation, approximate the Coulomb interaction matrix-element:

$$V_e(k,k') \cong \begin{cases} \mathrm{V_e}; & |\xi_{\vec{k}}|,\ |\xi_{\vec{k}'}| < \varepsilon_\mathrm{F} \\ 0; & |\xi_{\vec{k}}|,\ |\xi_{\vec{k}'}| > \varepsilon_\mathrm{F} \end{cases} \tag{D.30}$$

Then the integral equation (D.27) transforms into a simple algebraic equation

$$\mathrm{U_C} + \frac{N(0) \cdot V_\mathrm{e}}{2} \int\limits_{-\varepsilon_\mathrm{F}}^{\varepsilon_\mathrm{F}} \frac{\mathrm{d}\xi}{\sqrt{\xi^2 + \varphi_\mathrm{C}^2}} \cdot \theta\left(\sqrt{\xi^2 + \varphi_\mathrm{C}^2} - \hbar\omega_\mathrm{C}\right) \mathrm{U_C} = V_\mathrm{e} \tag{D.31}$$

Considering that $\varphi_\mathrm{C} \sim k_\mathrm{B} T_\mathrm{C} \ll \hbar\omega_\mathrm{C}$, the integral with respect to ξ is in fact taken from ω_C to ε_F, and neglecting φ_C^2 under the square root we get:

$$\int_{0}^{\varepsilon_F} \frac{d\xi}{\xi}\theta(\xi - \hbar\omega_C) = \int_{\hbar\omega_C}^{\varepsilon_F} \frac{d\xi}{\xi} = \ln\frac{\varepsilon_F}{\hbar\omega_C}$$

Thus we have the following expression for the pseudopotential

$$U_C = \frac{V_e}{1 + N(0)V_e \ln\dfrac{\varepsilon_F}{\hbar\omega_C}} \tag{D.32}$$

which turns out somewhat weakened in comparison with the Coulomb potential V_e owing to the inequality $\varepsilon_F \gg \hbar\omega_C$. This pseudopotential is often referred to as 'intermediate', because an effective reduced Coulomb potential μ^* is expressed through it.

Calculation of the 'Impurity Scattering'[1] Part $\sum_{\text{Scat}}(\vec{k}, \omega)$

With this aim in view let us average equation (D.4) over the Fermi surface S_F:

$$\sum_{\text{Scat}}(\omega) = \left(\int_{S_F} \frac{d^2k}{\hbar v_{\vec{k}}}\right)^{-1} \cdot \int_{S_F} \frac{d^2k}{\hbar v_{\vec{k}}} \cdot \int_{S_F} \frac{d^2k'}{\hbar v_{k'}} \cdot \int d\xi_{\vec{k}'} \cdot \frac{|\tilde{w}(\vec{k} - \vec{k}_1)|^2}{v_0}$$
$$\times S_0(\vec{k} - \vec{k}') \cdot \frac{Z(\vec{k}', \omega) \cdot \hbar\omega\tau_0 - \varphi(\vec{k}', \omega) \cdot \tau_1}{|Z(\vec{k}', \omega)\hbar\omega|^2 - \varphi^2(\vec{k}', \omega) - \xi^2_{\vec{k}'}} \tag{D.33}$$

Here, as in the calculation of $\sum_{\text{Ph}}(\vec{k}, \omega)$, Eq. (D.6) is used; $\varepsilon^2(\vec{k}, \omega)$ is replaced by $\xi^2_{\vec{k}'}$; and allowance is made for the fact that the integral containing $\xi_{\vec{k}} \cdot \tau_3$ vanishes. Next, assuming again that the angular dependence of $\varphi(\vec{k}, \omega)$ and $Z(\vec{k}, \omega)$ is weaker than for the other cofactors, we replace them again by the averaged expressions $Z(\omega)$ and $\varphi(\omega)$, averaged over the Fermi surface, and introduce the abbreviation

$$W = \left(\int_{S_F} \frac{d^2k}{\hbar v_{\vec{k}}}\right)^{-1} \cdot \int_{S_F} \frac{d^2k}{\hbar v_{\vec{k}}} \cdot \int_{S_F} \frac{d^2k'}{\hbar v_{k'}} \cdot \frac{|\tilde{w}(\vec{k} - \vec{k}')|^2}{v_0} S_0(\vec{k} - \vec{k}') \tag{D.34}$$

Then Eq. (D.33) takes the form:

$$\sum_{\text{Scat}}(\omega) = \int d\xi \cdot W \cdot \frac{Z(\omega) \cdot \hbar\omega \cdot \tau_0 - \varphi(\omega)\tau_1}{Z^2(\omega)(\hbar\omega)^2 - \varphi^2(\omega) - \xi^2} \tag{D.35}$$

Assuming

1) A precise definition of the term 'impurity scattering' in the present context is given below.

$$\sum_{\text{Scat}}(\omega) = -Z_{\text{Scat}}(\omega)\hbar\omega \cdot \tau_0 + \varphi_{\text{Scat}}(\omega)\tau_1$$

and equating the coefficients of the matrices τ_0 and τ_1 in the left- and right-hand sides of Eq. (D.35), we obtain:

$$Z_{\text{Scat}}(\omega) = -W \int \mathrm{d}\xi \frac{Z(\omega)}{Z^2(\omega)(\hbar\omega)^2 - \varphi^2(\omega) - \xi^2} \tag{D.36}$$

$$\varphi_{\text{Scat}}(\omega) = -W \int \mathrm{d}\xi \frac{\varphi(\omega)}{Z^2(\omega)(\hbar\omega)^2 - \varphi^2(\omega) - \xi^2} \tag{D.37}$$

It is seen from these relationships that

$$\frac{Z_{\text{Scat}}(\omega)}{Z(\omega)} = \frac{\varphi_{\text{Scat}}(\omega)}{\varphi(\omega)} = -\eta(\omega) + 1 \tag{D.38}$$

Because

$$Z(\omega) = Z_{\text{Ph}}(\omega) + Z_{\text{C}}(\omega) + Z_{\text{Scat}}(\omega)$$

$$\varphi(\omega) = \varphi_{\text{Ph}}(\omega) + \varphi_{\text{C}}(\omega) + \varphi_{\text{Scat}}(\omega)$$

then it follows from (D.38) that

$$Z'(\omega) = Z_{\text{Ph}}(\omega) + Z_{\text{C}}(\omega) = Z(\omega) - Z_{\text{Scat}}(\omega) = \eta(\omega) \cdot Z(\omega)$$

$$\varphi'(\omega) = \varphi_{\text{Ph}}(\omega) + \varphi_{\text{C}}(\omega) = \varphi(\omega) - Z_{\text{Scat}}(\omega) = \eta(\omega) \cdot \varphi(\omega)$$

This means that the *gap function*

$$\Delta(\omega) = \frac{\varphi(\omega)}{Z(\omega)} = \frac{\eta(\omega) \cdot \varphi(\omega)}{\eta(\omega) \cdot Z(\omega)} = \frac{\varphi_{\text{Ph}}(\omega) + \varphi_{\text{C}}(\omega)}{Z_{\text{Ph}}(\omega) + Z_{\text{C}}(\omega)} = \Delta'(\omega) \tag{D.39}$$

is determined *only by the electron-phonon and Coulomb interactions*, whereas the 'elastic electron impurity scattering' from the disordered position of the ions does *not* contribute to $\Delta(\omega)$ (this is known as *Anderson's theorem*)[2)]. Thus the gap function $\Delta(\omega)$, and hence the transition temperature, are described only by the previously derived equations (D.12), (D.13) and (D.29).

To find $Z_{\text{Scat}}(\omega)$ and $\varphi_{\text{Scat}}(\omega)$ we evaluate the integrals on the r.h.s. of (D.36) and (D.37):

2) Note that Anderson's theorem does not apply to *inelastic* scattering or to *elastic spin-flip scattering*, i.e. here we consider as usual exclusively *elastic potential scattering*. Therefore we have used the *primed* symbol 'impurity scattering' in the heading instead of the more general term *scattering*.

$$\int_{-\infty}^{+\infty} \frac{d\xi}{\xi^2 + [\varphi^2(\omega) - Z^2(\omega)(\hbar\omega)^2]} = \frac{\pi}{\sqrt{\varphi^2(\omega) - Z^2(\omega)(\hbar\omega)^2}}$$

Now take into account that the integrand on the r.h.s. of (D.34) depends only on the difference $(\vec{k} - \vec{k}')$ of the arguments. Then

$$W = \frac{1}{2\pi}\frac{\hbar}{\tau} \tag{D.40}$$

where

$$\frac{1}{\tau} = \frac{k_F}{(2\pi)^2\hbar^3}\frac{1}{v_0}\int |w(\theta)|^2 S(\theta)d\Omega$$

is the electron-ion collision time. Here the following definition has been used:

$$f(\theta) = f\left(k_F\sqrt{1-\cos\theta}\right) = f\left(k_F \sin\frac{\theta}{2}\right)$$

Introducing the gap function $\Delta = \frac{\varphi}{Z}$, we find

$$Z_{\text{Scat}}(\omega) = \frac{1}{2}\frac{\hbar}{\tau}\frac{1}{\sqrt{\Delta^2(\omega) - \hbar^2\omega^2}} \tag{D.41}$$

$$\varphi_{\text{Scat}}(\omega) = \frac{1}{2}\frac{\hbar}{\tau}\frac{\Delta(\omega)}{\sqrt{\Delta^2(\omega) - \hbar^2\omega^2}} \tag{D.42}$$

The function $\eta(\omega)$ is determined by the relationship (D.38),

$$\eta^{-1}(\omega) = 1 - \frac{1}{2}\frac{\hbar}{\tau}\frac{Z(\omega)}{\sqrt{\Delta^2(\omega) - (\hbar\omega)^2}} \tag{D.43}$$

Assuming $Z(\omega) \approx 1$ in this expression, we obtain $\eta(\omega)$ in the form reported by Abrikosov and Gorkov (1958, 1960).

According to the above-mentioned findings

$$Z'(\omega) = Z_{\text{Ph}}(\omega);\; Z'(\omega)\Delta(\omega) = Z_{\text{Ph}}(\omega)\Delta_{\text{Ph}}(\omega) + \varphi_C$$

where $Z_{\text{Ph}}(\omega)$, $\Delta_{\text{Ph}}(\omega)$ and φ_C are specified by the expressions (D.12), (D.13) and (D.29), respectively. Written in an explicit form they give finally the following set of equations for $Z'(\omega)$ and $\Delta(\omega)$:

$$[1 - Z'(\omega)]\hbar\omega = \int_{-\infty}^{+\infty} \mathrm{d}Z'\, K_{\mathrm{Ph}}(Z', \omega) \cdot \mathrm{Re}\frac{Z'}{\sqrt{Z'^2 - \Delta^2(Z')}} \operatorname{sign} Z' \tag{D.44}$$

$$\begin{aligned} Z'(\omega)\Delta(\omega) = {} & \int_{-\infty}^{+\infty} \mathrm{d}Z'\, K_{\mathrm{Ph}}(Z', \omega) \cdot \mathrm{Re}\frac{\Delta(Z')}{\sqrt{Z'^2 - \Delta^2(Z')}} \operatorname{sign} Z' \\ & - \mathrm{U_C} N(0) \int_{0}^{\hbar\omega_C} \mathrm{d}Z' \tanh\frac{\beta Z'}{2} \cdot \mathrm{Re}\frac{\Delta(Z')}{\sqrt{Z'^2 - \Delta^2(Z')}} \end{aligned} \tag{D.45}$$

References

I. Abarenkov, V. Heine, *Phil. Mag.* 9 (1964) 451.

A. A. Abrikosov, 'Fundamentals of the Theory of Metals', Nauka, Moscow, 1987 (in Russian); English translation: North Holland, Amsterdam, 1988.

A. A. Abrikosov, L. P. Gorkov, *ZhETP* 34 (1958) 1438 (in Russian); translated in *Sov. Phys. JETP* 12 (1961) 243.

A. A. Abrikosov, L. P. Gorkov, *ZhETP* 39 (1960) 1781 (in Russian); translated in *Sov. Phys. JETP* 12 (1961) 1243.

A. I. Akhieser, V. G. Barjakhtar, S. V. Peletminskij, 'Spin Waves', Nauka, Moscow, 1967 (in Russian); English translation: North Holland, Amsterdam, 1968.

P. B. Allen, R. C. Dynes, *Phys. Rev. B* 12 (1975) 905.

P. W. Anderson, B. J. Halperin, C. J. Varma, *Phil. Mag.* 25 (1972) 1.

A. E. O. Animalu, V. Heine, *Phil. Mag.* 12 (1966) 1249.

P. J. Anthony, A. C. Anderson, *Phys. Rev. B* 14 (1976) 5198.

N. W. Ashcroft, *Phys. Rev. C* 1 (1968) 232.

B. G. Austin, V. Heine, L.-J. Sham, *Phys. Rev.* 127 (1962) 276.

R. Balescu, 'Equilibrium and Nonequilibrium Statist. Mechanics', John Wiley, New York, 1975.

V. G. Barjakhtar, V. N. Krivoruchko, D. A. Yablonskij, 'Green's Functions in the Theory of Magnetism', Naukowa Dumka, Kiev, 1984 (in Russian).

H. Beck, H. J. Güntherodt, 'Glassy Metals II: Atomic Structure and Dynamics, Electronic Structure, Magnetic Properties', Vol. 1, Springer, Berlin, 1983 (see Güntherodt and Beck).

H. Beck, H. J. Güntherodt, 'Glassy Metals III: Amorphization Techniques, Catalysis, Electronic and Ionic Structure', Springer, Berlin, 1994.

D. K. Belashchenko, 'Structure of Liquid and Amorphous Metals', Metallurgia, Moscow, 1986 (in Russian).

G. Bergmann, 'Amorphous Metals and their Superconductivity', Phys. Reports (*Phys. Lett. C*) 27 (1976) 159.

J. D. Bernal, *Nature* 183 (1959) 141.

J. D. Bernal, *Nature* 185 (1960) 68.

J. D. Bernal, in: 'Liquids: Structure, Properties and Solid Interactions', ed. by T. J. Hugel, Elsevier, Amsterdam, 1963.

K. Binder, A. P. Young, *Rev. Mod. Phys.* 58 (1986) 801.

J. L. Black, *Phys. Rev. B* 17 (1978) 2740.

J. L. Black, 'Low-Energy Excitations in Metallic Glasses', in: 'Glassy Metals I', ed. by H. R. Güntherodt and H. Beck, Springer, Berlin, 1981, p. 167.

J. L. Black, B. L. Gyorffy, *Phys. Rev. Lett.* 41 (1978) 1595.

V. B. Bobrov, S. A. Trigger, *Solid State Commun.* 56 (1985a) 21.

V. B. Bobrov, S. A. Trigger, *Solid State Commun.* 56 (1985b) 29.

N. N. Bogoliubov, V. V. Tolmachev, D. V. Shirkov, 'A New Method in the Theory of Superconductivity', Izdat. Acad. Nauk USSR, Moscow, 1958 (in Russian); English translation: Consultants Bureau, New York, 1959; abbreviated version: *Fortschr. Physik* 6 (1958) 605.

A. M. Bratkowskii, V. G. Vaks, A. V. Trefilov, *Zh. Eksp. Teor. Fiz.* 86 (1984) 1241 (in Russian); translated in *Sov. Phys. JETP* 59 (1984) 1245.

E. G. Brovman, Yu. M. Kagan, *Usp. Fiz. Nauk* 112 (1974) 369 (in Russian); translated in *Sov. Phys. Uspekhi* 17 (1974) 125.

E. G. Brovman, Yu. M. Kagan, A. Kholas, *Fiz. Tverdogo Tela* 12 (1970) 1001 (in Russian); translated in *Sov. Phys. Solid State* 12 (1970) 786.

E. G. Brovman, Yu. M. Kagan, A. Kholas, *Zh. Eksp. Teor. Fiz.* 61 (1971) 2429 (in Russian); translated in *Sov. Phys. JETP* 34 (1971) 1300.

E. G. Brovman, Yu. M. Kagan, A. Kholas, *Zh. Eksp. Teor. Fiz.* 62 (1972) 1492 (in Russian); translated in *Sov. Phys. JETP* 35 (1972) 783.

U. Buchenau, M. Prager, N. Nücker, A. Y. Dianaux, N. W. Ahmad, W. A. Philipps, *Phys. Rev. B* 34 (1986) 5565.

G. S. Cargill III, 'Structure of Metallic Alloy Glasses', in: 'Solid State Physics', Vol. 30, ed. by H. Ehrenreich, F. Seitz and D. Turnbull, Academic Press, New York, 1975, p. 227.

D. M. Ceperley, B. J. Alder, *Phys. Rev. B* 36 (1987) 2092.

S. Chakravarty, J. H. Rose, D. Wood, N. W. Ashcroft, *Phys. Rev. B* 24 (1981) 1624.

C. L. Chien, R. Hasegawa, *Phys. Rev. B* 16 (1977) 2115, 3024.

M. H. Cohen, *J. Phys. et Radium* 23 (1962) 1249.

M. H. Cohen, V. Heine, 'The Fitting of Pseudopotentials to Experimental Data and their Subsequent Application', in: 'Solid State Physics', Vol. 24, ed. by H. Ehrenreich, F. Seitz and D. Turnbull, Academic Press, New York, 1970, p. 32.

C. A. Croxton, 'Liquid State Physics: A Statistical Mechanical Introduction', Cambridge University Press, Cambridge, 1974.

L. F. Cugliandolo, G. Lozano, *Phys. Rev. Lett.* 80 (1998) 4979.

L. F. Cugliandolo, J. Kurchan, G. Parisi, *J. de Physique I* 11 (1994) 1641.

L. Cui, N. H. Chen, S. J. Jeon, I. F. Silvera, *Phys. Rev. Lett.* 72 (1994) 3048.

L. B. Davies, P. J. Grundy, *Phys. Stat. Solidi (a)* 8 (1971) 189.

L. B. Davies, P. J. Grundy, *J. Non-cryst. Solids* 11 (1972) 179.

A. S. Davydov, 'Quantum Mechanics', Fizmatgiz, Moscow, 1963; translated, edited and with additions by D. ter Haar, Pergamon Press, Oxford, 1976.

W. Döring, 'Introduction into Quantum Mechanics', Vandenhoek & Ruprecht, Göttingen, 1962 (in German).

O. V. Dolgov, A. G. Maximov, *Trans. FIAN* 148 (1983) 3 (in Russian).

J. M. Dubois, P. H. Gaskell, G. Le Caer, *Proc. R. Soc. London Ser. A* 402 (1985) 323.

F. J. Dyson, *Phys. Rev.* 102 (1956) 1217, 1230.

G. M. Eliashberg, *ZhETP* 38 (1960) 966; ibid. 39 (1960) 1437 (in Russian); translated into English in *Sov. Phys. JETP* 11 (1960) 696 and 12 (1961) 1000.

R. Evans, D. A. Greenwood (eds) 'Liquid Metals '76', IOP, Bristol, 1977.

R. Evans, R. Kumaravadivel, *J. Phys. C* 9 (1976) 1891.

W. Felsch, *Z. Phys.* 219 (1969) 280.

W. Felsch, *Z. Angew. Physik* 29 (1970a) 217.

W. Felsch, *Z. Angew. Physik* 30 (1970b) 278.

A. .L Fetter, J. D. Walecka, 'Quantum Theory of Many-Particle Systems', MacGraw-Hill, San Francisco, 1971.

J. L. Finney, *Proc. R. Soc. London Ser. A* 319 (1970) 479.

I. Z. Fisher, 'Statistical Theory of Liquids', Fizmatgiz, Moscow, 1961.

M. Folmer, 'Kinetics of New Phase Formation', Nauka, Moscow, 1986 (in Russian).

I. N. Frantsevich, F. F. Voronov, S. A. Bakuta, 'Elasticity Constants and Elasticity Modules for Metal and Non-Metal Solids', Naukova Dumka, Moscow, 1982 (in Russian).

S. Franz, M. Mezard, G. Parisi, L. Peliti, *Phys. Rev. Lett.* 81 (1998) 1758.

T. Fujiwara, H. S. Chen, Y. Waseda, *Z. Naturforsch.* 37a (1982) 611.

Yu. M. Galperin, V. G. Karpov, V. I. Kozub, *Adv. Phys.* 38 (1989) 669.

P. Gaskell, 'Models for the Structure of Amorphous Solids', in: 'Material Science and Technology', ed. by R. W. Cahn, P. Haasen, E. J. Kramer, Vol. 9: 'Glasses and Amorphous Materials' (Volume Editor J. Zarzycki), VCH, Weinheim, 1991, p. 493.

D. J. W. Geldart, S. H. Vosko, *J. Phys. Soc. Japan* 20 (1965) 20.

D. J. W. Geldart, S. H. Vosko, *Can. J. Phys.* 44 (1966) 2137.

I. M. Gell-Mann, K. A. Brueckner, *Phys. Rev.* 106 (1957) 364.

D. D. Gilman, H. J. Leamy (eds) 'Metallic Glasses', American Society for Metals, Me-

tals Park, OH, 1978; Russian translation: Metallurgiya, Moscow, 1984.
V. L. GINZBURG, D. A. KIRZHNITZ, *Phys. Rep. (Phys. Lett. C)* 4 (1972) 343.
V. L. GINZBURG, D. A. KIRZHNITZ, 'Problems of High-temperature Superconductivity', Nauka, Moscow, 1977, in Russian.
M. D. GIRARDEAU, *J. Math. Phys.* 12 (1971) 165.
W. GÖTZE, L. SJÖGREN, *Rep. Progr. Phys.* 55 (1992) 241.
B. GOLDING, J. E. GRAEBNER, *Phys. Rev. Lett.* 37 (1976) 852.
B. GOLDING, J. E. GRAEBNER, R. J. SCHULTZ, *Phys. Rev. B* 14 (1967) 1660.
B. GOLDING, J. E. GRAEBNER, R. J. SCHULTZ, B. I. HALPERIN, *Phys. Rev. Lett.* 30 (1973) 223.
B. GOLDING, J. E. GRAEBNER, A. B. KANE, *Phys. Rev. Lett.* 37 (1976) 1248.
G. S. GREST, S. R. NAGEL, A. RAHMAN, *Phys. Rev. Lett.* 49 (1982) 1271.
A. I. GUBANOV, *Fiz. Tverdogo Tela* 2 (1960) 502 (in Russian); translated in *Sov. Phys. Solid State* 2 (1960) 468.
H.-J. GÜNTHERODT, H. BECK, 'Glassy Metals I: Ionic Structure, Electronic Transport, and Crystallization', Springer, Berlin, 1981; Vol. II, III: see Beck and Güntherodt.
B. A. GURSKI, Z. A. GURSKI, *Ukr. Fiz. Zhurnal* 21 (1976a) 1609 (in Russian).
Z. A. GURSKI, B. A. GURSKI, *Fiz. Metallov i Metallovedenie* 46 (1976b) 903 (in Russian).
Z. A. GURSKI, B. A. GURSKI, *Ukr. Fiz. Zhurnal* 22 (1977) 796 (in Russian).
Z. A. GURSKI, G. L. KRASKO, *Doklady Akad. Nauk USSR* 197 (1971) 1980 (in Russian); translated in *Sov. Phys. Doklady* 16 (1971) 298.
J. HAFNER, *J. Phys. F: Metal Phys.* 6 (1976) 243.
J. HAFNER, *Phys. Rev. B* 15 (1977a) 617.
J. HAFNER, *Phys. Rev. A* 16 (1977b) 351.
J. HAFNER, 'From Hamiltonians to Phase Diagrams – The Electronic and Statistical Treatment of sp-bonded Metals', Springer, Berlin, 1987.
B. I. HALPERIN, *Ann. N.Y. Acad. Sci.* 279 (1976) 173.
J. HAMMERBERG, N. W. ASHCROFT, *Phys. Rev. B* 9 (1974) 409.
K. HANDRICH, S. KOBE, 'Amorphous Ferro- and Ferrimagnets', Akademieverlag, Berlin, 1980 (in German); Russian translation: Mir, Moscow, 1982.
S. HANDSCHUH, J. LANDES, U. KOBLER, CH. SAUER, G. KISTERS, A. FUSS, W. ZINN, *J. Magn. Magn. Mater.* 119 (1993) 254.
W. A. HARRISON, 'Pseudopotentials in the Theory of Metals', Benjamin, New York, 1966.
A. HARRISON, 'Electronic Structure and the Properties of Solids', Freeman, San Fransisco, 1980.
M. HASEGAWA, *J. Phys. F: Metal Phys.* 6 (1976) 649.
R. HASEGAWA, R. LEVY (eds), 'Amorphous Magnetism', Plenum Press, New York, 1977.
C. HAUSLEITNER, J. HAFNER, *J. de Physique IV (Colloque)* 1-C5 (1991) 25.
C. HAUSLEITNER, J. HAFNER, *Phys. Rev. B* 47 (1993) 5689.
L. VON HEIMENDAHL, *J. Phys. F: Metal Phys.* 9 (1979) 161.
V. HEINE, M. L. COHEN, D. WEAIRE, 'The Pseudopotential Concept', in: 'Solid State Physics', Vol. 24, ed. by H. Ehrenreich, F. Seitz and D. Turnbull, Academic Press, New York, 1970, p. 1.
C. HERRING, *Phys. Rev.* 57 (1940) 1169.
H. R. Hilzinger, *J. Magn. Magn. Mater.* 83 (1990) 370.
H. O. HOOPER, A. M. DE GRAAF (eds) 'Amorphous Magnetism', Plenum Press, New York, 1973.
J. HUBBARD, *Proc. R. Soc. London A* 243 (1957) 364.
S. HUNKLINGER, W. ARNOLD, S. STEIN, R. NAVA, K. DRANSFELD, *Phys. Lett.* 42A (1972) 253.
T. ICHIKAWA, *Phys. Stat. Solidi (a)* 19 (1973) 707.
S. V. IORDANSKY, A. M. FINKEL'SHTEIN, *Zh. Eksp. Teor. Fiz.* 62 (1972) 473 (in Russian); translated in *Sov. Phys. JETP* 35 (1972) 215.
S. N. ISHMAEV, S. L. ISAKOV, I. P. SADIKOV, E. SVAB, L. KOSZEGI, A. LOVAS, GY. MESZAROS, *J. Non-Cryst. Solids* 94 (1987) 11.
J. Z. JÄCKLE, *Z. Physik* 257 (1972) 2740.
J. Z. JÄCKLE, L. Piche, W. Arnold, S. Hunklinger, *J. Non-Cryst. Solids* 20 (1976) 365.
W. L. JOHNSON, in: 'Topics of Applied Physics' Vol. 46, Springer, Berlin Heidelberg, 1981, p. 191.

I. R. Juchnowsky, Z. A. Gurski, 'Quantum-Statistical Theory of Disordered Systems', Naukowa dumka, Kiev, 1991 (in Russian).
Yu. M. Kagan, V. V. Pushkarev, A. Kholas, *Zh. Eksp. Teor. Fiz.* 73 (1977) 967 (in Russian); translated in *Sov. Phys. JETP* 46 (1977) 511.
S. D. Kaim, N. P. Kovalenko, E. V. Vasiliu, *J. Phys. Studies* 1 (1997) 589.
Y. Kakehashi, T. Uchida, M. Yu, *Phys. Rev. B* 56 (1997) 8807.
T. Kaneyoshi, 'Introduction to Amorphous Magnets', World Scientific, Singapore, 1992.
A. E. Karakozov, E. G. Maximov, S. A. Mashkov, *ZhETP* 68 (1975) 1937 (in Russian); translated in *Sov. Phys. JETP* 41 (1976) 971.
E. I. Kcharkov, V. I. Lysov, V. E. Fedorov, 'Physics of Liquid Metals', Vyshcha shkola, Kiev, 1979 (in Russian).
N. K. Kikoin, 'Physical Constant Tables: Reference Book', Atomizdat, Moscow, 1976 (in Russian).
Ch. Kittel, 'Quantum Theory of Solids', Wiley, New York, 1963.
Ch. Kittel, 'Introduction to Solid State Physics', Wiley, New York, 1971.
P. Kizler, PhD thesis, University of Stuttgart, 1988.
P. Kizler, *Phys. Rev. Lett.* 67 (1991) 3555.
P. Kizler, *Phys. Rev. B* 48 (1993) 12488.
P. Kizler, E. Hertlein, P. Vargas, S. Steeb, *J. Physique* C8 (1986) 1019.
P. Kizler, P. Lamparter, S. Steeb, *Z. Naturforsch.* 43a (1988) 1047.
M. W. Klein, R. Brout, *Phys. Rev.* 132 (1963) 2412.
W. Kob, J. L. Barrat, *Phys. Rev. Lett.* 78 (1997) 4581.
N. P. Kovalenko, I. Z. Fisher, *Usp. Fiz. Nauk* 108 (1972) 209 (in Russian); translated in *Sov. Phys. Uspekhi* 15 (1973) 592.
N. P. Kovalenko, Yu. P. Krasny, *Physica B* 162 (1990) 115.
N. P. Kovalenko, L. M. Kuz'mina, *Fizika zhidkogo sostoyania* 9 (1981) 34.
N. P. Kovalenko, Yu. P. Krasny, G. Weck, *Phys. Stat. Sol. (b)* 166 (1971) 117.
N. P. Kovalenko, Yu. P. Krasny, S. A. Trigger, 'Statisticheskaia Theoria Zhidkih Metallov', Nauka, Moscow, 1990.
N. P. Kovalenko, Yu. P. Krasny, V. V. Mikho, *J. Molec. Liquids* 58 (1993) 45.
Yu. P. Krasny, V. P. Onishchenko, *Ukr. Phys. J.* 17 (1972) 1705.
Yu. P. Krasny, N. P. Kovalenko, *Fizika zhidkogo sostoyania* 9 (1981) 34.
Yu. P. Krasny, N. P. Kovalenko, *Physica B* 162 (1990) 128.
Yu. P. Krasny, N. P. Kovalenko, V. V. Mikho, *Physica B* 162 (1990a) 162.
Yu. P. Krasny, N. P. Kovalenko, V. F. Zarev, *Ukr. Phys. J.* 35 (1990b) 41.
Yu. P. Krasny, N. P. Kovalenko, V. V. Mikho, *J. Molec. Liquids* 58 (1993) 45.
Yu. P. Krasny, N. P. Kovalenko, V. V. Mikho, E. P. Gurnitskaya, *Physica B* 183 (1993a) 115; (1993b) 123; (1993c) 130.
Yu. P. Krasny, N. P. Kovalenko, V. A. Tesis, *Mol. Phys. Rep.* 11 (1995) 143.
Yu. P. Krasny, N. P. Kovalenko, V. A. Tesis, *J. Non-cryst. Solids* 205–207 (1996) 669.
Yu. P. Krasny, N. P. Kovalenko, U. Krey, *Physica B* 240 (1997) 173.
Yu. P. Krasny, N. P. Kovalenko, J. Krawczyk, *Physica B* 264 (1998) 92.
U. Krauss, U. Krey, *J. Magn. Magn. Mater.* 98 (1991) L1.
U. Krey, *Z. Physik B* 31 (1978) 247.
U. Krey, *Z. Physik B* 38 (1980) 243.
U. Krey, *Z. Physik B* 42 (1981) 231.
U. Krey, *J. de Physique Lettres* 46 (1985) L845.
U. Krey, H. Ostermeier, J. Zweck, *Phys. Stat. Solidi (b)* 144 (1987) 203.
U. Krey, S. Krompiewski, U. Krauss, *J. Magn. Magn. Mater.* 86 (1990) 85.
U. Krey, U. Krauss, S. Krompiewski, *J. Magn. Magn. Mater.* 103 (1992) 37.
S. Krompiewski, U. Krauss, U. Krey, *Phys. Rev. B* 39 (1989a) 2819.
S. Krompiewski, U. Krauss, U. Krey, *Physica B* 61 (1989b) 219.
S. Krompiewski, U. Krauss, U. Krey, *J. Magn. Magn. Mater.* 103 (1992) 31.
R. Kumarvaadivel, R. Evans, *J. Phys. C* 9 (1976) 3877.
C. Laermans, *Phys. Rev. Lett.* 42 (1979) 256.
P. Lamparter, S. Steeb, 'Structure of Metallic Glasses, Amorphous and Molten Alloys', in: 'Material Science and Technology', ed. by R. W. Cahn, P. Haasen, E. J. Kramer, Vol. 1: 'Structure of Solids' (Volume Editor V. Gerold), VCH, Weinheim, 1993, p. 217.
L. D. Landau, E. M. Lifshitz, 'Quantum Mechanics. Nonrelativistic Theory', Fizmatgiz, Moscow, 1963, Vol. III of the Course in Theoretical Physics (in Russian); translated by J. S. Sykes and J. S. Bell from the 3rd edn,

revised and enlarged with the assistance of L. P. Pitaevskii. Pergamon Press, Oxford, 1977.

L. D. Landau, E. M. Lifshitz, 'Statistical Physics', part 2, Nauka, Moscow, 1976 (in Russian) translated version: see Lifshitz and Pitaevskii.

P. K. Leung, G. Wright, *Phil. Mag.* 30 (1974) 185–995.

E. M. Lifshitz, Yu. Kagan, *ZhETP* 62 (1972) 385 (in Russian); translated in: *Sov. Phys. JETP* 35 (1972) 206.

E. M. Lifshitz, L. P. Pitaevskii, 'Physical Kinetics', Nauka, Moscow, 1978 (in Russian); English translation: Pergamon Press, Oxford, 1981.

E. M. Lifshitz, L. P. Pitaevskii, 'Statistical Physics', part 2: 'Theory of the Condensed State' (Vol. 9 of the 'Course on Theoretical Physics' by L. D. Landau and E. M. Lifshitz), Pergamon Press, Oxford, 1980.

J. Lindhard, *Kon. Danske Videnssk. Selsk., mat.-fys. medd.* 28 (1954) 57.

P. Lloyd, C. A. Sholl, *J. Phys. C* 1 (1968) 1620.

R. Lorenz, J. Hafner, *J. Magn. Magn. Mater.* 139 (1995) 209.

L. F. Lou, *Solid State Commun.* 19 (1976) 335.

S. G. Louie, S. Froyen, M. L. Cohen, *Phys. Rev. B* 26 (1982) 1738.

F. E. Luborsky (ed.) 'Amorphous Metallic Alloys', Butterworths, London, 1983; Russian translation: Metallurgia, Moscow, 1987.

H. Mao, R. J. Hemley, *Rev. Mod. Phys.* 66 (1994) 671.

N. H. March, M. P. Tosi, 'Atomic Dynamics in Liquids', Macmillan, London, 1976.

E. G. Maximov, *ZhETP* 57 (1969) 1660 (in Russian); translated in *Sov. Phys. JETP* 30 (1970) 197.

R. M. Mazo, *J. Chem. Phys.* 39 (1963) 1214.

W. L. McMillan, *Phys. Rev.* 167 (1967) 331.

A. B. Migdal, *ZhETP* 34 (1958) 1438 (in Russian); translated in *Sov. Phys. JETP* 34 (1958) 996.

H. A. Mook, C. C. Tsuei, *Phys. Rev. B* 16 (1977) 2184.

H. A. Mook, N. Wakayabashi, D. Pan, *Phys. Rev. Lett.* 34 (1975) 1029.

K. Moorjani, J. M. Coey, 'Magnetic Glasses', Elsevier, Amsterdam, 1984.

P. Morel, P. W. Anderson, *Phys. Rev.* 125 (1962) 1263.

E. G. Moroni, G. Kresse, J. Hafner, *Phys. Rev. B* 56 (1997) 15629.

V. L. Moruzzi, P. M. Marcus, P. C. Pattnaik, *Phys. Rev. B* 37 (1988) 8803.

V. L. Moruzzi, P. M. Marcus, J. Kübler, *Phys. Rev. B* 39 (1989) 6957.

N. F. Mott, E. A. Davis, 'Electronic Processes in Non-crystalline Materials', Clarendon Press, Oxford, 1979.

J. N. Murrell, S. F. A. Kettle, J. M. Tedder, 'Valence Theory', J. Wiley & Sons, London, 1965.

U. Mussel, H. Rieger, W. Kob, J. L. Barrat, *Phys. Rev. Lett.* 81 (1998) 930.

S. R. Nagel, J. Tauc, *Phys. Rev. Lett.* 35 (1975) 380.

V. V. Nemoshkalenko, A. V. Romanova, A. G. Il'insky *et al.*, 'Amorphous Metal Alloys', Naukova Dumka, Kiev, 1987, in Russian.

P. Nozieres, D. Pines, *Phys. Rev.* 111 (1958) 442.

D. G. Onn, 'Thermal Properties of Amorphous Metallic Alloys', in: 'Amorphous Metallic Alloys', ed. by F. E. Luborsky, Butterworths, London, 1983, p. 453; Russian translation: Metallurgia, Moscow, 1987.

H. Ostermeier, PhD thesis (in German), University of Regensburg, 1988, unpublished.

H. Ostermeier, U. Krey, *Mat. Sci. Eng.* 99 (1988) 273.

A. Paintner, F. Süss, U. Krey, *J. Magn. Magn. Mater.* 154 (1996) 107.

G. Kh. Panova, A. A. Chernoplekov, N. A. Shikov, B. I. Savel'ev, M. N. Khlopkin, *Zh. Eksp. Teor. Fiz.* 88 (1985) 1012 (in Russian); translated in *Sov. Phys. JETP* 61 (1985) 595.

D. A. Papaconstantopoulos, 'Handbook of the Band Structure of Elemental Solids', Plenum Press, New York, 1986.

M. C. Payne, M. P. Teter, D. C. Allen, T. A. Arias, J. D. Joannopoulos, *Rev. Mod. Phys.* 64 (1992) 1045.

J. K. Percus, G. J. Yevick, *Phys. Rev.* 110 (1958) 1.

W. A. Phillips, *J. Low-Temp. Phys.* 7 (1972) 351.

W. A. Phillips, *J. Non-Cryst. Solids* 31 (1978) 267.

W. A. Phillips (ed.) 'Amorphous Solids: Low Temperature Properties', Springer, Berlin, 1980.

D. Pines, P. Nozieres, 'The Theory of Quantum Liquids', Benjamin, New York, 1966.

D.E. Polk, *Acta Met.* 20 (1972) 485.

S.J. Poon, 'Superconducting Properties of Amorphous Metallic Alloys', in: 'Amorphous Metallic Alloys', ed. by F.E. Luborsky, Butterworths, London, 1983; Russian translation: Metallurgia, Moscow, 1987.

L. Pusztai, *Z. Naturforsch.* 46a (1991) 69.

J. Ram, Y. Singh, *J. Chem. Phys.* 66 (1977) 924.

H. Rieger, *Physica A* 224 (1996) 267.

G.S. Rushbrooke, M. Silbert, *Molec. Phys.* 12 (1967) 505.

G.S. Rushbrooke, P.J. Wood, *Molec. Phys.* 1 (1958) 257.

K. Samwer, H.J. Fecht, W.L. Johnson, 'Amorphization in Metallic Systems', in: 'Glassy Metals III', ed. by H. Beck and H.-J. Güntherodt, Springer, Berlin, 1994, p. 5.

D.J. Scalapino, 'The Electron-Phonon Interaction and Strong-Coupling Superconductivity', in: 'Superconductivity', ed. by R.D. Parks, Dekker Inc., New York, 1969, p. 449.

S.K. Sinha, J. Ram, Y. Singh, *J. Chem. Phys.* 66 (1977) 5013.

J.C. Slater, G.F. Koster, *Phys. Rev.* 94 (1954) 1498.

V.I. Smirnov, 'Course of High Mathematics', Vol. I, Fizmatgiz, Moscow, 1956 (in Russian).

R.B. Stephens, *Phys. Rev. B* 8 (1973) 2896.

D.J. Stevenson, N.W. Ashcroft, *Phys. Rev. A* 9 (1974) 782.

I.-B. Suck, H. Rudin, 'Vibrational Dynamics of Metallic Glasses Studied by Neutron Inelastic Scattering', in: 'Glassy Metals II', ed. by H. Beck and H.-J. Güntherodt, Springer, Berlin, 1983, p. 217.

I.-B. Suck, H. Rudin, H.-J. Güntherodt, H. Beck, *Phys. Rev. Lett.* 50 (1983) 49.

N. Troullier, J.L. Martins, *Phys. Rev.* B43 (1991) 1993; 8861.

S.V. Tyablikov, 'Methods of Quantum Theory of Magnetism', Nauka, Moscow, 1965 (in Russian); English translation: Plenum Press, New York, 1967.

T. Uchida, Y. Kakehashi, *J. Appl. Phys.* 81 (1997) 3859.

T. Uchida, Y. Kakehashi, *J. Magn. Magn. Mater.* 177–181 (1998) 83.

I.A. Vakarchuk, I.F. Margolych, 'Magnon Spectrum of Two-component Amorphous Ferromagnets', Preprint ITP-87-2p, Kiev, 1987 (in Russian).

I.A. Vakarchuk, V.M. Tkachuk, *Teor. Mat. Fiz.* 79 (1989) 446 (in Russian).

I.A. Vakarchuk, Yu.K. Rudavski, G.V. Ponedilok, 'Free Energy of Amorphous Ferromagnets with Heisenberg Exchange Interaction', Preprint ITP-81-45p, Kiev, 1981 (in Russian).

I.A. Vakarchuk, Yu.K. Rudavski, G.V. Ponedilok, *Phys. Stat. Sol. (b)* 128 (1985a) 231.

I.A. Vakarchuk, Yu.K. Rudavski, G.V. Ponedilok, B.V. Monianchin, 'Thermodynamical Properties of Amorphous Ferromagnets with Liquid-like Structure', Preprint ITP-85-10p, Kiev, 1985b (in Russian).

I.A. Vakarchuk, V.M. Migal, V.M. Tkachuk, *Teor. Mat. Fiz.* 75 (1988) 306.

V.G. Vaks, A.I. Larkin, S.A. Pikin, *ZhETP* 3 (1961) 1541 (in Russian).

V.G. Vaks, A.I. Larkin, S.A. Pikin, *ZhETP* 53 (1967) 281 (in Russian); translated in *Sov. Phys. JETP* 26 (1968) 188.

V.G. Vaks, S.P. Kravchuk, A.B. Trevilov, *Fiz. Tverdogo Tela* 19 (1977) 271 (in Russian); translated in *Sov. Phys. Solid State* 19 (1970) 740.

D. Vanderbilt, *Phys. Rev. B* 41 (1990) 7892.

P. Vashista, K.S. Singwi, *Phys. Rev. B* 6 (1972) 875.

S.V. Vonsovskij, Ju.A. Izyumov, E.Z. Kurmaev, 'Superconductivity of Transition Metals and Compounds', Nauka, Moscow, 1977 (in Russian); English translation: Springer, Berlin, 1982.

J.D. Weeks, D. Chandler, H.C. Anderson, *J. Chem. Phys.* 54 (1971) 5237.

M. Yu, Y. Kakehashi, *J. Phys. CM: Cond. Matt.* 8 (1996) 5071.

J. Zarzycki, 'Special Methods of Obtaining Glasses and Amorphous Materials', in: 'Material Science and Technology', ed. by R.W. Cahn, P. Haasen, E.J. Kramer, Vol. 9: 'Glasses and Amorphous Materials' (Volume Editor J. Zarzycki), VCH, Weinheim, 1991, p. 91.

R.C. Zeller, R.O. Pohl, *Phys. Rev. B* 4 (1971) 2029.

J.M. Ziman, 'Models of Disorder', Cambridge University Press, London, 1979.

Index

a
ab initio molecular dynamics 70
acoustic absorption 78
acoustic sound waves 108
additional low-energy vibrational excitation 79
additive pair potentials 70
adiabatic approximation 15, 185
– Born-Oppenheimer 130
aging 2, 38, 80, 81, 215
alkaline metals 51, 53
ambiguity of the pseudopotential 24
amorphous solids 1
amorphous Y or Gd substrate 174
analytic continuation 152
Anderson's theorem 266
anharmonic excitation 82
anharmonic term 113
annealed approximation 242
annealed averaging 38
annealing procedures 2
annihilation operator 113, 114
anomalous low-temperature behaviour 77 ff.
anomalous part of the Green's function 190
anomalous thermal physics of glasses 129
anticommutation rules 17, 181
Ashcroft pseudopotential 106, 201, 204, 205, 211
asperomagnet 176
asymmetry energy 84
atomic force microscopy 80
atomic properties of metallic glasses 77
attenuation of phonon excitations 109

b
ball bearing model 8
band magnetism 11
band-theoretic approach 171 ff.
barrier height 84, 91
BCS theory 192
Bloch's law 136, 137, 164, 168, 169, 171
blocks (in diagram theory) 230 ff., 237
Bohr magneton 139, 220
Boltzmann approximation 63
bonding or antibonding states 70
Born approximation 150
Born-Oppenheimer approximation 15, 130
Bose commutation relations 105
Bose operators 157
Brillouin function 141
Brillouin light scattering 137

c
Cauchy part 131 *see also* principal part
characteristic time of the tunneling nucelation of the H_2 molecule 66
chronological ordering operator 221
closed loops 42
Cohen-Abarenkov-Heine pseudopotential 48
collective quantum states of groups of particles 68
collinear magnetism 178
compensating background charges 16
compensation point 171
compensation temperature 170
completely orthogonalized plane waves (COPW) 22 ff.
composition range 215
compressibility 57
conditional distribution functions 55
configurational averaging 139
confluence processes (of phonons) 115, 116
continued-fraction algorithm 73
continued-fraction technique 168
contractions 224 ff.
contrast in the size of the atoms 149
core states 19

correlation energy 30, 37
corrosion properties 215
coupling 192, 195, 196
– intermediate 192, 195
– strong 192, 196 ff.
– weak 192
creation and annihilation operators 33
creation operator 113, 114
crystal symmetry 57
Curie point 144

d
Debye contribution 78
Debye formula 99
decoupling 151 ff.
deep eutectics 2
deformation potential 85
degeneration temperature 16
delocalized electronic states 19
demixing transition 2
dense random packing of hard spheres 3
densely packed frozen liquids 81
diagrammatic expansion 41
diagrammatic rules 41 ff., 234, 239
– for the Eliashberg theory 247 ff.
– for the Heisenberg ferromagnet 234
diagrammatic technique by Vaks *et al.* 141, 150, 227
diagrams 39, 41 ff.
– bubble diagram 45
– connected 39, 218, 238
– disconnected 39, 237
– indexed 232
– irreducible 151
– topologically discernible 232
– topologically equivalent 232
diatomic ordering 59, 62, 65 ff.
dielectric function 36, 37
– Geldart-Vosko form 54, 61
see also permittivity
differential cross-section 4
diffraction experiment 4
direct Coulomb repulsion 47
direct process (Hartree contribution) 32, 36
dispersion curves 108, 109, 122, 133, 153, 154
– for longitudinal vibrational excitations 133
– for transversal vibrational excitations 133
distribution of barriers 90
double-differential cross-section 150
double-well potential 84
drawback of model pseudopotentials 28
DRPSS model 8
dynamical structure functions 133
Dyson equation 251 ff.
Dyson's time-ordering symbol 39

e
effective dispersion curves 133
effective Hamiltonian 15, 25, 162
– of non-interacting bosons 162
effective ion-ion interaction 56
effective pair interaction 56
effective pair interaction potential 53, 55, 57
eigenfrequencies 130
eigenvectors 130
Einstein's summing convention 72, 173, 176
electron diffraction 4
electron liquid 29
electron microscopy 80
electron self-energy 255
electron-electron interaction 29
electronic structure 172 ff.
electron-ion interaction 29, 41, 180 ff.
electron-ion potential 58, 59
electron-ion system 15
electron-phonon coupling constant 192, 211
electron-phonon interaction 179 ff.
Eliashberg equations 179 ff., 209
– in binary alloys 209
Eliashberg function 189 ff., 261
Eliashberg theory 204
energy relaxation 11
Euler constant 194
EXAFS 11
excess density of states 80
exchange diagram 33
exchange integral 139
exchange interaction 135
exchange interaction radius 142
exchange process 31 ff., 36

f
Fermi distribution 30
Fermi's Golden Rule 93, 112
ferrimagnetic 170
ferromagnetic resonance 137
fluctuation corrections 166
free energy 164
free path length of sonic phonons 125 ff.
Friedel oscillation 48, 75
frozen liquid 140, 153
– hard-sphere model 153
frozen-liquid model 102 ff., 161, 167, 182

frozen-liquid picture 8
frustration 92

g
gap function 188, 209
Geldart-Vosko formula 44
generalized coordinate 84
Gibbs-Bogolyubov variational principle 54
glassiness 84
grand canonical thermodynamic potential 37
Green's functions 41, 132, 150ff., 163, 168, 182ff.
– anomalous 184, 190
– causal 150, 152, 187, 238ff.
– Nambu 184
– retarded 150, 152, 155, 256
Green's function formalism 131
Green's lines 237
Green's operator 131
Grüneisen coefficient 77ff.

h
H_2 molecule 67
Hamiltonian in the second-quantized form 17
Hamiltonian of vibrating atoms 112
hard core 57
hard-sphere models 3, 82, 148
hard-sphere structure factor 106, 201
harmonic approximation 241
Hartree process 32, 36
Hartree-Fock approximation 30, 31
heat capacity 94, 110, 164
– at moderately low temperatures 110ff.
heat conductivity 119, 121, 123
– temperature dependence 122
Heisenberg model 11, 135, 178, 220
Heisenberg operators 182
high-temperature superconducting materials 179
Holstein-Primakoff representation 157
Holstein-Primakoff transformation 138
Hubbard's Coulomb repulsive energy 177
Hund's rule exchange 177
hydrogenation 172ff., 178

i
imaginary time 150, 182
indirect electron-electron interaction 262
indirect interactions 47, 59ff.
indirect interionic interaction 61
industrial applications 215
inelastic double-differential cross-section 133
inelastic neutron scattering 133, 137, 162
inelastic neutron scattering cross-section 168
inelastic scattering cross-section 162
interacting electron gas 34
interaction lines 237
interaction representation 183ff., 238
interactions 111ff., 180ff.
– between electrons and phonons 111
– between phonons 111
– between phonons and tunneling states 111
interference function 9
interionic interaction potential 54, 59
interionic pair potential 61ff.
intermediate densities 36
intermediate pseudopotential 265
internal frequencies 42
internal momenta 42
internal stresses 215
internuclei distances 63
inter-particle potential 57
interproton separations 67
ion electron interaction potential 60
ion subsystem 16
ion-ion interaction 48
irreducible correlation functions 245
isotopes 4
isotropic spin-glass 176
itinerant approach 11
itinerant description 178
itinerant ferromagnetism 172
itinerant magnetism 171ff., 178
– of amorphous metals 13

k
kinetic equations 115ff., 124
kinetic phenomena 111
Kirkwood's superposition approximation 55

l
Landé factor 139, 220
laser glazing 2
Lennard-Jones glass 79
Lennard-Jones potential 11, 51, 69, 72
lifetime of an ion in a well 66
lifetime of metastable metallic hydrogen 68, 69
linear response function 58
Lindhard function 43

liquid metals 54 ff.
local directions 171
local magnitudes 171
local polarization vectors 176
localization of the electrons on ionic pairs 66
logarithmic divergence 34
longitudinal sound waves 208
long-range Coulomb interaction 34

m

magnetic interaction radius 149
magnetic moment density 164
magneto-volume effects 172
magnon excitations 163, 168
– energy spectrum 163
majority-spin d-band 172
many-body formalism 216 ff.
many-ion tunneling mechanism 66
many-ionic tunneling nucleation 68
many-particle interaction 58
many-particle tunneling 59
Matsubara frequencies 41, 151, 152, 185, 207, 253
– Fermion 185, 253
McMillan formula 197, 198, 209
mean field approximation 153
see also molecular field approximation; self-consistent field approximation
mean field theory 174
mean free path of a phonon 93
melt spinning 1
metallic hydrogen 58 ff., 61 ff.
metalloid 1, 71 ff.
metalloid atoms 52
metal-metalloid alloys 136
metal-metalloid amorphous alloys 52
metastable metallic hydrogen 59, 67, 68
– vestiges of the molecular phase 68
minority-spin d-band 172
mode coupling theory 216
model formation 2
model pseudopotentials 25 ff.
– Ashcroft's pseudopotential 27, 54
– Cohen-Abarenkov-Heine pseudopotential 27
– cut-off function 28
– drawback of model pseudopotentials 28
– empty-core pseudopotential 27
– Krasko-Gurski pseudopotential 54
– pointwise-ion pseudopotential 27
– pseudopotential of Abarenko and Heine 26
– pseudopotential of Animalu and Heine 51
– screened pseudopotential 28
– Shaw's pseudopotential 27
– ultrasoft pseudopotentials 28
molecular dynamics 10, 70, 79
molecular field (Pierre Weiss) 174
molecular field approximation 136, 144
see also mean field approximation; self-consistent field approximation
molecular field models 135
molecular hydrogen phase 59
momentum representation 57
– for the interaction potentials 17
Moruzzi *et al.* 175
Mössbauer spectroscopy 174
most diverging terms 34, 36

n

Nagel-Tauc criterion 76
Nambu operator s 181, 184, 205, 206, 251
neutralizing background 185
neutron diffraction 4
non-collinear magnetism 178
non-collinear spin configuration 174
non-linear response function 58
non-local pseudopotential theory 7, 69
non-pair forces 52
non-pair interaction 57
non-spherical contribution 69
non-uniqueness of the pseudopotential 21
n-particle polarization Green's function 44
nucleation of the H_2 molecule 59
numerical factor 233
numerical studies 130

o

occupation numbers 114
occupied states 173
online calculation of the electronic state 70
optic sound waves 108
orthogonality to the wave functions of delocalized states 23
orthogonalized plane waves (OPW) 19, 20

p

pair distribution function 103
pair forces 52
pair interactions 57, 59, 63 ff.
pair potential 6, 7, 61 ff.
parabolic potential 82

paramagnetic state 173
partial densities of states 73 ff.
partial distribution function 5, 7
partial structure factors 208
partial summation 45
Pauli matrices 176, 181
Pauli principle 32, 173
Percus-Yevick approximation 8, 106, 201
Percus-Yevick equation 56, 57
Percus-Yevick radial distribution function 9
permittivity (dielectric function) 240
phase-coherent tunneling 96
phonon echo 96
phonon excitations 168
phonon mean free path 94
phonon spectral function 194
phonons 82, 123, 124
– thermal 123, 124
– ultra-sound 123, 124
planar arrangement of four ions 67
plateau 123
polarization Green's function 41, 44
polarization of phonons 132
polarization operator 42, 45, 241 ff.
– of Geldart and Vosko 48
Polk networks 81
polyvalent metals 52
positive charge background 180
potential landscape 62 ff.
– by a group of ions 64
principal part 131
see also Cauchy part
projection operator 27, 131
pseudopotential 20, 180
pseudopotential method 15
pseudopotential theory 6, 18
pseudowave function 24

q

quantum tunneling 86
quartz 79
quasi-cristalline Hamiltonian 106
quasi-equilibrium configurations 187
quasi-equilibrium position 182
quasi-magnon approximation 162
quasi-magnon dispersion curves 153, 154
quasi-magnon gas 162
quasi-magnon scattering 164
quasi-phonon approximation 106
quasi-phonon dispersion curves 153
quasi-phonon excitations 107, 122
quasi-phonons 113, 114, 207 ff., 242 ff.
quenched system 81

r

radial distribution 7
radial distribution function (RDF) 3 ff., 9, 12, 57
random phase approximation (RPA) 43
rapid quenching 80
reciprocal space 57
reduced radial distribution 5, 8
renormalization of the exchange 168
renormalized interaction 54
resolvent formalism 131
resolvent matrix 73
roton minimum 106, 109 ff, 122 ff., 130, 133, 138
roton region 162
roton-like anomaly 80
rotons 121

s

saturation of the ultrasound resonance attenuation 95
second-quantized representation 181
self-averaging property 81
self-consistent effective magnetic field 174
self-energy 251
semi-empirical calculation 70
semi-empirical rules 73
silicate glasses 79, 81
skeleton diagrams 229 ff.
Slater determinant 29
Slater-Koster 72 ff., 174
soft phonons 106
soft quasi-phonons 121
soft-core model 5, 6
softening 200
– of the phonon spectrum 179
soft-sphere model 8, 167
solid metallic hydrogen 58
sound absorption 123, 128
sound attenuation 95
sound velocities 95, 106 ff.
sound waves 210
sound-wave absorption 125
spectral density of the electron-phonon interaction 261
speromagnet 176
spin frustration 92
spin glasses 83, 92
spin-glass properties 178
spin-wave dispersion law 137, 150
spin-wave energy spectrum 152
spin-wave excitations (magnons) 171
spin-wave stiffness 137, 153, 156, 164, 166 ff., 171

split maximum 8
splitting processes 115, 116
stability of amorphous alloys 75
– of the amorphous state 2
steric aspects 3
strong-coupling equations 194
structural disorder 136
structural frustration 92
structural models of glassy metals 7
structure factor 199, 243
structure function 3, 5, 7, 9
sublattice magnetization 170
superconducting metals 179 ff.
superconductivity of simple metals 202
– parameters for the crystalline and amorphous states 202 ff.
superionic conductors 83
superposition approximation 146, 165

t

temperature scattering matrix 184, 221
theory of liquids 82
thermal conduction 94 ff.
thermal conductivity 121
thermodynamic Green's function 41
thermodynamic perturbation theory 58, 158, 162, 164, 206
thermodynamic potential 37, 40, 41, 45, 46, 51
thermodynamic time 238
three-body correlations 11
three-ion interaction in a cluster of four ions 65
three-ion interactions 59, 63 ff.
three-particle interaction 53, 56, 57
three-proton interaction 63
tight-binding formalism 72 ff.
time-ordering operator 217 ff., 241
tracing operation 37
transducers 215
transition from metallic hydrogen into a molecular phase 66
transition metal (TM)-metalloid (M) amorphous alloys 69
transition temperature 191
transversal sound waves 208
tree diagrams 142
triplets of ions 67
tunneling excitations 82 ff.
tunneling levels 63, 80
tunneling states 91
tunneling transition 65
two-level systems 84
two-particle interaction 53, 57

u

Umklapp processes 117, 199
uncontracted 229

v

vertex 41
vertex function 240
vertices 143
vibrational contributions 77
vibrational oscillations 38
vibrational rotons 138
volume enhancement (by hydrogenation) 172
volume expansion 173
Voronoi polyhedron 172, 174, 175

w

Wick's theorem 39, 141
Wick's theory 230
– for spin operators 230
width of the excitations 133
Wigner-Seitz cell 172

x

X-ray diffraction 4

y

Yukawa potential 145

z

Zeeman energy 157
zero pressure condition 28